Medical Virology 8

A Continuation Order Plan is available for this series. A continuation order will bring delivery of each new volume immediately upon publication. Volumes are billed only upon actual shipment. For further information please contact the publisher. Plenum began publication of volumes in this series effective with Volume 8.

Medical Virology 8

Edited by
Luis M. de la Maza
and
Ellena M. Peterson
University of California, Irvine Medical Center
Orange, California

SPRINGER SCIENCE+BUSINESS MEDIA, LLC

Library of Congress Card Catalog Number 89-657524
ISBN 978-1-4899-0893-3 ISBN 978-1-4899-0891-9 (eBook)
DOI 10.1007/978-1-4899-0891-9

Proceedings of the 1988 International Symposium on Medical Virology, held September 22–24, 1988, in San Francisco, California

Originally published by Plenum Press, New York in 1989
Softcover reprint of the hardcover 1st edition 1989

FOREWORD

No other area of biology has grown as fast and become as relevant over the last decade as virology. It is with no little amount of amazement, that the more we learn about fundamental biological questions and mechanisms of diseases, the more obvious it becomes that viruses permeate all facets of our lives. While on one hand viruses are known to cause acute and chronic, mild and fatal, focal and generalized diseases, on the other hand, they are used as tools for gaining an understanding of the structure and function of higher organisms, and as vehicles for carrying protective or curative therapies. The wide scope of approaches to different biological and medical virological questions was well represented by the speakers that participated in this year's Symposium. While the epidemic by the human immunodeficiency virus type 1 continues to spread without hope for much relief in sight, intriguing questions and answers in the area of diagnostics, clinical manifestations and therapeutical approaches to viral infections are unveiled daily. Let us hope, that with the increasing awareness by our society of the role played by viruses, not only as causative agents of diseases, but also as models for better understanding basic biological principles, more efforts and resources are placed into their study.

Luis M. de la Maza
Ellena M. Peterson

Irvine, California
March, 1989

ACKNOWLEDGEMENTS

To organize an international meeting is a task that requires the efforts of many individuals, and thus it would be very difficult to recognize all those that have contributed to this year's undertaking. We want, however, to at least name those individuals that in a most unselfish manner, put forth the time and effort to make this Symposium a reality. Our thanks go to all the speakers for their excellent lectures and the chapters that they contribute for this book, and to Drs. Thomas C. Cesario and Edwin E. Lennette for chairing the sessions. We want to acknowledge also the participants for their willingness to share their knowledge and their questions and by so doing making the meeting a stimulating intellectual experience.

Throughout the year the staff of the Division of Medical Microbiology at the University of California Irvine, provided us with continuous support and particular mention should be made of Marie Pezzlo and Sandra Aarnaes, who shared their efforts with us almost on a daily basis. The completion of these Proceedings would not have been possible without the secretarial support of Penny Richardson and the assistance of Mirtha Nodal and Lillian Dalgleish in preparing the manuscripts. The Plenum Publishing Co. helped us with their editorial support throughout this undertaking. We are particularly grateful to Melanie Yelity and Gregory Safford for their support.

We want also to express our appreciation to the following organizations that provided the financial support to make the Symposium a reality:

Bayer AG/Miles, Inc.
Bethesda Research Laboratories
Bion Enterprises, LTD
Bio-Rad Laboratories
Coulter Immunology
Eastman Pharmaceuticals
Genentech, Inc.
Gull Laboratories, Inc.
Hoffman-La Roche, Inc.
HyClone Laboratories, Inc.
Lilly Research Laboratories
Merck Sharp & Dohme
Monsanto Company
Ortho Diagnostics Systems, Inc.
Pharmacia LKB Biotechnology, Inc.
The Rupp and Bowman Company
Sandoz Research Institute
Schering Corporation
SmithKline & French Laboratories
Syva Company
Triton Biosciences, Inc.
The Upjohn Company
Virion (U.S.), Inc.
ViroMED Laboratories, Inc.
Whittaker Bioproducts
Wyeth Laboratories

CONTENTS

NEWER APPROACHES TO PERINATAL HERPES SIMPLEX VIRUS TESTING

ANN L. WARFORD AND EVAN STEINBERG

Regional Virology Laboratory and Department of Pediatrics, Southern California Kaiser Permanente, North Hollywood and Los Angeles, California, USA

INTRODUCTION

The prevalence of genital herpes simplex virus (HSV) infections in the United States as estimated from "first office visits" has increased 9-fold over the period from 1966 to 1984 (Centers for Disease Control, 1986). The incidence of neonatal HSV infections has also increased, as reported from King County, Washington, from 2.6 to 28.2 per 100,000 live births over the period from 1966 to 1982 (Sullivan-Bolyai et al. 1983). In Southern California we have experienced similar increases from 2 cases in 1982 (8.2/100,000 live births) to 6 and 7 cases in 1985 and 1986 (25.5/100,000). The mortality associated with neonatal HSV infections has decreased with antiviral therapy from 80% to 15% of the cases with central nervous system (CNS) and to 50% of cases with disseminated disease (Stagno and Whitley, 1985). One third to one-half of these treated infants, however, have neurologic sequelae which may appear as late as 1 to 2 years after treatment (Whitley et al. 1980, 1986).

Since babies are usually colonized with HSV after rupture of membranes or during vaginal delivery, if those women shedding HSV at delivery could be identified, prevention of most neonatal infection is theoretically possible (Binkin et al. 1984; Corey and Spear, 1986). To minimize the risk of transmission of HSV by women with a history or symptoms of genital herpes in themselves or a sexual partner, the practice developed of culturing such "high risk" pregnant women regularly between week 32 and delivery. Only those women with no lesions and a negative culture within the week of delivery were delivered vaginally (American Academy of Pediatrics, 1980). Unfortunately, 70% of neonates with HSV infection are born to women with no history or symptoms of genital HSV (Whitley et al.

1980). While 5% or less of pregnant women have a history of lesions or other symptoms of genital herpes or a partner with this disease (Binkin et al. 1984; Prober et al. 1988), serologic studies of antibodies to HSV-2 specific glycoprotein G in pregnant women suggest that 18 to 34% have been infected with HSV-2 (Frenkel et al. 1987; Prober et al. 1988). Genital cultures from unselected groups of pregnant women have confirmed shedding of HSV-2 at delivery in women who are seropositive but lacking clinical history or symptoms of genital herpes. In addition, asymptomatic or undiagnosed primary maternal HSV-2 infection may occur in the perinatal period (Prober et al. 1988; Yeager and Arvin, 1984).

But, even in women identifiable by history, antepartum cultures failed to predict asymptomatic shedding of HSV at delivery and resulted in unnecessary cesarean deliveries with a 6-fold higher risk of maternal mortality and with a complication rate of 28% (Arvin et al. 1986; Binkin et al. 1984). Nonetheless, the most recent American College of Obstetrics and Gynecology recommendations are to perform viral cultures on all pregnant women with suspicious lesion(s) or a "high risk" history and to have at least one negative culture before, and no lesions present, at vaginal delivery for such women. Cesarean delivery is indicated for all patients with a lesion (American College of Obstetricians and Gynecologists, Committee on: Obstetrics: Maternal and Fetal Medicine, 1987).

In developing an alternate approach to minimize neonatal HSV disease at Southern California Kaiser-Permanente Medical Group, the following projection (Table 1) was used:

Assuming 30,000 (approximately 24,500 to 27,500 in 1981-87) deliveries in 1988 and 15% cesarean births per obstetrical indication, 25,500 vaginal deliveries could be anticipated. If 40% of these women have historical or serologic evidence of HVS-2 infection, then 10,200 women would be at risk of genital HSV. With 1.4% (Arvin et al. 1986) of these high risk women shedding HSV asymptomatically at the time of delivery, 143 women would be expected to have a positive HSV culture at delivery. The attack rate in the infant delivered vaginally to a mother with active recurrent genital HSV is estimated to be 5% (Binkin et al. 1984; Corey et al. 1983) which would yield seven cases of neonatal HSV. Six to seven cases of neonatal HSV were identified in the Southern California Kaiser-Permanente region annually during 1984-1986.

One could propose replacing weekly culturing of "high risk" women with cultures taken at delivery of either all mothers or all infants in an attempt to identify and treat neonatal exposure to HSV. Since previous studies have reported delays of five to seven days between onset of the symptoms of neonatal HSV and diagnosis and treatment of these cases

TABLE 1. PROJECTION OF NEONATAL HSV INFECTION WITHIN SOUTHERN CALIFORNIA PERMANENTE FOR ONE YEAR

30,000	Deliveries per year
- 4,000	30,000 x 15% non-HSV cesarean deliveries
25,500	Vaginal deliveries
x 40%	Women with a history of serologic evidence of HSV-2 infection
10,200	Women at risk of HSV infection
x 1.4%	Asymptomatic shedding of HSV at delivery
143	Asymptomatic mothers with HSV positive viral cultures
x 5%	Risk of HSV disease to infant
7	Cases of neonatal HSV per year

(Sullivan-Bolyai et al. 1986; Whitley et al. 1980), a better outcome with lower rates of mortality and late sequelae might be expected with more rapid detection of infants exposed to HSV.

Culturing infants at delivery has the advantage of directly and rapidly identifying birth exposure, but there are several problems with this approach. Samples from conjunctiva, nose, oropharynx, and rectum are easily obtained, but in some cases of neonatal HSV disease only CSF or brain tissue yielded HSV on culture (Sullivan-Bolyai et al. 1983; Whitley et al. 1980). Prober et al. (1988) reported that surface cultures obtained at delivery were negative from 13 of 16 infants whose mothers had positive delivery cultures. One of these 13 infants subsequently developed HSV meningo-encephalitis. Thus, surface cultures of neonates at birth, as with surveillance cultures of "high risk" women, fail to identify all infants at risk. Animal models of neonatal HSV (Overall et al. 1984) suggest that HSV replication begins in mucous membranes 18-24 hours after viral challenge. Cultures obtained from infants before this time might reflect transient maternal HSV contamination rather than neonatal infection. While cultures of neonates at 24 hours of age might be more specific, the current practice of early discharge of babies make them impractical.

The proposed alternative to weekly antepartum cultures of only "high risk" patients is, therefore, to obtain cervical and vulvar samples during labor from all women delivering vaginally and those women with rupture of membranes six hours or more before abdominal delivery. Infants of women with positive cultures would be recalled for evaluation. This would include a thorough clinical evaluation and cultures of conjunctivae, oropharynx, and lesion site(s) if any.

This proposed alternative protocol would benefit from laboratory diagnostic tests that are both less expensive and more rapid than standard seven-day culture. The clinical incubation period for HSV is five to seven days. The onset of neonatal symptoms is usually seen between five to 17 days (Stagno and Whitley, 1985; Whitley and Hutto, 1985). In order to provide earlier intervention, laboratory detection of HSV should be shorter than this five day incubation interval or two days for maternal sample testing followed by two days for infant testing.

The cost aspects of the current and proposed protocols are outlined in Table 2. Using the current practice of weekly herpes cultures of 1,500 patients (30,000 x 5% high risk patients) starting at week 32 to 34 until delivery at 38 to 42 weeks, the average number of cultures per patient is eight, yielding approximately 12,000 cultures. The 1987 average direct laboratory cost was approximately $15 per herpes culture. The 12,000 cultures of high risk patients cost $180,000. According to the estimates of Binkin et al. (1984), each excess cesarean delivery due to positive antepartum culture(s) costs $2,300. These deliveries also have a complication rate of 28% and require four extra days hospital stay at $300.00 per day. With a 1.4% positive culture rate for the week before delivery from 1,500 high risk women 21 cesarean deliveries would occur. Such deliveries cost $2,300 each plus $7,060 for complication or a total of $55,360. Therefore, the cost of the current protocol is at least $235,360, not including the costs of 6-7 cases of neonatal HSV. The actual cesarean delivery rate approaches 20% of 30,000 deliveries or 24,000 vaginal deliveries. In order to have a cost equivalent program using the proposed testing protocol on all 24,000 women having vaginal deliveries, the previous protocol cost of $235,360 would now be divided by 24,000 yielding $10 per patient for direct laboratory cost.

Although rapid direct specimen tests for HSV antigen are available including fluorescent antibody (FA), immunoperoxidase (IP), enzyme immunoassay (EIA), and latex (LA), these methods are not as sensitive and specific as viral isolation (VI) even with samples obtained from lesions. Moseley et al. (1981) demonstrated the difference in the sensitivity of both VI and DFA using a polyclonal antibody according to the stage of the lesions sampled (Table 3). The DFA test sensitivity varied from 29% with ulcers in recurrent disease to a maximum sensitivity of 80% with vesicular lesions from patients with primary HSV. In this same study lower sensitivity was demonstrated in cervical lesions as compared to vulvar -

TABLE 2. COST COMPARISON OF CURRENT AND PROPOSED OBSTETRIC HSV PROTOCOL

CURRENT PRACTICE		PROPOSED PROCEDURE	
Annual deliveries	30,000		30,000
OB with history symptoms of HSV	x5%	Actual cesarean rate 20%	- 6,000
High risk OB	1,500	Vaginal deliveries	24,000
Avg. No. HSV cultures (32 to 40±2 weeks)	x8	Previous lab and hospital costs	$235,360
Total No. cultures	12,000		
Avg. lab cost/culture	x $15	Divided by vaginal deliveries	24,000
Direct lab HSV costs	$180,000 (subtotal)		
Plus hospital costs cesarean deliveries		Cost equivalent test program per HSV test	$10
Positive HSV culture week of delivery	1,500		
	x 1.4%		
Inc. cesareans	21		
Inc. surgery cost at $2,300 each	$48,300		
Complication cost at 28% x $1,200	$ 7,060		
Cesarean costs	$55,360 (subtotal)		
Total of lab and hospital costs	$235,360		

48% versus 71%. Results similar to DFA were seen with the immunoperoxidase labeled polyclonal antibody (Moseley et al. 1981). In 1987, Lafferty et al. (Table 4) reported a similar correlation of sensitivity with duration or stage of genital lesions using a monoclonal DFA test. Ten percent of samples from crusted lesions were positive for HSV as contrasted with 87% of vesicle samples with DFA. Culture was only 2-7% more sensitive in the samples from patients with recurrent lesions. Culture sensitivity decreased with the age of the lesions from an isolation rate of 65-70% with fresh lesions to only 15% sensitivity with lesions over 120 hours old (Lafferty et al. 1987).

Higher titers of HSV have been reported from vesicles than ulcerated lesions and the duration of shedding is 11.4 days in initial infections

TABLE 3. SENSITIVITY OF VIRAL ISOLATION, INDIRECT IMMUNOPEROXIDASE, AND DIRECT FLUORESCENT ANTIBODY TEST ACCORDING TO CLINICAL STAGE OF HSV LESIONS.

Initial Disease Patients (No.)	VI Positive (%)	DFA Positive (%)
Vesicle (10)	100%	80%
Pustule (9)	89%	67%
Ulcer (22)	82%	41%
Recurrent Disease Patients (No.)		
Vesicle (20)	90%	75%
Pustule (3)	67%	33%
Ulcer (7)	43%	29%

Moseley et al. 1981.

but only 3.1 days in recurrences in females (Corey et al. 1983; Corey and Spear, 1986; Guinan et al. 1981). The peak viral titer determined by Guinan et al. (1981) in recurrent lesions was $10^{3.2}$ plaque forming units (PFU) on day 2 after lesion onset (Figure 1). The titer of HSV isolated from asymptomatic patients has been 10 to 1,000 times lower (10^1 to 10^3 tissue culture infection doses ($TCID_{50}$) as compared to titers from lesions of 10^3 to 10^6 $TCID_{50}$ (Corey et al. 1983; Corey and Spear, 1986).

TABLE 4. COMPARISON OF VI AND DFA TESTING BY LESION STAGE IN RECURRENT GENITAL INFECTION[a]

Lesion Stage (No.)	VI Positive (%)	DFA Positive (%)
Vesicle (30)	70	87
Pustule (24)	67	67
Ulcer (47)	32	30
Crust (29)	17	10

[a]Includes only those specimens with adequate cell counts (more than 20 exfoliated cells). Lafferty et al. 1987.

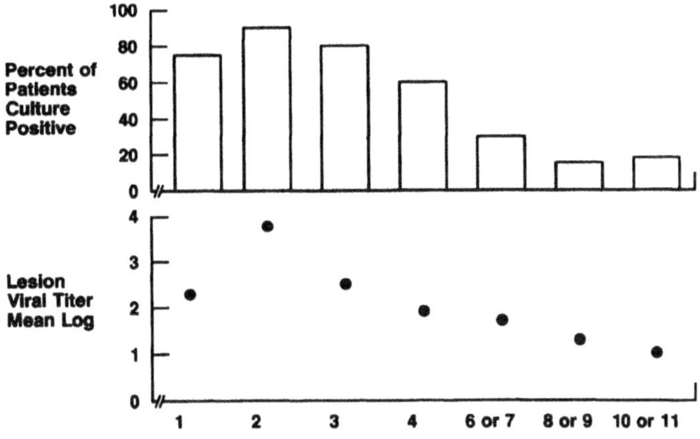

Figure 1. Frequency of positive cultures and the virus titers lesions during illness in women with recurrent HSV infection. Guinan et al. 1981.

In evaluating a commercial HSV antigen EIA test, we also found dramatically lower sensitivity with cervical specimens from predominantly asymptomatic obstetric patients (Warford et al. 1984; 1986). The EIA sensitivity with cervical specimens was 21.7%, while sensitivities of 41.9% to 100% were obtained from other symptomatic sites (Warford et al. 1984) (Table 5). The EIA sensitivity correlated with the titer of HSV as reflected by days to detection of CPE with only 8.3% of day 4 CPE specimens EIA positive and 71.4% of specimens with CPE detectable in 24 hours EIA positive (Table 6).

The EIA positive/culture negative specimens were also a major concern since reporting of these results would increase the cesarean delivery rate by 1.7%, although none of the infants from these women developed symptomatic HSV infection (Warford et al. 1986).

False positive DFA results in neonates have also been reported (Sever, 1983; Sullivan-Bolyai et al. 1986). Sullivan-Bolyai et al. (1986) had 5/36 DFA positive/culture negative oropharyngeal samples from neonates with other diagnoses which were placed on antiviral therapy.

Both significantly higher false positive and false negative results were obtained in our hands with the HSV latex antigen test. Even with genital lesion samples (n=100), only 34.7% of culture positives were also latex positive with 16.6% false positives in the reagent control well further reducing the utility of the test for direct specimens (Warford, unpublished data).

TABLE 5. SOURCE OF SPECIMENS FOR POSITIVE DIRECT ELISA AND HSV CULTURE RESULTS

SPECIMEN SOURCE	No. (%) of specimens with the following result: ELISA + CULTURE -	ELISA + CULTURE +	ELISA - CULTURE +
Labia-vulva	1(0.8)	55(41.9)	75(57.3)
Cervix	22 (36.7)[a]	13(21.7)	25(41.6)
Penus-scrotum	0	30(65.2)	16(34.8)
"Genital"	1(3.8)	17(65.4)	8(30.8)
Vagina	0	16(64.0)	9(36.0)
Buttocks	0	15(100.0)	0
Rectal-anal	0	2(66.7)	1(33.3)
Chin	0	0	1(100.0)
Knee	0	0	1(100.0)
Urine[b]	1(100.0)	0	0
Liver tissue[c]	0	1(100.0)	0

[a]Three patients had previous positive HSV cultures within 10 days
[b]Urine specimen showed no CPE for 28 days; same results when reported
[c]Disseminated neonatal HSV infection

Warford et al. 1984

TABLE 6. NUMBER OF DAYS TO DETECTION OF CPE FOR HSV-POSITIVE CULTURES FOUND POSITIVE OR NEGATIVE BY DIRECT ELISA

Days to detection	No. (%) of specimens with the following result: Positive by Direct ELISA	Negative by Direct ELISA
1	95(71.4)	38(28.6)
2	55(47.4)	61(52.6)
3	3(10.3)	26(89.7)
4	1(8.3)	11(91.7)
5	1(25.0)	3(75.0)
7	0	1(100.0)

Warford et al. 1984

Hsiung and Landry (1986) have pointed out that at least 10^3 viral particles are required to detect HSV antigen by immunologic tests (FA, EIA, IP) in contrast with cell culture which theoretically requires only one infectious viral particle (Table 7). DNA hybridization offers superior specificity (100%) but again requires 10^3 HSV PFU, which as mentioned above, is near the peak titer obtained from recurrent lesions. A sensitivity of only 78% was reported by Redfield et al. (1983) using a 32p labeled DNA probe in genital lesion samples. Also, both the cost and technical requirements of this type of assay at present preclude use in the diagnostic laboratory.

While standard tube culture remains the most sensitive and specific method of detecting HSV, seven days are required to obtain maximum sensitivity which has been variously reported as 53% from recurrent lesions to 100% from primary herpetic vesicles (Corey and Spear, 1986; Lafferty et al. 1987; Moseley et al. 1981). Culture has been used in conjunction with HSV pre-CPE antigen detection, primarily with IP, but also FA and EIA as a method of amplifying virus concentration to a detectable level (Fayram et al. 1983; Morgan and Smith, 1984; Sever, 1983). Gleaves et al. (1985) found only 16 hours preincubation were required to obtain 100% FA detection of HSV from fresh dermal and mucosal specimens in centritrifuged shell vial monolayers. The spin amplified EIA test (SATEIA) was reported to be 97.6% sensitive after 48 hours incubation in a study using predominantly female genital specimens (Michalski et al. 1986). However, Zhao et al. (1987) demonstrated that 48 hours were insufficient to detect low virus titers (<50 PFU) (Table 8) which would be expected in an asymp-

TABLE 7. TIME REQUIRED FOR DIAGNOSIS OF HSV INFECTION

Estimated number of virus articles/ml	Method of detection	Virion/viral component	Time Required for reporting results
1-10	Cell culture isolation	Virus infectivity	1-2 days
10^6-10^7	Electron microscopic examination	Virus particles	30 min
10^3-10^6	Immunologic assays[a]	Viral antigens	24-48 hr
10^3-10^6	Nucleic acid hybridization	Viral genomes	24 hr

[a] Including IF., IP., ELISA., etc. Hsiung and Landry, 1986

TABLE 8. EFFECT OF VIRUS CONCENTRATION OF SPEED OF DETECTION OF HSV IN CLINICAL SPECIMENS BY CPE AND IP STAINING IN THREE CELL CULTURE SYSTEMS

Virus titer per 0.1 ml	No. tested	Cell culture	Cumulative no. (%) positive at time: 24 h p.i. CPE	24 h p.i. ABC	48 h p.i. CPE	48 h p.i. ABC
<50 PFU	17	ML	13(76)	5(88)	15(88)	15(88)
		MRC-5	6(35)	13(76)	10(59)	12(71)
		Vero	0(0)	8(47)	7(41)	10(59)
>50 PFU	15	ML	15(100)	15(100)	-[a]	-
		MRC-5	15(100)	15(100)	-	-
		Vero	5(33)	15(100)	15(100)	15(100)

[a]All cultures were positive by CPE and ABC. Zhao et al. 1987

tomatic obstetric population. Further enhancement of pre-CPE FA antigen detection was obtained by West et al. (Abs. Ann. Mtg. Am. Soc. Microbiol., 1987, C-150:348) in shell vials with the use of medium containing dexamethasone (DEX) (10^{-5}M) and dimethyl sulfoxide (DMSO) (Abs. Ann. Mtg. Am. Soc. Microbiol., 1987, C-150:348).

We evaluated the use of these amplification methods, centrifugation, DEX medium and three commercial EIA tests, for the detection of HSV in obstetric samples obtained before and during labor and in neonatal samples. We first used the SAT-EIA (Ortho Diagnostics, Carpinteria, CA) with 1,814 pregnant female genital specimens with centrifugation at 2,500 xg for 30 min in primary rabbit kidney (PRK) flat-bottom tubes followed by the Ortho 4.5 h EIA on culture lysates at 16 or 36 hours. These results, shown in Table 9, were disappointing with only 59.7% sensitivity at 16 h and 86.4% at 36 h. This sensitivity was similar to our cell culture (TC), previously described, in which 53.7% of isolates were reported at 16-18 hours and 84.9% at 36 hours (Warford et al. 1984).

Both the modified 2.5 hour Ortho EIA with a new substrate system as well as the Fairleigh Dickinson (FDL) EIA tests were next compared to TC using PRK after 36 h with 154 female genital and neonatal specimens (Table 10). The modified Ortho SAT-EIA detected 71 of 73 culture positives for a sensitivity of 97.3% as compared to only 51 of 73 (69.9%) detected with the FDL.

The effects of the DEX medium and type of cells were compared using the modified Ortho SAT-EIA at 36 hours (Table 11). In the DEX-MRC-5 cells 71 of 80 (88.8%) TC positives were identified while 74 of 80 (92.5%) CPE

TABLE 9. COMPARISON OF CULTURE WITH THE SPIN-AMPLIFIED 4.5 HOUR ORTHO EIA AFTER 16 AND 36 HOURS INCUBATION WITHOUT DEXAMETHASONE

EIA Result	No. of specimens with typical HSV CPE in culture: Positive	Negative	% Sensitivity	% Specificity
16 Hours				
Positive	46	2	59.7%	99.8%
Negative	31	943		
36 Hours				
Positive	57	0	86.4%	100.0%
Negative	9	720		

TABLE 10. COMPARISON OF CULTURE WITH THE SPIN-AMPLIFIED EIA AFTER 36 HOURS USING FDL OR MODIFIED ORTHO KIT IN PRK

EIA Result	No. of specimens with typical HSV CPE in culture: Positive	Negative	% Sensitivity	% Specificity
FDL				
Positive	51	0	69.9%	100.0%
Negative	22	81		
Ortho (mod.)				
Positive	71	0	97.3%	100.0%
Negative	2	81		

positives were detected in PRK cells. Although five positives were found in PRK alone as compared to two in DEX-MRC-5 this difference was not statistically significant (p=0.42). Actually, the average OD of the positive EIA values was higher, 2.83, in DEX-MRC-5 than in PRK with an average positive OD of 2.19. Using two cell types, one PRK and one DEX-MRC-5, the SAT-EIA identified 147 of 153 (96.1%) of CPE positives with pre-delivery female genital specimens (Tables 10 and 11).

TABLE 11. COMPARISON OF CULTURE WITH THE SPIN-AMPLIFIED ORTHO (MODIFIED) EIA AFTER 36 HOURS WITH MRC-5 AND PRK IN DEXAMETHASONE MEDIUM

EIA Result	No. of specimens with typical HSV CPE in culture: Positive	Negative	% Sensitivity	% Specificity
MRC-5				
Positive	71	0	88.8%	100.0%
Negative	9	204		
PRK				
Positive	74	0[a]	92.5%	100.0%
Negative	4	204		
Both cells				
Positive	76	0[a]	95.0%	100.0%
Negative	4	204		

[a] Nonreproducible positive in 1 of 2 duplicate wells.

When the DEX medium pretreatment of MRC-5 cells was evaluated in seven-day TC with 2,604 genital samples, the difference in results was greatest with the low titer specimens which required 3-6 days for observation of CPE (Table 12). In these cultures the CPE was detected 1 to 3

TABLE 12. COMPARISON OF AMOUNT OF CPE AND TIME TO DETECTION IN MRC-5 WITH AND WITHOUT DEXAMETHASONE MEDIUM

	NO. OF CULTURES (%) WITH CPE: Equal	DEX First[a]	DEX[b]	MEM First[a]	MEM[b]	Total
CPE DAY						
1 - 2	238(69.0%)	28(8.1%)	66(19.1%)	2(0.6%)	11(3.2%)	345
3 - 6	15(38.5%)	14(35.9%)	5(12.850	3(7.7%)	2(5.1%)	39
TOTAL	253(65.9%)	42(10.9%)	71(18.5%)	5(1.3%)	13(3.4%)	384

[a]CPE detected 1 to 3 days earlier or other cell tube negative at day 7
[b]1+ to 2+ greater CPE reported.

days earlier from the DEX tube in 35.9% of 39 positives compared to 7.7% witout DEX. Greater CPE (1 to 2+) was reported from 12.8% of 39 isolates with DEX as compared to only 5.1% without DEX. Altogether, in TC 42 of 384 (10.9%) isolates were reported first from the DEX tubes and the CPE was greater in 71 of 384 (18.5%). These results compared to only 1.3% and 3.4% respectively, without DEX. After using the DEX pretreated MRC-5 for 120 days, we noted an increase in 16-18 hour isolates from 51.6 to 63.2% (Table 13). The difference was 86.5% with DEX at 36-38 hours while 84% were reported at this time without DEX.

A total of 712 neonatal samples obtained from conjunctiva, throat, CSF and skin sites during the first month of life (average age 10.6 days) were also tested by both the 36 h SAT-EIA and TC incubated and examined for 28 days. The SAT-EIA and TC had 100% agreement, all negative, if rectal samples were excluded. Due to three of 123 early non-reproducibly positive rectal SAT-EIA results, rectals were not further tested by the spin amplified EIA method. One of the stated limitations of the HSV antigen EIA is occurrence of false positives due to high concentrations of *Staphylococcus aureus* with affinity for the Fc portion of the capture antibody. Cross reactions with other organisms have not been reported. We do not use or recommend the SAT-EIA alone for neonatal samples since the sensitivity of this method is not determined.

When we evaluated the 16-hour shell vial FA method of Gleaves et al. (1985) using neonatal specimens, two CSF specimens were reported as positive. Neither infant, however, had any positive viral cultures nor developed serologic evidence of HSV although both infants had meningitis (unpublished data). One possible cause of false positive fluorescence in these shell-vial cultures is lipofuscin, an autofluorescent age pigment of cells. These pigment granules resulting from lipid peroxidation accumulate in cultivated diploid cells with age, oxygen or temperature variations over time (Thaw et al. 1984). The appearance of lipofuscin can be confused with the cytoplasmic HSV fluorescence although the color is not apple green (Figures 2a,b). Another possible cause of nonspecific fluorescence in the shell vial FA test is the interference by cell proteins that can bind certain portions of antibodies and give false positive results discussed by Sever et al. (1983).

In November 1987 we started a pilot study at one of the nine hospital in the Southern California region to evaluate the proposed protocol of sampling all asymptomatic pregnant women at delivery with the SAT-EIA and TC. The results at eight months are as follows (Table 14):

TABLE 13. COMPARISON OF TIME TO DETECTION OF CPE FOR 120 DAYS WITH MEM MEDIUM AND 120 DAYS WITH DEXAMETHASON MEDIUM[a]

Day	Number of CPE Positive Cultures Reported (Cumulative %) MEM Medium	Dexamethasone Medium
1	614(51.6%)	827(63.2%)
2	386(84.0%)	305(86.5%)
3	125(94.5%)	124(96.0%)
4	42(98.0%)	30(98.3%)
5	12(99.0%)	13(99.3%)
6	8(99.7%)	5(99.7%)
7	3(100.0%)	4(100.0%)
Total	1190	1308

[a] Dexamethasone in MRC-5 only

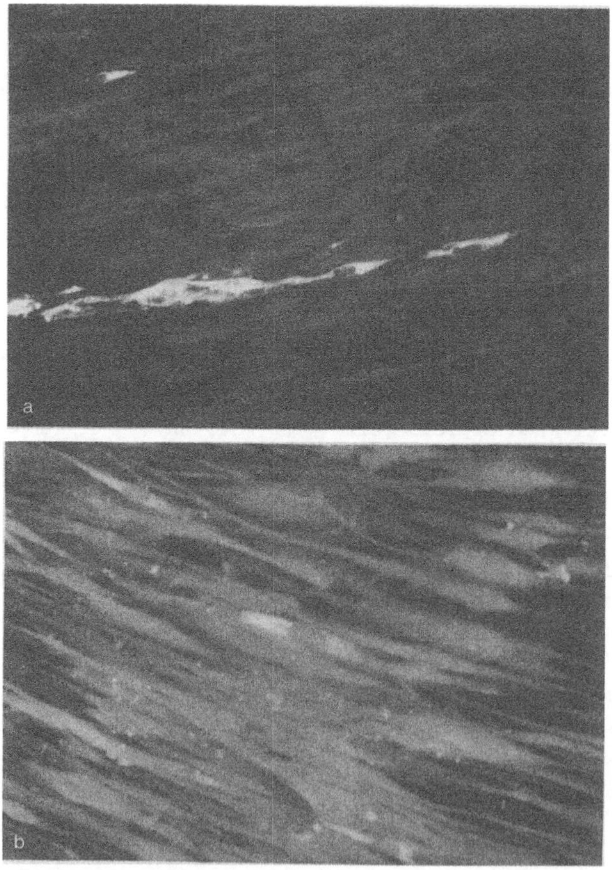

Figure 2. Diploid cells a) HSV b) Lipofuscin

TABLE 14. COMPARISON OF CULTURE TO SPIN-AMPLIFIED ORTHO EIA (MODIFIED) AFTER 36 HOURS WITH SPECIMENS OBTAINED AT DELIVERY

	No. of specimens with typical HSV CPE in culture POSITIVE	NEGATIVE	% Sensitivity	% Sensitivity
EIA Result	Positive	Negative		
Positive	22	1[a]		
Negative	1[b,c]	3,073	95.7%	99%

[a]Spin culture only positive due to gross bacterial contamination.
[b]TC only positive at day 4 in one of 3 TC tubes inoculated, patient in false labor, no delivery within 4 days.
[c]One additional TC CPE positive due to *Molluscum contagiosum.*

A swab of the cervix and labia was collected from the 3,098 asymptomatic women at the time of delivery. Twenty-two of the 23 (95.7%) CPE positives were detected by SAT-EIA in 36 h using PRK and DEX-MRC-5. The CPE positive and SAT-EIA negative sample was obtained from a patient in false labor, who did not deliver within four days of culture. This specimen developed CPE in one of three TC tubes at day four. The other two tubes did not develop CPE by day seven. The false positive SAT-EIA culture had bacterial contamination which produced grossly visible turbidity in all tubes. One additional sample reported as CPE positive was repeatedly negative by both EIA and the Syva culture confirmation monoclonal FA test and lost CPE on subpassage as is typical of *Molluscum contagiosum* in cell culture (Dennis et al. 1985).

The spin-amplified culture EIA method in DEX medium had the following advantages:

(1) Direct laboratory costs were reduced by using an automated detection method rather than the labor-intensive microscopic examination of culture tubes for seven days. Labor costs were further reduced since the technicians employed for the operation of the pre-programmed Beckman Biomek robotic workstation (Figure 3) have salaries lower by 50% as compared to licensed trained virologists required for cell culture.

(2) Use of an EIA test provides an objective numeric result with duplicate values and control samples to monitor reproducibility and precision.

Figure 3. Beckman Biomek robotic workstation

(3) The detection and identification steps for HSV are combined into a single test. The previous procedure in obstetric patients was to later identify or exclude HSV from all culture tubes exhibiting CPE by FA or EIA. Culture cytopathic effects include those from the viral effects of HSV, enteroviruses, multifocal varicella zoster virus, cytomegalovirus, adenoviruses and *Molluscum contagiosum*, as well as degeneration or toxicity from trichomonads, bacterial and fungal contamination, unknown substances in genital samples and bad lots of cells. This separate confirmation step also prolonged the time to final reporting of isolates.

(4) The additional costs of repeated or subpassaged cultures, which averaged 3-5% during the previous year due to many of the culture problems above, are eliminated with the SAT-EIA. The exception noted in this trial period was the need for repeat cultures of those with gross bacterial contamination.

(5) The SAT-EIA assists the early diagnosis and treatment of infants at risk of HSV infection by reducing the turnaround time for testing of both mother and infant to 4 days, which is before the average onset of neonatal HSV infections acquired during vaginal delivery.

The pilot study is continuing in asymptomatic women at delivery, but our preliminary results have found that the 36-hour, spin-amplified culture with the modified Ortho HSV antigen EIA and DEX pretreatment of MCR-5 detected 147 of 153(96.1%) HSV isolates from obstetrical genital samples and 22 of 23(95.7%) labor and delivery sample HSV positives with specificities of 100% and 99.9% respectively, in these two populations. The observed rate of asymptomatic shedding of HSV at delivery was 0.26% lower than projected, and in agreement with figures reported elsewhere (Prober et al. 1988).

REFERENCES

American Academy of Pediatrics, Committee on Infectious Disease, Committee on Fetus Newborn (1980) Perinatal herpes simplex virus infections. Pediatrics. 66:147-149.

American College of Obstetricians and Gynecologists, Committee on: Obstetrics: Maternal and Fetal Medicine (1987) Perinatal herpes simplex virus infections. Am Coll Ob Gyn, Washington, D.C.

Arvin, AM, PA Hensleigh, CG Prober, DS Au, LL Yasukawa, AE Wittek, PE Pallumbo, SE Paryan, AS Yeager (1986) Failure of antepartum maternal cultures to predict the infant's risk of exposure to herpes simplex virus at delivery. N Engl J Med 315:796-800.

Binkin NJ, JP Koplan, W Cates (1984) Preventing neonatal herpes. The value of weekly viral cultures in pregnant women with recurrent genital herpes. JAMA 251(21)2816-2821.

Center for Disease Control (1986) Genital herpes infection --United States, 1966-1984. MMWR. 35(24)402-403.

Corey L, HG Adams, ZA Brown, KK Holmes (1983) Genital herpes simplex virus infections: clinical manifestations, course, and complications. Ann Intern Med 908:958-972.

Dennis J, LS Oshiro, JW Buntner (1985) Molluscum contagiosum, another sexually transmitted disease: its impact on the clinical virology laboratory. J Infec Dis 151:376.

Fayram SL, S Aarnaes, LM de la Maza (1983) Comparison of Cultureset to a conventional tissue culture-fluorescent antibody technique for isolation and identification of herpes simplex virus. J Clin Microbiol 18:215-216.

Frenkel LM, JP Shen, E Garratty, YJ Bryson. 1987. A prospective study of pregnant women to determine prevalence and risk factors of asymptomatic infection with herpes simplex virus (HSV) type 2 and the risk of shedding virus at delivery. Ped Res 21:325A.

Gleaves CA, DJ Wilson, JAD Wold, TF Smith (1985) Detection and serotyping of herpes simplex virus in MRC-5 cells by use of centrifugation and monoclonal antibodies 16 h postinoculation. J Clin Microbio 21:29-32.

Guinan ME, J MacCalman, ER Kern, JC Overall, SL Spruance (1981) The course of untreated recurrent genital herpes simplex infection in 27 women. N Engl J Med 304:759-763.

Hsiung GD, M Landry (1986) Rapid Viral Diagnosis. Clin Microbiol News 8:173-177.

Lafferty WE, S Krofft, M Remington, R Giddines, C Winter, A Cent, L Corey (1987) Diagnosis of herpes simplex virus by direct immunofluorescence and viral isolation from samples of external genital lesions in a high prevalence population. J Clin Microbiol 25:323-326.

Michalski FJ, M Shaikh, F Sahraie, S Desai, L Verano, J Valla, BHA Neni (1986) Enzyme-linked immunosorbent assay spin amplification technique for herpes simplex virus antigen detection. J Clin Microbiol 24:310-311.

Morgan MA, TF Smith (1984) Evaluation of an enzyme-linked immunosorbent assay for the detection of herpes simplex virus antigen. J Clin Microbiol 19:730-32.
Moseley RD, L Corey, D Benjamin, C Winter, ML Remington (1981) Comparison of viral isolation, direct immunofluorescence, and indirect immunoperoxidase techniques for detection of genital herpes simplex virus infection. J Clin Microbiol 13:913-918.
Overall JC, RJ Whitley, AS Yeager, GH McCraken and JD Nelson (1984) Prophylactic or anticipatory antiviral therapy for newborns exposed to herpes simplex infection. Ped Infect Dis 3:193-195.
Prober, CG, PA Hensleigh, FD Boucher, LL Yasukawa, DS Au, AM Arvin (1988) Use of routine viral cultures at delivery to identify neonates exposed to herpes simplex virus. N Engl J Med 318:887-891.
Prober CG, WM Sullender, LL Yasukawa, DS Au, AS Yeager, AM Arvin (1987) Low risk of herpes simplex virus infections in neonates exposed to the virus at the time of vaginal delivery to mothers with recurrent genital herpes simplex virus infections. N Engl J Med 316:240-244.
Redfield DC, DD Richman, S Albanil, MN Oxman, GM Wahl (1983) Detection of herpes simplex virus in clinical specimens by DNA hybridization. Diag Microbiol Infect Dis 1:117-128.
Sever JL (1983) New tissue culture-fluorescent method speeds detection of herpes simplex virus. JAMA 250:3045.
Stagno S, RJ Whitley (1985) Herpesvirus infections of pregnancy. Part II: herpes simplex virus and varicella-zoster virus infections. N Engl J Med 313: 1327-1330.
Sullivan-Bolyai JZ; HF Hull, C Wilson, L Corey (1983) Neonatal herpes simplex virus infection in King County, Washington JAMA 250:3059-3062.
Sullivan-Bolyai JZ, HF Huff, C Wilson, AL Smith, L Corey (1986) Presentation of neonatal herpes simplex virus infections: implications for a change in therapeutic strategy. Ped Infect Dis 5:309-314.
Thaw HH, VP Collins, UT Brunk (1984) Influence of oxygen tension, prooxidants and antioxidants on the formation of lipid peroxidation products (lipofusacin) in individual cultivated cells. Mech Age Develop 24:211-223
Warford AL RA LEvy, KA Kekrut (1984) Evaluation of a commercial enzyme-linked immunosorbent assay for detection of herpes simplex virus antigen. J Clin Microbiol. 20:490-493.
Warford AL RA Levy, KA Rekrut, E Steinberg (1986) Herpes simplex virus testing of an obstetric population with an antigen enzyme-linked immunosorbent assay. Am J Obstet Gynecol. 154:21-28.
Whitley RJ, AJ Nahmias, S Soong, GG Galasso, CL Fleming, CA Alford (1980) Vidarabine therapy of neonatal herpes simplex virus infection. Pediatrics 66:495-501.
Whitley RJ, A Arvin, L Corey, D Powell, S Plotkin, S Starr, C Alford, J Connor, AJ Nahmias, SJ Soone, The NIAID Collaborative Antiviral Study Group (1986) Vidarabine versus acyclovir therapy of neonatal herpes simplex virus infection. Ped Res 20:323A.
Whitley RJ, AJ Nahmias, AM Visintine, CL Fleming, CA Alford (1980) The natural history of herpes simplex virus infection of mother and newborn. Pediatrics 66:489-494.
Whitley RJ, C. Hutto (1985) Neonatal herpes simplex virus infections. Ped Rev 7:119-126.
Yeager AS, AM Arvin (1984) Reasons for the absence of a history of recurrent genital infections in mothers of neonates infected with herpes simplex virus. Pediatrics 73:188-193.
Zhao L, ML Landry, ES Balkovic, GD Hsiung (1987) Impact of cell culture sensitivity and virus concentration on rapid detection of herpes simplex virus by cytopathic effects and immunoperoxidase staining. J Clin Microbiol 25:1401-1405.

DISCUSSION

Peterson E (University of California Irvine Medical Center, Orange, CA):

Nice talk Ann. I'm just a little concerned about the false-positive spinal fluids in the shell vials from the neonates. Do you stain your shell vials with both monoclonals separately, HSV-1 and HSV-2?

Warford A:

Yes, we do.

Peterson E:

Were they positive on both?

Warford A:

No, they were positive on 1. But in all fairness to Syva, and they demonstrated to us, they took cells and they shocked them, taking them in an out of the incubator and refrigerator, aging them, and they showed us that you can get this phenomenon even in your negative control. So if this is our problem, it's a function of having commercial cells rather than the product.

Peterson E:

Ann, when you went back and took another look at those particular cover slips, the slide you showed was more of a dull yellow, nuclear fluorescence.

Warford A:

It didn't look that way to me. The majority of us felt that they were positive and looked like herpes. In fact, I think we sent a couple of them off to you for EM. What really convinced us that they were false positives was when we started to see this in our negative control vials later on down the road. We sent the specimens off to Syva, they could not get them to be positive. Then, of course we did long-term serologic studies on the infants and they never developed antibody. Not, at least, in the first 6-8-10 week period.

Sieger, Chuck (Viromed):

Nice talk Ann. One question I do have. Do you think it's a function of dexamethasone or perhaps the age of the culture that give you the best isolation rate in view of the article that was published by Tom Smith and now Curt here at the meeting.

Warford A:

I'm not saying that your cells are bad Chuck. I'm just saying that being dependent on commercial cells, I know there's nothing in life that's fair. I don't think it's fair to have studies out of Mayo Clinic or where Curt is now, up in Washington, where people get to make their own cells and have a hands-on knowledge of which cells they're using. How many hours and minutes old they are, what kind of split they've had, etc.. and really baby the cells along, and now do these studies and present their data. I feel that it's just not real world. I mean, this is life in the trenches. You get commercial cells in and of course, we prescreen them now. We learned our lesson. We look at them and we decide which ones look like hell and dump them. But, still, I don't think you're going to get the same data out of somebody who makes their own cells. I think this is a problem with Hsiung's data also. She says that primary rabbit kidney is exquisitely better. But if she had to take cells from three different companies every week, we tried the first three years in my lab, that's the kind of study I wanted to do. I wanted to find the definitive cell line to isolate herpes. We would do a year's study and not find any significant difference at the end of the year between one cell line and another cell line with the same patient specimen. I have to say that I really believe that fresh cells are going to make a much bigger difference than anything I can do. On the other hand, I have to look at what I can do to improve our laboratory diagnosis. I agree with you. I think if I can have really optimal quality cells, the dexamethasone probably might not make as big a difference as it does in our study.

Burnette G (Burnette Labs, Buena Park, CA):

The number of cases that you have studied is really impressive. Considering small laboratories, and actually a physician neonatologist, I would like you to comment on this line of thought. To take the amniotic fluid in a high risk case of pregnancy and do *in situ* hybridization using herpes-1 and herpes-2 DNA probes, and somehow if the probe is positive, then it's not worth doing the C-section, the fetus is already contaminated. If the probe turns out to be negative, obviously with a high risk patient, the idea would be to do the C-section to avoid contamination through the birth canal. In summary, the amniotic fluid cells seems to be a good source, not only to do probes for herpes, but also papilloma virus and other kinds of infection. I would like you to comment on that.

Warford A:

I'll comment, but this is just opinion. I have no data to substantiate my opinion. I can just say that it's going to be a whole new ball game when we have good probes and PCR, obviously. Right now, if we were to follow your recommendation, we need to realize that in the first place, there are a lot of women in prolonged labor that appear to develop their lesions at a later point in time. Secondly, to do this test takes several hours. If in fact the amniotic fluid is contaminated, which they've shown it's not always to be, it's been our experience at Kaiser that if there is a significant length of time between the time the amniotic fluid is contaminated and the women delivers, in the neighborhood of 2-4-6-8 hours, that you're going to have a negative outcome. So, you're going to need a very fast test if you do this. That's just a personal comment.

LABORATORY TESTS FOR HUMAN T-LYMPHOTROPIC VIRUS TYPE I

DANA GALLO

California Department of Health Service, Division of Laboratories, Viral and Rickettsial Disease Laboratory, 2151 Berkeley Way, Berkeley, CA 94704 (USA)

INTRODUCTION

In 1978, the first human retrovirus was isolated from the lymphocytes of a patient in the United States with a mature T-cell malignancy (Poiesz et al. 1980). This virus is now known as human T-lymphotropic virus type I (HTLV-I). In 1980, serologic tests revealed that almost 100% of sera collected from patients with adult T-cell leukemia (ATL) in Japan were positive for HTLV-I antibody (Kalyanaraman et al. 1982a; Robert-Guroff et al. 1982). Subsequently, other areas endemic for HTLV-I infection were found, including the Caribbean Islands (Catovsky et al. 1982; Gallo et al. 1983a), the southeastern United States (Blayney et al. 1983), parts of South America (Gallo et al. 1983a; Robert-Guroff, 1984), parts of southern Italy (Manzari et al. 1984) and many locations in Africa (Hunsmann et al. 1983); Saxinger et al. 1984). HTLV-I infection in theses areas is associated with 50-70% of all lymphoid leukemias of adulthood.

A simian T-cell lymphotropic virus (STLV-I) possesses greater than 95% nucleotide sequence homology with HTLV-I and is particularly prevalent in African green monkeys (Hunsmann et al. 1983; Miyoshi et al. 1982). It has been postulated that HTLV-I originated in Africa and was brought to other endemic areas by commercial and slave trades dating back to the sixteenth century (Gallo et al. 1983b).

DISEASE

ATL cases in Japan are noted to have 1) onset in adulthood; 2) leukemic cells with T-cell properties; 3) leukemic cells with pleomorphic cellular morphology and nuclear irregularity; 4) hepatosplenomegaly and lymphadenopathy; and 5) hypercalcemia. The disease in Caribbean blacks has been shown to be the same as ATL in Japan (Sarngadharan et al. 1987). In the United States, HTLV-I associated lymphoma cases may more frequently have hypercalcemia and bony abnormalities than those found in Japan and the Caribbean. Different stages of the infection have been defined: 1) acute ATL - aggressive non-Hodgkin's lymphoma with frequent skin involvement, hypercalcemia, and advanced stage of presentation with median survival less than six months; 2) transition stage-lymphoma, maculopapular rash, nuclear irregularity, but no hypercalcemia (Ratner et al. 1987); 3) chronic ATL - white cell count above 10,000, cough, and skin disease; 4) smoldering ATL cutaneous involvement, and 0.5% to 3% circulating leukemic cells with highly convoluted nuclei (condition may remain stationary for many years but may transform into the acute form of the disease); and 5) healthy carrier state.

In acute, chronic, and smouldering ATL, monoclonal integration of HTLV-I proviral DNA is present, but in the healthy carrier there is no detectable proviral DNA. The quantity of proviral DNA in peripheral blood cells from smoldering ATL patients was found to be less than that from acute or chronic ATL patients. It has been suggested that detection of proviral DNA would be useful for diagnosis of ATL and for classification of T-cell malignancies (Yamaguchi et al. 1984). An intermediate stage has been described, where polyclonal integration of HTLV-I proviral DNA is found, accompanied by only a few atypical lymphoid cells in the peripheral blood along with various clinical complaints not associated with ATL but indicative of immunodeficiency (Yamaguchi et al. 1988a).

Tropical spastic paraparesis (TSP), a spinal cord disease prevalent in Jamaica, Colombia, Martinique, Trinidad, South India, and South and West Africa, has been associated with HTLV-I. Various antibody studies have revealed HTLV-I seropositivity rates between 60 and 70% in these patients, which is much higher than in the normal population (Bartholomew et al. 1986; Gessain et al. 1985; Rodgers-Johnson et al. 1985), but the antibody correlation with TSP is not as high as it is with ATL. HTLV-I antibody is found in the cerebrospinal fluid (CSF) of TSP patients with HTLV-I antibody in their sera. HTLV-I associated myelopathy (HAM) described in Japan (Osame et al. 1986) appears to be clinically identical to TSP (Roman and Osame, 1988).

TRANSMISSION

Before screening of blood donors for HTLV-I antibody was begun, it was estimated that in Kyushu, Japan, an endemic area, 40,000 units of HTLV-I positive blood were transfused annually, with a seroconversion rate of 62% in seronegative individuals transfused with positive blood. No ATL has been reported in these patients, but the disease has a long incubation period (Sandler, 1986). Transmission of HTLV-I by blood transfusion occurs via infected cells. Seroconversion has been documented in recipients of whole blood or blood components containing cells but not in recipients of fresh-frozen plasma from positive donors (Okochi et al. 1984).

Transmission from husband to wife apparently occurs more readily than from wife to husband. In one Japanese study, the chance that a seronegative wife of a seropositive husband would convert to positive was estimated to be 60.8% but the chance that a negative husband of a positive wife would convert was only 0.4% (Kajiyama et al. 1986). An HTLV-I serosurvey of 697 Dutch homosexuals revealed five positive individuals. The human immunodeficiency virus type-1 (HIV-1) positivity rate in this group was 34%. HTLV-I was imported into the Netherland homosexual population after HIV-1, and it was concluded that HTLV-I was spreading more slowly than HIV-1 (Goudsmit et al. 1987). In Trinidad, 15% of sera from homosexual men were positive for HTLV-I antibody, but 40% of the sera were positive for HIV-1 antibody, suggesting that HIV-1 is more efficiently transmitted in this group, in view of the fact that this infection has been more recently introduced to Trinidad that HTLV-I (Bartholomew et al. 1987a, 1987b). In a study of homosexual men in Trinidad, more patients progressed to AIDS in a given time period if infected with both HIV-1 and HTLV-I than if infected only with HIV-1, and it was proposed that the HTLV-I transactivating transcriptional gene (tax) may accelerate the cytopathic effects associated with HIV-1 replication in cells (Bartholomew et al. 1987a, 1987b).

Transmission from mother to infant is relatively common. In contrast, none of 82 children of a positive father and a negative mother was HTLV-I positive. It has been shown that mother-to-child transmission is established by 24 months after birth. Intrauterine infection has rarely been documented, and virtually all transmission from mother to infant appears to occur through infected breast milk. It is estimated that approximately 200 children per year become infected in the Nagasaki Prefecture (population 1.6 million), an HTLV-I endemic area in southern Japan (Sugiyama et al. 1986).

In one study, mothers with HTLV-I antigen-bearing cells possessed HTLV-I antibody titers of 1:256,000 or greater as measured by the gel particle agglutination test. Sixty percent of the children of these mothers were also antigen carriers. No carrier children were found from mothers with antibody titers less than 1:4,000. In an intervention program, in which carrier mothers refrained from breast feeding, none of 47 children followed for at least 12 months developed HTLV-I antibody or antigen-positive cells. The authors stated that refraining from breastfeeding by carrier mothers is the only practical and currently available measure to break the endemic cycle of HTLV-I infection (Hino et al. 1987).

It is assumed that there are about one million healthy carriers in Japan (Kobayashi et al. 1988). The most prevalent antibody-positive group other than ATL patients is family members of ATL patients (Robert-Guroff et al. 1983). In a survey for HTLV-I antibody prevalence in Georgia and Florida, four black patients with lymphoma were HTLV-I positive, as were five family members. A small survey of 95 samples from healthy blacks in Georgia revealed an antibody prevalence of 2.1% (Blayney et al. 1983).

PROTEINS

The envelope precursor of HTLV-I, gp62 or gp68, depending on the cell line, is processed into gp46 and p20, an the gag precursor p53 is processed to p15, p19, and p24 (Hattori et al. 1984). The tax gene, p40X, is critical for transformation (Lee et al. 1984). Activation of this gene leads to subsequent transformation of the infected cells, and it also activates the genes for both interleukin 2 (IL2) and the IL2 receptor. Although IL2 production is only observed in the initial period after activation, the expression of the IL2 receptor remains. The reverse transcriptase protein has a molecular weight of 95 kDa.

Human T-lymphotropic virus type II (HTLV-II), isolated from the cells of a patient with a T-cell variant of hairy cell leukemia (Kalyanaraman et al. 1982b), has an overall nucleotide sequence homology of about 40% with HTLV-I. The precursor envelope protein, gp67, is processed to gp52 and the transmembrane gp21. The nonglycosylated precursor protein is p53. The tax gene is p38X. The most immunogenic gag antigen, p24, is highly cross-reactive with HTLV-I, as is gp52.

ANTIGEN DETECTION

Direct examination of freshly biopsied tissues has not been found to be very efficient for detection of HTLV-I or -II. It appears that fresh tumor tissues do not express mature virus particles or even virus structural antigens. Nucleic acid hybridization by Southern blotting of tumor tissue has been suggested (Essex, 1985). The polymerase chain reaction method has been shown to be more sensitive than Southern blotting for HTLV-I detection in peripheral blood lymphocytes (Bhagavati et al. 1988).

ISOLATION

The discovery of T-cell growth factor, IL2 (Morgan et al. 1976) made possible the long-term culture of T-lymphocytes and the subsequent isolation of HTLV-I an HTLV-II. When a patient's T-cells are infected with one of these agents, they sometimes are transformed and grow indefinitely *in vitro*. These T-cell lines are predominantly OKT4+, have partial or complete independence of exogenous IL2, and express high levels of IL2 receptors. Studies of isolates from various countries have shown that the majority are identical.

Most HTLV-I or -II infected T-cells do not become transformed. Co-cultivation of the patient's T-cells with cells from cord blood or bone marrow has been more successful than with peripheral blood lymphocytes, and cell-free transmission to these cells has rarely been successful. The co-cultivated cells are propagated in the presence of IL2 and expression of virus can be detected by the reverse transcriptase (RT) test or by electron microscopy (EM). Both these techniques are relatively insensitive, however. The RT method is based upon detection of the enzyme in viral particles present in culture fluids. Unlike HIV-1, HTLV-I and -II cultures do not usually produce large quantities of viral particles in the supernatant fluid. EM is a very time-consuming procedure and requires approximately 10^6 viral particles per ml for reliable detection.

A short-term culture method for detecting HTLV-I antigen in patient's cells has been used in Japan to determine carrier state in HTLV-I antibody positive individuals (Sugiyama et al. 1986). Cells were separated by Ficoll, the T-lymphocytes were isolated by rosette formation with neuraminidase-treated sheep red blood cells, and the cells were cultured in 96-well plates with IL2 for up to two weeks. Antigen expressing cells were detected by immunofluorescence (IF) using human serum or a monoclonal antibody against p24 of HTLV-I. In our experience, specific fluorescence in lymphocyte cultures is very difficult to determine. Nonspecific binding of the detection serum and the fluorescein conjugate to cells containing Fc receptors for IgG, and binding of IgG-producing lymphocytes in the cultures with the antispecies globulin conjugate, often mask specific reactions.

SEROLOGIC TESTS

Several serologic methods have been used for detection of antibody to HTLV-I.

Membrane Fluorescence

Live, infected cells were incubated with the patient's serum, washed by centrifugation, and stained with a fluorescein-conjugated anti-human globulin conjugate. Cytoplasmic membrane fluorescence constituted a positive reaction (Essex et al. 1983).

Fixed-Cell Immunofluorescence (IF)

The serum was incubated on HTLV-I infected cells which had been dried on the slide and fixed in acetone. The antigen-antibody reaction was detected by an anti-human globulin conjugate (Hinuma et al. 1981).

Enzyme Immunoassay (EIA)

Both laboratory and commercially prepared EIA tests employing whole virus lysates and p24 antigen have been described. Some of the methods included an inhibition test employing HTLV-I or HTLV-II infected cell lysates for specificity determinations (Robert-Guroff et al. 1983, 1986).

Solid-Phase Radioimmunoassay

Microtiter plates were coated with disrupted HTLV-I or with a negative control, a 1:10 dilution of serum was incubated with both antigens, the plates were washed, and rabbit anti-human IgG was added. After incubation and washing, iodinated protein A was added, the plates were washed, and the wells were counted in a gamma counter. Specimens were considered positive if binding to HTLV-I could be decreased by addition of extracts of HTLV-I producing cells but not with uninfected cells (Blayney et al. 1983).

Radioimmune Precipitation (RIP)

Purified p24 from the HTLV-I cell line, MT2, was labeled with ^{125}I and incubated with the patient's serum overnight. The antigen-antibody complex was collected on *Staphylococcus aureus* cells, and the radioactivity in the precipitate was determined (Jason et al. 1985).

Radioimmunoasssay-Polyacrylamide Gel Electrophoresis (RIP-PAGE)

Infected cells were grown in the presence of ^{35}S-cysteine or methionine, the antigen was disrupted, lysed, and the suspension was clarified by ultracentrifugation. The proteins were incubated with the patient's serum and the complex was collected with protein A-Sepharose. After washing, the precipitate was eluted and the proteins were separated by PAGE. The gel was incubated on x-ray film and the radioactive bands were determined (Yamamoto et al. 1983).

Western Blot

Partially purified virus was disrupted and the proteins separated by PAGE and transferred to nitrocellulose paper, which was cut into strips. Each strip was incubated with the serum, and the reaction with the various proteins was detected by EIA (Tsang et al. 1986).

Gel Particle Agglutination (PA)

Particles coated with partially purified viral antigens were mixed with an equal volume of serially diluted serum and incubated 2 hr at room temperature. An agglutination titer of 1:16 or higher was considered positive (Ikeda et al. 1984).

One study compared PA, EIA, and IF for measurement of HTLV-I antibody in 2,316 serum samples from Japanese blood donors. Discrepant sera were tested by RIP-PAGE. Twenty-six sera were positive by all three methods, two were positive by PA and IF and also by RIP. Eleven were positive only by PA, and three of these were RIP positive. Eight were positive only by EIA and none of these were reactive by RIP. Thus, PA was apparently more sensitive than IF and IF was more sensitive than EIA for detecting antibody. One of the PA positive/IF negative samples was shown to possess IgM antibody to HTLV-I. Three were eight nonspecific reactions by PA and by EIA and none by IF, as determined by RIP. The authors considered RIP the most sensitive and specific method for the three for detecting HTLV-I antibody (Kobayashi et al. 1988).

No difference was found in antibody response between ATL patients and healthy carriers using IF, RIP-PAGE and a RIP method employing ^{125}I-labeled gp68. The ability of the sera to precipitate gp68 correlated with IF titer, and healthy carrier sera, which had higher IF titers than ATL sera, usually precipitated gp68 more efficiently than did ATL sera. Three sera from ATL patients were negative in the RIP assay but positive by IF. By RIP-PAGE, gp68, p53, p24, p19, and p15 could be detected. If IF titers were less than 1:40, only gp68 was present, but sera with higher IF titers also precipitated the core proteins (Yamamoto et al. 1983).

These serologic methods have also been used to screen sera from individuals with various diseases of unknown origin, and the results have often been difficult to interpret. In 1983, it was reported that using the membrane fluorescence test, HTLV-I antibody was found in sera of approximately 25% of the AIDS patients tested (Essex et al. 1983). In 1985, 30% of the sera from multiple sclerosis (MS) patients were found to react with p24 antigen of HTLV-I, and it was suggested that HTLV-I or a relative may cause MS (Koprowski et al. 1985). A more recent study found no correlation of HTLV-I antibody with MS or with AIDS (Karpas et al. 1987). High titers of IgM antibody to HTLV-I were found in sera from patients with systemic lupus erythematosus (Phillips et al. 1986). Another study indicated that these reactions were probably nonspecific (Yamaguchi et al. 1988b).

Sera from 11 Japanese patients with hairy cell leukemia and an anti-hairy cell serum were reacted by Western blot with HTLV-I and -II. The patients' sera reacted with structural proteins of HTLV-I and/or HTLV-II but the anti-hairy cell serum reacted only with HTLV-II. The authors suggested that because these reactions were weak, they may be due to infection by an unrecognized retrovirus capable of cross-reaction with HTLV-I and -II. All patients but one were from outside ATL endemic areas (Katayama et al. 1987).

There was a report that 6 of 12 sera from patients with a rare lymphoproliferative disease of granular lymphocytes (LDGL) reacted by Western blot with p19 or p24 (Starkebaum et al. 198. Another group (Pandolfi et al. 1987) studied sera from 27 patients with LDGL employing two commercial EIA tests, one commercial Western blot test, and a Western blot they developed. In seven cases, antibodies were found with at least one test. Three sera were positive by one commercial EIA. A serum drawn in 1982 from one patient was positive but his 1987 serum was negative. Only one serum was positive by the other commercial EIA method. None of the five sera tested positive by the commercial Western blot, but all reacted with p19 or, in one case, with both p19 and p24, when the authors' blot method was employed. They concluded that serological testing for HTLV-I is not yet well standardized and results of individual tests should therefore be interpreted with great caution.

It is apparent that each of the serologic tests has some drawbacks. The membrane fluorescence test requires more cells than the IF method and is more difficult to read. There are many dead cells in lymphocyte suspension cultures and dead cells stain nonspecifically in the membrane immunofluorescence test. Only cytoplasmic membrane staining is scored as specific. In addition, since these cells are suspended in the mounting medium, they tend to move, which makes reading difficult.

Many of the transformed HTLV-I cell lines are 100% infected and, thus, there is no comparable uninfected cell line for use as a nonspecific control in the IF methods. One line, MT-1, which posses only 2-5% infected cells, has been used for IF, but this line must be screened before use, since any drop in the number of infected cells would make reading laborious. Alternate methods described include: 1) screening out negative sera on a 100% infected line and then checking the specificity of the reactors on MT-1 (Okochi et al. 1984), and 2) mixing infected cells with cells from an unrelated, uninfected T-cell line. Sera that react nonspecifically can be absorbed with an uninfected cell pellet and retested (Gallo et al. 1988), but this method does not remove nonspecific staining in all instances.

The EIA method was reported to be more sensitive than the solid phase RIA (Blayney et al. 1983), but the EIA has problems with specificity. Confirmation tests employing competition methods improve the reliability of positive reactions, but these methods are not available with commercial kits. It has been suggested that EIA antigens of whole disrupted virus contain little p24, an antigen which is broadly cross-reactive with HTLV-II, and that EIA tests employing the p24 antigen are more sensitive, especially if the virus is an HTLV-I relative (Marx, 1987).

The IgG heavy chain position is around 50 kDa and cellular proteins tend to be sticky in PAGE at around 45 kDa. Nonspecific bands in these areas are common in the RIP-PAGE method, which may mask specific reactions. In low-titered sera, only the envelope protein gp68 may be precipitated. The radioactive label ^{35}S has a half-life of around 88 days, which necessitates frequent preparation and standardization of antigen, thus effectively limiting this method to the research laboratory.

Western blot antigens are usually prepared from concentrated and sucrose density gradient purified virus from cell culture supernatant fluids. It was found that gp46 was only loosely attached to the viral and cell surface and tended to be shed into the culture medium. After purification on density gradients, only small amounts of this protein were present. The envelope protein gp68 was found only on the cell surface and not at all on free viral particles in the culture medium (Schneider et al. 1984). Thus, HTLV-I Western blot antigens mainly consist of core proteins.

Sera from ATL patients often do not display as many band reactions as sera from healthy antibody-positive individuals (Hattori et al. 1985). A uniform set of criteria for interpretation of HTLV-I Western blot reactions have not yet been established, and reactions with core bands only may be nonspecific.

Over a four year period, starting in 1983, we tested 3,349 sera for HTLV-I antibody by the IF method (Gallo et al. 1988). The MT2 strain of HTLV-I and the Mo-T strain of HTLV-II were used for IF slide preparation. One part infected cells was mixed with 3 parts uninfected T-cells (H9), which served as a nonspecific (NS) control. A positive serum reacted with 25% of the cells. If a serum stained all the cells, it was absorbed with an H9 cell pellet to remove the NS and the test was repeated. Specimens that were NS after absorption were dropped from the study. Specimens reactive with HTLV-I were tested on HTLV-II slides. A serum was regarded as positive if it reacted with both antigens.

The specimens were from homosexual and bisexual men, encephalitis and meningitis cases, sera collected for an AIDS serosurvey from juvenile halls, prenatal and sexually transmitted disease (STD) clinics and from marriage license applicants, and specimens from blood donors and from intravenous (IV) drug abusers. Seven sera from the STD clinic, one from an encephalitis patient, one premarital specimen and 232 of 940 sera from IV drug abusers were positive.

One hundred ninety-eight of the IV drug abuser sera displayed diffuse, cytoplasmic staining and these specimens were also positive on the HTLV-II infected slides. Most sera reacted more brightly with the HTLV-II antigen. An additional thirty-four sera stained particulate cytoplasmic membrane antigen on the HTLV-I slides. Ten of these 34 did not react with HTLV-II. Three were 31 EIA negative/IF positive specimens. Nineteen of the discrepant sera displayed diffuse and 12 exhibited particulate staining. Twenty displayed O.D. readings greater than 70% of the EIA positive cutoff and 28 reacted in the HTLV-I Western blot, often with only one band, p24. There were no EIA positive/IF negative sera. The HTLV-II antigen appeared to be more sensitive than the HTLV-I antigen by Western blot, as well as by IF. Many sera which reacted with only one band by Western blot with HTLV-I reacted with multiple bands when the HTLV-II antigen was employed. One specimen was negative with HTLV-I but reactive with the HTLV-II antigen.

When the Western blot antigen was heated at 65^{0}C for 30 minutes to disrupt it, several sera reacted with only the p19 protein and two different commercial p19 mouse monoclonal antibodies reacted with the p24 protein. When the lysate was boiled, however, the specimens reacts with the p24 protein and the monoclonals reacted with the p19, as they should. Thus, we decided to use the boiling procedure. It may be important to know how the antigen was disrupted when interpreting HTLV-I Western blot results from various laboratories.

The 10 sera that reacted by IF with the HTLV-I antigen only displayed very low EIA ratios and were negative by RIP-PAGE. One reacted with p19 of HTLV-I by Western blot, one with both p19 and p24, and three reacted with p24 of HTLV-II only. These results were interpreted as inconclusive.

Sera which fluoresced brightly by IF, reacted with several bands by Western blot, and gave high EIA ratios, precipitated multiple proteins by RIP-PAGE, principally gp68, p40X, p33, p28, and p24 of HTLV-I, and gp52, p38X, p33, p28, and p24 of the HTLV-II antigen. Most IF positive/EIA negative sera reacted only with gp68 with HTLV-I and were either negative with HTLV-II or reacted with p33 only. The reaction with the envelope protein gp68 by RIP-PAGE was helpful in verifying the specificity of Western blot p24 only reactions.

We absorbed some of the positive sera with HTLV-I and -II infected cell pellets and repeated the IF test, in an attempt to determine if the reactions were more closely related to HTLV-I or to HTLV-II. Six different staining patterns were found: 1) diffuse staining that was reduced by HTLV-I absorption; 2) diffuse staining that was not; 3) particulate staining accompanied by HTLV-II reactions which was reduced by HTLV-I absorption; 4) particulate staining which was not reduced; 5) particulate staining of HTLV-I only that was reduced; and 6) that which was not. It would appear that some of these reactions may be due to some other, not recognized, retrovirus(s).

In this study, the IF test was more sensitive than the EIA method used, but any positive reaction should be verified by another test. It has been suggested that a serum must react with at least two different gene products by blot to be considered positive for HTLV-I antibody by this method (Dodd and Gallo, 1988). In our experience, many positives will be missed using this criterion. IF positive sera which reacted with p24 only with the HTLV-I antigen often reacted with multiple bands when the HTLV-II antigen was employed. These specimens also precipitated the gp68 envelope protein by RIP-PAGE.

At this time, we are using the IF, EIA, Western blot, and RIP-PAGE for HTLV-I antibody determinations. Sera are screened by the HTLV-I IF and the EIA methods. IF reactors are tested on the HTLV-II slides, and are scored as positive if they stain with both antigens. If the IF and EIA results do not agree, Western blot is performed. A Western blot reaction with p19 or p24 only, when accompanied by an IF positive result, is considered specific. The RIP-PAGE is performed on IF positive/Western blot negative sera. Specimens that react by IF with the HTLV-I antigen only are reported as inconclusive.

Virtually 100% of ATL sera are positive for HTLV-I antibody and HTLV-I has been isolated or detected in the lymphocytes of many of these patients. The significance of HTLV-I antibody in IV drug abusers has yet to be determined. Most of the studies on this group are being performed with commercially available EIA kits, and the cutoffs for these tests have been established using ATL sera. Whether these cutoffs are appropriate for use in screening IV drug abuser sera remains to be seen. It is difficult, if not impossible, to establish sensitivity and specificity values for serologic tests run on a group of individuals without disease. Seroprevalence studies are in progress in many areas of this country, and it is evident that a significant portion of IV drug abusers have been exposed to HTLV-I or HTLV-II, or to some other agent that shares antigens with HTLV-I and -II. Virus isolates must be obtained and characterized to determine if these positive serologic results are due to HTLV-I or to cross-reactions with some other, perhaps benign, agent(s).

REFERENCES

Bhagavati A, Bartholomew C, Cleghorn F, Charles W, Ratan P, Roberts L, Maharaz K, Jankey N, Daisley H, Hanchard B, Blattner W (1986) HTLV-I and tropical spastic paraparesis. Lancet 2:99-100.

Bartholomew C, Blattner W, Cleghorn F (1987a) Progression to AIDS in homosexual men co-infected with HIV and HTLV-I in Trinidad. Lancet 2:1469.

Bartholomew C, Saxinger WC, Clark JW, Gail M, Dudgeon A, Mahabir B, Hull-Drysdale B, Cleghorn F, Gallo RC, Blattner WA (1987b) Transmission of HTLV-I and HIV among homosexual men in Trinidad. JAMA 257:2604-2608.

Blayney DW, Blattner WA, Robert-Guroff M, Jaffe E, Fisher R, Bunn PA, Patton MG, Rarick HR, Gallo RC (1983) The human T-cell leukemia-lymphoma virus in the Southeastern United States. JAMA 250:1048-1052.

Catovsky D, Greaves MF, Rose M, Galton DAG, Goolden AWG, McCluskey DR, White JM, Lampert I, Bourikas G, Ireland R, Brownell AI, Bridges JM, Blattner WA, Gallo RC (1982) Adult T-cell lymphoma-leukemia in blacks from the West Indies. Lancet 1:639-642.

Dodd R, Gallo D (1988) Detection of HIV virus and viral antigen/testing for other retroviruses. In: Hausler WJ, Getchell JP (eds), Third Consensus Conference on HIV Testing, Association of State and Territorial Public Health Laboratory Directors, University of Iowa, Iowa City, Iowa, pp 11-13.

Essex M, McLane MF, Lee TH, Falk L, Howe CWS, Mullins JI, Cabradilla C, Francis DP (1983) Antibodies to cell membrane antigens associated with human T-cell leukemia virus in patients with AIDS. Science 220:859-862.

Essex M (1985) Human T-cell leukemia viruses. In: Lennette EH, Balows A, Hausler WJ Jr., Shadomy (eds) Manual of Clinical Microbiology, American Society for Microbiology, Washington, D.C., pp 840-844.

Gallo D, Hoffman MN, Cossen Ck, Diggs JL, Hurst JW, Penning LM (1988) Comparison of immunofluorescence, enzyme immunoassay, and Western blot methods for detection of antibody to human T-cell leukemia virus type I. J Clin Microbiol 26:1487-1491.

Gallo RC, Kalyanaraman VS, Sarngadharan MG, Sliski A, Vonderheid Ec, Maeda M, Nakao Y, Yamada K, Ito Y, Gutensohn N, Murphy S, Bunn PA, Jr., Catovsky D, Greaves MF, Blayney W, Jarrett WFH, zur Hausen H, Seliqmann M, Brouet JC, Haynes BF, Jeqasothy BV, Jaffe E, Cossman J, Broder S, Fisher RI, Golde DW, Robert-Guroff M (1983a) Association of the human type-C retrovirus with a subset of adult T-cell cancers. Cancer Res 43:3892-3899.

Gallo RC, Sliski A, Wong-Staal F (1983b) Origin of human T-cell leukemia-lymphoma virus. Lancet 2:962-963.

Gessain A, Barin F, Vernant JC, Maurs L, Gout O, Calender A, de The G (1985) Antibodies to human T-lymphotropic virus type I in patients with tropical spastic paraparesis. Lancet 2:407-410.

Goudsmit J, de Wolf F, van de Wiel B, Smit L, Bakker M, Albrecht-van Lent N, Coutinho RA (1987) Spread of human T-cell leukemia virus (HTLV-I) in the Dutch homosexual community. J Med Virol 23:115-121.

Hattori S, Kiyokawa T, Imagawa K-I, Shimuzu F, Hashimura E, Seiki M, Yoshida M (1984) Identification of gag and env gene products of human T-cell leukemia virus (HTLV). Virology 136:338-347.

Hattori T, Robert-Guroff M, Chosa T, Matsuoka M, Yamaguchi K, Ishii T, Gallo RC, Takatsuki K (1985) Natural antibodies in sera from Japanese individuals infected with HTLV-I do not recognize HTLV-III. Blood 66:645-647.

Hino S, Doi H, Yoshikuni H, Sugiyama H, Ishimaru T, Yamabe T, Tsuji Y, Miyamoto T (1987) HTLV-I carrier mothers with high-titer antibody are at high risk as a source of infection. Jpn J Canc Res (Gann) 78:1156-1158.

Hinuma Y, Nagata K, Hanaoka M, Nakai M, Matsumoto T, Kinoshita K, Shirakawa S, Miyoshi I (1981) Adult T-cell leukemia: Antigen in an ATL cell line and detection of antibodies to the antigen in human sera. Proc Natl Acad Sci USA 78:6476-6480.

Hunsmann G, Schneider J, Schmitt J, Yamamoto (1983) Detection of serum antibodies to adult T-cell leukemia virus and non-human primates and in people from Africa. Int J Canc 32:329-332.

Ikeda M, Fujino R, Matsui T, Yoshida T, Komoda H, Imai J (1984) A new agglutination test for serum antibodies to adult T-cell leukemia virus. Gann 75:845-848.

Jason J, McDougal JS, Cabradilla C, Kalyanaraman VS, Evatt BL (1985) Human T-cell leukemia virus (HTLV-I) p24 antibody in New York City blood product recipients. Am J Hematol 20:129-137.

Kajiyama W, Kashiwaji S, Ikematsu H, Hayashi J, Nomura H, Okochi K (1986) Intrafamilial transmission of adult T-cell leukemia virus. J Infect Dis 154:852-857.

Kalyanaraman VS, Sarngadharan MG, Nakao Y, Ito Y, Aoki T, Gallo RC (1982a) Natural antibodies to the structural core protein (p24) of the human T-cell leukemia (lymphoma) retrovirus found in sera of leukemia patients in Japan. Proc Natl Acad Sci USA 79:1653:1657.

Kalyanaraman VS, Sarngadharan MG, Robert-Guroff M, Miyoshi I, Blayney D, Golde D, Gallo RC (1982b) A new subtype of human T-cell leukemia virus (human T-cell leukemia virus-II) associated with a T-cell variant of hairy cell leukemia. Science 218:571-573.

Karpas A, Malik K, Lida J (1987) Studies of human retroviruses in relation to adult T-cell leukemia, acquired immune deficiency syndrome,and multiple sclerosis. Arch Virol 95:237-249.
Katayama I, Maruyama K, Fukushima T, Nishibe Y, Kayano H (1987) Cross-reacting antibodies to human T-cell Leukemia virus-I and -II in Japanese patients with hairy cell leukemia. Leukemia 1:401-404.
Kobayashi S, Yoshida T, Hiroshige Y, Matsui T, Yamamoto N (1988) Comparative studies of commercially available particle agglutination assay and enzyme-lined immunosorbent assay for screening of human T-cell leukemia virus type I antibodies in blood donors. J Clin Microbiol 26:308-312.
Koprowski H, DeFreitas EC, Harper ME (1985) Multiple sclerosis and human T-cell lymphotropic retroviruses. Nature 318:154-160.
Lee TH, Coligan JE, Sodroski JA, Haseltine WA, Salahuddin SZ, Wong-Staal F, Gallo RC, Essex M (1984) Antigens encoded by the 3'-terminal region of human T-cell leukemia virus: Evidence for a functional gene. Science 220:57-61.
Manzari V, Fazio VM, Marinotti S, Gradilone A, Collalti E, Fattorossi A, Stoppacciaro A, Verani P, Frati L (1984) Human T-cell leukemia/lymphoma virus (HTLV-I) DNA: Detection in Italy in a lymphoma and in a Kaposi sarcoma patient. Int J Canc 34:891-892.
Marx JL (1987) Leukemia virus linked to nerve disease. Science 236:1059-1061.
Miyoshi I, Yoshimoto S, Fujishita M, Hirokuni T, Kubonishi I, Niiya K, Minezawa M, (1982) Natural adult T-cell leukemia virus infection in Japanese monkeys. Lancet 2:658-659.
Morgan D, Ruscetti FW, Gallo RC (1976) Selective *in vitro* growth in T-lymphocytes from normal human bone marrow. Science 193:1007-1008.
Okochi K, Sato H, Hinuma Y (1984) A retrospective study on transmission of adult T-cell leukemia virus by blood transfusion: Seroconversation in recipients. Vox Sang 46:245-253.
Osame M, Usuku K, Izumo S, Ijichi N, Amitani H, Igata A, Matsumoto M, Tara M (1986) HTLV-I associated myelopathy, a new clinical entity. Lancet 1:1031-1032.
Pandolfi F, Shriver K, Scarselli E, Fiorelli V, Semenzato G, Zambello R, Chisesi T, de Rossi G, Strong DM (1987) HTLV-I antibodies and lymphoproliferative disease of granular lymphocytes. Lancet 2:1527.
Phillips PE, Johnston SSL, Runge LA, Moore JL, Poiesz BJ (1986) High IgM antibody to human T-lymphotropic virus type I in systemic lupus erythematosus. J Clin Immunol 6:234-241.
Poiesz BJ, Ruscetti FW, Gazsdar AF, Bunn PA, Minna JD, Gallo RC (1980) Detection and isolation of type-C retrovirus from fresh and cultured lymphocytes of a patient with cutaneous T-cell lymphoma. Proc Natl Acad Sci USA 77:7415-7419.
Ratner L, Griffith RC, Marselle L, Hoh M, Wong-Staal F, Saxinger C (1987) A lymphoproliferative disorder caused by human T-lymphotropic virus type I. Am J Med 83:953-958.
Robert-Guroff M, (1984) Human T-cell leukemia virus in immigrants to the United Kingdom, In: Gallo RC, Essex ME, Gross L (eds) Human T-Cell Leukemia/Lymphoma Virus. Cold Spring Harbor Laboratory, New York, pp 297-306.
Robert-Guroff M, Nakao Y, Notake K, Ito Y, Slishi A, Gallo RC (1982) Natural antibodies to human retrovirus HTLV in a cluster of Japanese patients with adult T-cell leukemia. Science 215:975-978.
Robert-Guroff M, Kalyanaraman VS, Blattner WA, Popovic M, Sarngadharan MG, Maeda M, Blayney D, Catovsky D, Bunn PA, Shibata A, Nakao Y, Ito Y, Aoki T, Gallo RC (1983) Evidence for human T-cell lymphoma-leukemia virus infection of family members of human T-cell lymphoma-leukemia virus positive T-cell leukemia-lymphoma patients. J Exp Med 157:248-258.
Robert-Guroff M, Weiss SH, Giron JA, Jennings AM, Ginzberg HM, Margolis IB, Blattner WA, Gallo RC (1986) Prevalence of antibodies to HTLV-I, -II, and -III in intravenous drug abusers from an AIDS endemic region. JAMA 255:3133-3137.
Rodgers-Johnson P, Gajdusek DC, Morgan O, Zaninovic V, Sarin PS, Graham DS (1985) HTLV-I and HTLV-III antibodies and tropical spastic paraparesis. Lancet 2:1247-1248.

Roman GC, Osame M (1988) Identity of HTLV-I-associated tropical spastic paraparesis and HTLV-I-associated myelopathy. Lancet 1:651.
Sandler SG, (1986) HTLV-I and -II: New risks for recipients of blood transfusions? JAMA 256:2245-2246.
Sandler SG (1987) HTLV-I and other retroviruses. In: Moore SB (ed) Transfusion-Transmitted Viral Diseases. American Association of Blood Banks, Arlington, Virginia, pp 19-35.
Sarngadharan MG, Orndorff SR, Markham PD (1987) Human T-lymphotropic retroviruses. In: de la Maza L, Peterson EM (ed), Medical Virology VI, Elsevier, pp 301-330.
Saxinger WC, Blattner WA, Levine PH, Clark J, Biggar R, Hoh M, Moghissi J, Jacobs P, Wilson L, Jacobson R, Crookes R, Strong M, Ansari AA, Dean AG, Nkrumah FK, Mourali N, Gallo RC (1984) Human T-cell leukemia virus (HTLV-I) antibodies in Africa. Science 225:1473-1476.
Schneider J, Yamamoto N, Hinuma Y, Hunsmann G (1984) Sera from adult T-cell leukemia patients react with envelope and core polypeptides of adult T-cell leukemia virus. Virology 132:1-11.
Starkebaum G, Kalyanaraman VS, Kidd PG, Loughan TP, Jr., Kadin ME, Singer JW, Ruscetti FW (1987) Serum reactivity to human T-cell leukemia/lymphoma virus type I proteins in patients with large granular lymphocytic leukemia. Lancet 1:596-598.
Sugiyama H, Doi H, Yamagushi K, Tsuji Y, Miyamoto T, Hino S (1986) Significance of postnatal mother-to-child transmission of human T-lymphotropic virus type I on the development of adult T-cell leukemia/lymphoma. J Med Virol 20:253-260.
Tsang VC, Hancock M, Wilson DF, Palmer DF, Whaley S, McDougal JS, Kennedy S (1986) Enzyme-linked immunoelectrotransfer blot technique (Western blot) for human T-lymphotropic virus type III/lymphadenopathy-associated virus (HTLV-III/LAV) antibodies. U.S. Department of Health and Human Services, Centers for Disease Control, Atlanta, Georgia, Immunology Series No. 15.
Yamaguchi K, Seiki M, Yoshida M, Nishimura H, Kawano F, Takatsuki K (1984) The detection of human T-cell leukemia virus proviral DNA and its application for classification and diagnosis of T-cell malignancy. Blood 63:1235-1240.
Yamaguchi K, Kiyokawa T, Nakada K, Yul LS, Asou N, Ishii T, Sanada I, Seiki M, Yoshida M, Matutes E, Catovsky D, Takatsuki K (1988a) Polyclonal integration of HTLV-I proviral DNA in lymphocytes from HTLV-I seropositive individuals: An intermediate state between the healthy carrier state and smouldering ATL. Brit J Hematol 68:169-174.
Yamaguchi K, Matutes E, Kiyokawa T, Nishimura Y, Ishii T, Takatsuki K, Catovsky D (1988b) Comparison of immunoperoxidase staining with indirect immunofluorescence, ELISA, and Western blotting assays for detecting anti-HTLV-I antibodies in systemic lupus erythematosus. J Clin Pathol 41:57-61.
Yamamoto N, Schneider J, Koyanagi Y, Hinuma Y, Hunsmann G (1983) Adult T-cell leukemia (ATL) virus-specific antibodies in ATL patients and healthy virus carriers. Int J Canc 32:281-287.

DISCUSSION

Lennette EH (Virion [U.S.], Inc., California Public Health Foundation, Berkeley, CA):

Dana, you tabulated geographic areas in which endemic foci of HTLV-1 infection have been found. Dr. Gold presented the same information on the map. Seroepidemiologic studies are only as good as the tools that are employed, and Dr. Gold mentioned the difficulties he had in interpreting the results of his tests. You mentioned all the difficulties including the lack of a good confirmatory test. So, the point I'm trying to make here is that many of the studies that have been reported on endemic foci have been based on the use of one test, usually ELISA, sometimes counter radioimmunoassays, seldom both. So how valid are these prevalence data? I would like to ask an epidemiologist in the group to respond. I'm not saying that it doesn't occur in these areas, but what is the true prevalence rate when you use only one test without any confirmation?

Gallo D:

I don't think that anyone can answer that right now because, as you just pointed out, we're still trying to nail down the good laboratory tests. In 1980 when they tested the ATL sera from Japan, virtually 100% were reported positive for HTLV-I antibody. Last week at the HTLV-I meeting at the FDA, it was reported that 72% of ATL patients are HTLV-I antibody positive. I think what's happening is that people are refining their tests and their confirmation methods and some of these original positives are dropping out. The figures from a few years ago probably have to be revised. The problem is it depends on what EIA test you use for screening. Some are more sensitive than others.

Lennette E:

The Japanese are great on the gelatin agglutination.

Gallo D:

It's fast.

Lennette E:

It's fast, and what is the specificity and sensitivity?

Gallo D:

Sensitivity is great, but specificity is poor.

Lennette E:

That falls in the same category as the latex test.

Burnette G (Burnette Labs, Buena Park, CA):

I'm going to make two comments. Number one, the Italian study of Dr. Purcella showed that in HIV positive drug abusers who are heterosexuals, their heterosexual partners were all negative when tested by serology. However, in their white cells, when tested with Dr. Gallo's HIV DNA probes, all the heterosexual partners had HIV in their lymphocytes, silent, without any production of antigens. Now, with the advent of the polymerase chain reaction, we will be using DNA probes for virus detection in the future. My second comment is that the Western blot can be very frustrating. In our experience, a Chiron kit that uses genetic engineered antigens, and uses the four most common antigenic components has brought a lot of clarification. Now we have only one core, one envelope and two surface antigens which are a lot easier to interpret than a huge number of bands. These are my comments and I would like you to give us your impression.

Gallo D:

I'm very impressed with PCR and it also scares me to death. When you have such a sensitive tool and no other diagnostic test to back it up, the results are difficult to interpret. We, in public health, like to have two diagnostic tests agree, especially in retrovirus serology. As for your comment about Western blot, we try to avoid doing Western blot because of the p24 only problem. We do EIA and immunofluorescence, and if those two agree, we're done. We perform Western blot only on discrepant sera. We do the same for HTLV-I.

Lennette E:

I propose a comment on PCR. There was a recent report in the Journal of Blood, I believe, on a patient who was thought to have Hodgkin's disease. The serologic tests to HTLV-1 were negative, they couldn't isolate the virus, but they did use the PCR and got a diagnosis. So, it is applicable. It's going to be a useful tool.

Gallo D:

The detection methods for HTLV-I isolation are frustrating. The RT is low, you can't use EM, and FA doesn't work for us, so we are sending cells fro our HTLV-I isolation attempts to John Sninski at Cetus for PCR. If his lab says the specimen is positive, we will hold the culture for months, looking for transformation. Without PCR we might not even know the virus was there.

THE ANALYSIS OF IMMUNITY TO SPECIFIC PROTEINS OF HERPES SIMPLEX AND VARICELLA ZOSTER VIRUS AND ITS CLINICAL APPLICATIONS

ANN M. ARVIN, WAYNE M. SULLENDER, FRANCOIS BOUCHER, PAMELA S. DIAZ, PAUL A. HENSLEIGH AND CHARLES G. PROBER

Infectious Disease Division, Department of Pediatrics and Department of Obstetrics and Gynecology, Stanford University School of Medicine, Stanford, California, USA

INTRODUCTION

Herpes viral glycoproteins are considered likely to be important targets of virus-specific immunity because they are expressed on the membranes of infected cells and because monoclonal antibodies to the surface glycoproteins of human herpes viruses often have neutralizing activity (Corey and Spear, 1986; Friedrichs and Grose, 1984; Keller et al. 1984; Rasmussen et al. 1985). Recent studies in viral immunology also demonstrate a host response to the nonglycosylated, internal or nucleoproteins of several viruses (Townsend et al. 1984; Yewdell et al. 1986). The investigation of immunity to proteins of both herpes simplex virus (HSV) and varicella-zoster virus (VZV) has progressed over the past few years in parallel with the expansion of knowledge about the molecular virology of these herpes viruses. These advances have allowed the mapping of gene sequences for many of the major viral proteins and their expression in plasmids and by transfection of mammalian cells (Davison and Scott, 1986; Pachl et al. 1987). Similarities between HSV and VZV have been defined at the molecular level with the identification of homologies between several gene sequences and, in the case of the HSV gB and VZV gp II, the demonstration of antigenic crossreactivity.

The initial studies of the humoral immune response to HSV and VZV proteins were based upon the immunoprecipitation of radiolabelled viral antigens from a detergent solubilized virion preparation or an extract of infected cells (Grose and Friedrichs, 1982; Zweerink and Corey, 1982).

This method successfully detects antibodies to viral proteins that are present in sufficient quantities in the antigen preparation and that have strong affinity for antigen binding. The antigen-antibody complexes are formed under non-denaturing conditions and the polyacrylamide gel electrophoresis of the bound complexes can be carried out under either reducing or non-reducing conditions. In the immunoblot, or Western blot method, viral proteins are separated by polyacrylamide gel electrophoresis and transferred to nitrocellulose, followed by direct reaction with the serum. This procedure enhances the detection of antibody binding to proteins that are present in lesser quantities but the viral proteins are denatured under the conditions required for the assay. Since protein-antibody interactions that are dependent upon conformational epitopes are eliminated by denaturing, only antibodies that react with linear epitopes of the viral proteins are detected by immunoblot. The application of immunoprecipitation or immunoblot methods to the analysis of immunoglobulin production against VZV proteins is further complicated by the difficulty of obtaining a cell free preparation of this virus. Human immunoglobulins bind non-specifically to cellular proteins which may have molecular weights that overlap with those of the viral proteins on the gel (Palumbo et al. 1984). Some HSV or VZV proteins and their precursors also have similar molecular weights which produce ambiguities in interpreting the patterns of specific antigen-antibody reactivity. If the reaction is not conformationally dependent, this problem can be addressed in the immunoblot procedure by first making a purified protein for transfer to nitrocellulose using immunoaffinity chromatography with a monoclonal antibody to the specific protein. The availability of monoclonal antibodies has provided another alternative method for the detection of antibodies to viral proteins in which the protein of interest is "captured" from an infected cell extract and bound to the solid phase by the monoclonal antibody (Cremer et al. 1985). This approach avoids the denaturing of the viral protein and allows the titration of the antibody response to a specific protein. Although less well developed than methods for the detection of immunoglobulins that react with HSV and VZV proteins, procedures for determining cell-mediated immunity to specific viral proteins have been established and are being applied to the study of the host response to these pathogens. Examples of these methods include the investigation of T-lymphocyte proliferation to purified proteins and synthetic peptides *in vitro* and the assessment of T-lymphocyte mediated cytotoxicity against specific proteins by infecting target cells with vaccinia recombinants that express a herpes viral protein (Zarling et al. 1986).

THE INVESTIGATION OF IMMUNITY TO HSV-2 GLYCOPROTEIN G (gG)

Background

Approximately ninety percent of symptomatic primary genital HSV infections and ninety-nine percent of recurrent genital HSV infections are caused by HSV-2 (Corey et al. 1983). The prevalence of asymptomatic HSV-2 infections is difficult to determine because the extensive antigenic homology between HSV-1 and HSV-2 leads to crossreactivity in serologic assays. The recent cloning of the genes of HSV-2 gG indicates that HSV-2 gG is much larger than its HSV-1 counterpart (McGeogh, et al. 1987) and type-specific epitopes of the HSV-2 gG polypeptide are identifiable using murine monoclonal antibodies (Roizman et al. 1984). The separation of HSV-2 gG from an infected cell extract by immunoaffinity chromatography with these monoclonal antibodies provided purified protein for use in an immunodot serologic assay for HSV-2 antibodies (Lee et al. 1985).

Description of Methods

In our method for detecting antibodies to HSV-2 gG (Sullender et al. 1988), a monoclonal antibody to HSV-2 gG was purified from ascites by fractionation with a staphylococcus A-Sepharose column, diluted 1:2,000 in phosphate buffered saline (PBS), added in aliquots to 100 µl/well to 96-well, flat bottom microtiter plates and incubated overnight at 4^0C. The plates were washed with PBS and coated with 1% normal mouse sera. A detergent solubilized extract of Hep-2 cells infected with HSV-2 or an uninfected cell control was added at a dilution with PBS containing 0.05% Tween 20, incubated for 1 h with 1% bovine serum albumin (BSA), washed and dried. Serial four-fold dilutions of the test sera were added to duplicate wells, beginning at a 1:16 dilution. The plates were incubated for 1 h at room temperature and washed. Bound antibody was detected using the Vectastain ABC kit. Biotinylated goat antihuman IgG was added for 30 min at room temperature. After washing, avidin DH-biotinylated horseradish peroxidase H was added for 30 min at room temperature, followed by washing with 0.1 M citrate buffer. The substrate solution, orthophenylene diamine in citrate buffer, was added. The enzyme reaction was stopped after 5 min with 2N H_2SO_4. The plates were read at O.D. 490 nm on an automated enzyme linked immunosorbent assay (ELISA) reader. HSV-1 and HSV-2 pooled sera controls were included in each assay. The specific O.D. of the test serum was calculated by subtracting the mean O.D. of duplicate control wells from the mean O.D. of antigen wells. The test serum was considered positive for HSV-2 gG antibody if the O.D. was equal to or greater than the O.D. of the HSV-2 control in the same assay. The presence of antibodies to HSV-2 gG in sera that were positive by the ELISA method was confirmed by reacting the same sera against HSV-2 gG protein by immunoblot assay. Immunoaffinity purified HSV-2 gG was made by coupling the monoclonal antibody to cyanogen bromide activated Sepharose 4B and incubating with a sonicated and detergent solubilized extract of HSV-2 infected fibroblasts or with an uninfected cell control extract. The bound protein was eluted with 3M potassium thiocyanate and concentrated by vacuum dialysis. The immunoaffinity purified antigen was electrophoresed using a 9% sodium dodecyl sulfate polyacrylamide gel cross-linked with N,N methylene-bisacrylamide and was transferred to nitrocellulose using a Trans-Blot chamber at $4^{0}C$ for 16 h with a constant 30-V power source. After transfer, the nitrocellulose strips were rinsed in PBS, pH 7.5, and incubated in 0.05% Tween-20 in 10 mM dibasic sodium phosphate and 0.9% NaCl, pH 7.5, for 30 min. Adjacent strips with HSV-2 gG antigen or the uninfected cell control were incubated with the test serum at a 1:20 dilution for 1 h. Each strip was washed and stained by the immunoperoxidase method.

Clinical Investigations of the Antibody Response to HSV-2 gG and its Potential Applications in Diagnostic Virology

We have investigated the immune response to HSV-2 gG in a series of clinical studies of otherwise healthy adult women with or without a history of symptomatic genital herpes. Infants exposed to maternal HSV infection at delivery and infants with neonatal herpes have also been tested for IgG antibodies to HSV-2 gG. Our experience, in keeping with the observations of other investigators (Ashley et al. 1985; Bernstein et al. 1984; Lee et al. 1985), suggests several clinical circumstances in which the measurement of antibodies to HSV-2 gG provides useful information.

Serologic Confirmation in Patients with Genital Herpes

HSV-2 gG antibodies are detectable by the ELISA method in serum from individuals with prior episodes of symptomatic genital herpes proved by viral culture. HSV-2 gG were detectable with the ELISA method at titers >1:16 in 98 of 102 sera (96%) from women with known past HSV-2 infection (Sullender et al. 1988). The mean O.D. of 1.00 + 0.23 SD was significantly above the mean O.D. of the HSV-1 and HSV seronegative control sera ($p < 0.001$, t test) (Figure 1.). Specific to antibody binding to HSV-2 gG was not demonstrated with any HSV-1 sera, obtained from children or from adults with serologically proved primary HSV-1 infections, or in sera that were negative by neutralization assay against HSV-1 and HSV-2.

The presence of antibodies to HSV-2 gG was confirmed by immunoblot using sera from subjects with culture-proved HSV-2 that were positive for HSV-2 gG antibody by the ELISA method. The fact that antibody reactivity was preserved under immunoblot conditions suggest that the HSV-2 gG antibodies recognize a linear epitope on the viral protein. The correlation between past HSV-2 infection and HSV-2 gG antibodies indicates that the HSV-2 gG ELISA could be used to confirm the diagnosis of genital herpes in a patient with a suggestive history who presented during an asymptomatic interval.

Diagnosis of Past HSV-2 Infection in Patients with no History of Genital Herpes

The assay for antibodies to HSV-2 gG provides an important marker of past HSV-2 infection based on HSV-2 immune status that is independent of the history of symptomatic genital herpes (Prober et al. 1988). Recent seroepidemiologic surveys, investigating the prevalence of antibodies to HSV-2 gG, show that most adults will not be aware of their past infections with HSV-2. For example, in a study of women in Atlanta, only one third of the subjects who had HSV-2 antibodies reported any history of recurrent genital lesions or known genital herpes (Nahmias et al. 1985). The serologic diagnosis of asymptomatic past HSV-2 infection can be used to analyse the risk of transmission of HSV-2 when one

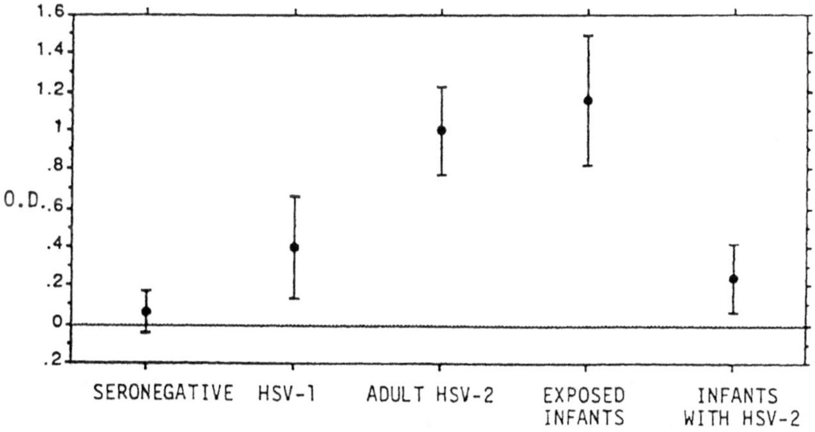

Figure 1. The detection of antibodies to HSV-2 gG by ELISA using a monoclonal antibody to HSV-2 gG. The O.D. values are shown as the mean plus one standard deviation from the mean. The populations tested included adults and children who were HSV-seronegative by indirect hemagglutination or neutralization (n-40), adults and children with past HSVV-1 (n=50), adults with HSV-2 (n-35) and infants tested within one week after the onset of neonatal HSV-2 infection (n=17).

partner has a history of symptomatic genital herpes and the other does not.

Using an immunoblot method to detect antibody binding to the gG band of an HSV-2 antigen preparation, Merta, et al. (1988) found that 10 to 38 asymptomatic exposed partners had HSV-2 antibodies at the initiation of a prospective study and that asymptomatic seroconversion to HSV-2 type specific antigen occurred in one subject. The capacity to investigate this population with a precise serologic method for HSV-2 infection demonstrated that the rate of transmission of genital HSV-2 is about 12% per year which is lower than suggested by earlier clinical observations. Given the high background frequency of past HSV-1 and HSV-2 infection among adult populations, the serologic distinction of HSV-1 from HSV-2 infection will also be necessary for clinical trials designed to evaluate the efficacy of HSV-2 vaccine preparations (Ashley et al. 1987).

Diagnosis of True Primary Genital HSV-2 Infection

Because of the problem of serologic crossreactivity between HSV-1 and HSV-2 antigens and the variable extent of disease caused by primary HSV-2 infection in different individuals, it is difficult to distinguish true primary genital HSV-2 infection from the first symptomatic episode in patients whose true primary infection was asymptomatic or unrecognized. Using immunoblot methods, Bernstein et al. (1984) and Mertz et al. (1988) found that many patients with a first clinically recognized episode of genital herpes had antibodies to HSV-2 proteins in serum taken at the onset of symptoms. In our experience, the HSV-2 gG ELISA method is valuable for the diagnosis of true primary HSV-2 infection by seroconversion, whether or not the individual has had past HSV-1 infection. Antibodies to HSV-2 gG were not detected until >28 days after the onset of genital herpes in women who had primary HSV-2 infection. Sera obtained from 21-30 days after the appearance of lesions showed faint reactivity when examined for HSV-2 gG antibodies by immunoblot but the O.D.s by HSV-2 gG ELISA were slightly below the O.D. for the HSV-2 pool serum control. The fact that HSV-2 gG antibodies were not detected until several weeks after the appearance of symptoms of genital herpes suggests that true primary genital HSV-2 infection can be defined by a positive culture in the absence of HSV-2 gG antibodies. Thus, in clinical practice, it should be possible to identify those patients most likely to benefit from antiviral therapy based upon a laboratory diagnosis of true primary genital herpes.

Applications of HSV-2 gG Antibody Testing to the Problem of Maternal HSV-2 Infection During Pregnancy

The determination of HSV-2 gG antibody status is particularly relevant to the problem of HSV-2 infection during pregnancy. Most neonatal HSV-infections result from asymptomatic maternal HSV shedding, as indicated by the fact that only one third of mothers of infants with neonatal herpes have signs of HSV infection at delivery for any history of genital herpes (Whitley et al. 1980; Yeager and Arvin, 1984). Under these circumstances, the serologic differentiation of past HSV-2 from past HSV-1 infection is necessary to identify pregnancies likely to be complicated by recurrent maternal HSV-2 infection. In our initial evaluation of the ELISA method as a screening tool to determine the prevalence of past HSV-2 infection, HSV-2 gG antibodies were found in sera from 21 of 110 pregnant women (19%) (Sullender et al. 1988). Only three of these women (14%) had a history suggesting past genital herpes; a fourth woman had no previous history of genital herpes but had a sexual partner with recurrent herpes. Thus, the failure of the maternal history to identify pregnancies potentially complicated by HSV-2 infection was confirmed by our observation that 81% of pregnant women with HSV-2 gG antibodies had no history of genital herpes or of exposure to genital herpes.

As part of a study designed to determine whether cultures done at delivery, regardless of the maternal history of genital herpes, would provide a better means of identifying infants at risk of neonatal HSV infection, we investigated whether maternal HSV-2 shedding at delivery correlated with serologic evidence of HSV-2 infection using the assay for antibodies to HSV-2 gG (Prober et al. 1988). HSV was isolated from 14 (0.20%) of 6,904 deliveries. The frequency of positive cultures was seven (0.22%) of 3,157 women from an upper socioeconomic population and seven (0.19%) of 3,747 women in a lower socioeconomic group. Only one (7%) of the 14 mothers identified as excreting HSV asymptomatically at delivery had a previous history compatible with genital herpes. By serologic analysis, 12 of the 14 women (86%) who were shedding HSV asymptomatically at delivery had evidence of previous infection and two women (14%) had evidence of a primary infection.

Twenty-one of 618 women (3.4%) in this study had a history of genital herpes based upon a review of their prenatal records. By comparison, the seroprevalence of HSV-2 gG antibodies was 17.7% (37/209) in this population which was equivalent to our initial finding of HSV-2 immunity

TABLE 1

Study Population	N	Number with HSV-2 gG titer > 1:16	Mean O.D. + SD
Pregnant women:			
Recurrent HSV-2	102	98 (96%)	1.00 + 0.23
Initial episode:			
5-14 d after onset	6	5 (83%)	
at delivery[a]	28	28 (100%)	0.96 + 0.23
Mothers of infants with HSV-2 infection	26	14 (54%)	0.49 + 0.26
Random screening	110	21 (19%)	
Infants:			
Of mothers with recurrent HSV-2 in pregnancy[b]	9	9 (100%)	1.40 + 0.21
Exposed to HSV-2 at delivery			
Known maternal history	25	23 (92%)	1.19 + 0.34
No maternal history	10	7 (70%)	1.05 + 0.32
With neonatal HSV-2			
< 1 week[c]	17	2 (12%)	0.25 + 0.12
2-6 weeks[c]	29	1 (4%)	0.29 + 0.13

[a] These samples were obtained from 1-7 months after the initial clinical episode of genital herpes.
[b] These samples were cord blood sera.
[c] Time of samples relative to onset of symptoms.

From Sullender et al. 1988.

among 19% of pregnant women (Sullender et al. 1988). Nine of the 37 women (24%) who were HSV-2 gG seropositive had a history of genital herpes. In contrast, only three of the 172 women (1.7%) who were HSV-2 gG seronegative had a history of genital herpes. The discrepancy between the number of women with serologic evidence of past infection and the number with compatible clinical histories was consistent with the observation that only one of 14 women who were shedding HSV at delivery had a past history of genital herpes.

The surveillance study proved that a minority of the infants at risk of neonatal HSV will be identified if concern about asymptomatic shedding of HSV at delivery is limited to women with past histories of genital HSV infection. Previous studies indicated that approximately one percent of women with known recurrent genital herpes shed the virus asymptomatically on the day of delivery (Arvin et al. 1986c). The observation of a 0.2% shedding rate in a population with a seropositivity rate of about 20%, supports the concept that HSV-2 gG seropositivity is a valid marker for identifying women at risk for asymptomatic shedding of HSV-2. If the frequency of shedding were predicted from the clinical history of genital herpes obtained from about three percent of the population, only two or three women would be expected to have asymptomatic shedding at delivery. As anticipated, the majority of exposed infants were born to mothers with subclinical past HSV-2 infections. The serologic screening of pregnant women using one of the methods for detecting antibodies to HSV-2 gG should improve the identification of infants that are at potential risk of HSV-2 exposure at delivery.

The HSV-2 gG antibody assay also provides a tool for investigating the consequences of primary genital HSV infection during pregnancy. Whether intrauterine infections are more likely to result from primary than recurrent maternal HSV remains unresolved. In a recent series, all of the cases of virologically proved intrauterine HSV infection were caused by HSV-2 (Hutto et al. 1987). Brown et al. (1988) have described significant fetal and neonatal morbidity associated with primary infection late in pregnancy. However, many women with an initial clinical episode of HSV-2 infection during pregnancy may be experiencing reactivation months or years after a clinically silent primary infection. In our experience five of six pregnant women who were tested within 5-14 days after the onset of the initial clinical episode of HSV-2 infection had HSV-2 gG antibodies (Table 1). By comparison with the kinetics of the HSV-2 gG antibody response in subjects after first episode primary HSV-2 infections, the detection of antibodies to HSV-2 gG shortly after the appearance of lesions suggested that these women had reactivation of a past asymptomatic HSV-2 infection. By serologic testing, only one woman had a true primary HSV-2 during pregnancy; her infant was born at term without sequelae. With a reliable laboratory method for diagnosing true primary infection, this morbidity might be reduced by giving antiviral therapy during pregnancy.

In a study in progress, we have preformed serologic screening for HSV-2 gG antibodies on 278 pregnant women using paired sera drawn at a mean interval of 20.4 weeks (Boucher et al. 1988). One woman acquired HSV-2 gG antibodies during pregnancy for a seroconversion rate of 0.36% or 0.9% per patient year. The prevalence of HSV-2 gG antibodies increased for 8.7% in the 15-20 year old age group to 27.8% among women older than 35 years. This data suggests that the acquisition of HSV-2 infection increases steadily during the child-bearing years and that asymptomatic seroconversion to HSV-2 among pregnant women does occur. By testing sera from pregnant women for HSV-2 gG antibodies at their first prenatal visit, in the third trimester, and at delivery, the outcome of pregnancies complicated by true primary HSV-2 infection can be determined.

Application to the Identification of Infants at Risk for Neonatal HSV-2 Infection

It is clear that primary genital HSV infections, some of which may be asymptomatic, sometimes occur near the time of delivery (Prober et al. 1988; Yeager and Arvin, 1984). The identification of an infant born to a mother with active HSV-2 infection who lacks HSV-2 gG antibodies, whether in her serum or in the cord blood, suggests that the maternal infection is primary. From the perspective of the risk to the infant, this is important information because the estimated attack rate for neonatal herpes is 50% if the maternal infection is primary compared to an attack rate of less than 8% with exposure to recurrent maternal infection (Nahmias et al. 1983; Prober et al. 1987). Our experience indicates that transplacentally acquired passive antibodies against HSV probably protect many infants from invasive infection despite exposure to HSV at delivery (Sullender et al. 1987). The fact that HSV-2 gG antibodies are transferred across the placenta was established among infants born to mothers who had at least one episode of recurrent HSV-2 infection during pregnancy (Table 1). The titer of HSV-2 G antibody in paired maternal and infant sera were equivalent in all cases.

We have investigated the transplacental transmission of HSV-2 gG antibodies among infants who were actually exposed to HSV at delivery in two clinical studies. In the first study, antibodies to HSV-2 gG were detected in 23 of 25 cord blood samples (92%) from exposed infants whose mothers had a known history genital herpes (Prober et al. 1987). None of these infants developed symptomatic HSV-2 infection. In the second study, which involved obtaining routine surveillance cultures at delivery without regard for maternal history, cord blood was available from 12 of 14 infants born to mothers who were shedding HSV (Prober et al. 1988). Eight (67%) of the 12 sera had demonstrable antibody to HSV-2 gG. Two of the four sera that were negative for antibody to HSV-2 gG were from infants born to women with evidence of asymptomatic primary infections at delivery. One of these two infants developed HSV-2 infection in the neonatal period. The other two cord blood samples which lacked HSV-2 gG antibody were from infants whose mothers had antibody to HSV-2 gG at delivery. Sera was available at 10 days postpartum form the two neonates whose cord blood samples were not obtained. Both infants had HSV-2 gG antibody.

The persistence of transplacentally acquired HSV-2 gG antibodies was determined by testing sera obtained between 28 and 125 days of age from the 13 infants who remained well. Antibody to HSV-2 gG was still present in five (65.5%) of eight samples obtained before 40 days of age (mean, 32 days) but none of seven samples obtained after 40 days of age (mean, 118 days) had HSV-2 gG antibody.

In contrast to the exposed infants who remained well, HSV-2 gG antibodies were detected in sera from only two of 17 infants (12%) tested within one week after the onset of symptoms of neonatal HSV-2 infection (Table 1; Figure 1). Seven infants with cutaneous infection and four infants with disseminated infection had no HSV-2 gG antibodies; two of six infants with HSV-2 encephalitis had HSV-2 gG antibodies at the time of diagnosis. The mean O.D. for sera obtained within one week after the onset of illness was significantly lower than the mean O.D. for sera from infants exposed to HSV-2 who did not develop symptoms ($p < 0.001$, t test) (Table 1; Figure 1). The serologic testing of mothers of infants with HSV-2 infection showed that only 14 of 26 mothers (54%) had HSV-2 gG antibodies in serum taken after the onset of symptoms in their infants but less than 28 days after delivery (Table 1). The mean O.D. for sera from mothers of infected infants was significantly less than the mean O.D. for sera from women with initial or recurrent HSV-2 during pregnancy ($p < 0.0001$, t test). Two of five mothers of infants with cutaneous HSV-2 and four of 10 mothers of infants with disseminated HSV-2 had HSV-2 gG antibodies (40%) compared to eight of 11 mothers of infants with HSV-2 encephalitis (72%).

The HSV-2 gG antibody assay was also applied to the investigation of the host response of infants with neonatal HSV-2 disease. Among 29 infants tested between 2-6 weeks after the onset of symptoms, one of 14 infants with encephalitis had HSV-2 gG antibodies while eight infants with cutaneous infection and seven infants with disseminated infection remained seronegative (Table 1). By 2.5-6 months, only two of six infants had HSV-2 gG antibodies > 1:16 although the O.D. for the late sample had increased in each case. HSV-2 gG antibodies were detected in six of eight sera from infants who were tested 6-18 months after neonatal HSV-2 infection; two infants remained seronegative despite severe HSV-2 encephalitis.

In summary, infants with HSV-2 infection who were tested during the first week of their illness and the mothers of these infants often had no detectable HSV-2 gG antibodies. In many cases, this circumstance probably resulted from primary maternal HSV infection. This data does not establish that HSV-2 gG antibodies have a protective role, per se. However, the absence of HSV-2 type specific antibodies in serum from exposed infants may help to identify a subpopulation of infants who are at increased risk for neonatal herpes.

THE INVESTIGATION OF IMMUNITY TO VZV PROTEINS

Background

With the prevention of the other common exanthems of childhood, such as measles, by immunization, VZV has emerged as an important cause of morbidity and mortality among otherwise healthy children as well as in immunocompromised patients (Preblud, 1981). The risk of serious morbidity is increased in adults who have primary VZV infection and varicella during pregnancy can produce the congenital varicella syndrome. Recently, these populations have benefitted from immunization with the live attenuated varicella vaccine (Brunell, et al. 1982; Gershon et al, 1984, 1988).

The cell-associated nature of VZV replication has complicated the characterization of the major proteins synthesized by VZV and the identification of the viral proteins that constitute important targets of the host immune response to VZV infection. Techniques based upon the availability of murine monoclonal antibodies to VZV, have facilitated the study of humoral and cellular immunity to specific VZV proteins (Grose and Litwin, 1988). The analysis of humoral and cellular immunity to VZV proteins provides a new approach for comparing the host response of normal and immunodeficient individuals to this virus and for evaluating natural and vaccine-induced immunity (Arvin, et al. 1986a; Brunell et al. 1987). For the purpose of making such comparisons, we have examined immunity to the gp I, gp III and p 170 proteins of VZV. Gp I is the VZV membrane glycoprotein made most abundantly in virally infected cells and elicits a major IgG antibody response after naturally acquired VZV infection (Davison et al. 1986; Grose, 1987). Gp III is a disulfide-linked protein of 140,000 mw in its native for that is also expressed on the membrane of VZV-infected cells. The p 170 is a protein that is expressed prominently over the nuclei of infected cells and is likely to be either an immediate early gene product (Shiraki and Hyman, 1987) or a nucleocapsid protein. The cellular immune response to this protein is of interest because it represents a non-membrane component of the virus.

Functionally, the antibodies produced to gp I and gp III may mediate viral neutralization. Murine monoclonal antibodies to these proteins often have neutralizing activity which is either complement dependent (gp I) or independent (gp III). Because of the cell-associated nature of VZV infection, the participation of antibodies to the VZV glycoproteins in antibody dependent cellular cytotoxicity may provide important host defense (Ito et al. 1985). The investigation of T-lymphocyte response to VZV proteins is of particular interest because of the evidence that cellular immunity plays a major role in the resolution of primary VZV infection (Arvin et al. 1987) and that diminished T-lymphocyte proliferation to VZV antigens correlates with an increased risk of VZV reactivation among elderly or immunocompromised patients (Miller et al. 1980).

Description of Methods

IgG antibodies to VZV proteins were measured using a solid phase capture assay in which monoclonal antibodies to VZV gp I, gp III or p 170 were diluted in PBS and added to flat-bottom microtiter plates (Arvin et al. 1986a). The gp I and gp III monoclonal antibodies were diluted 1:3,000 and the p 170 monoclonal antibody was diluted 1:10,000. After incubation for approximately 12 h, washing, and blocking with 1% mouse serum, the plates were dried and coated with a detergent solubilized extract of VZV infected melanoma cells, or an uninfected cell extract as a control, and incubated overnight. The plates were washed and 3% BSA in PBS was added to block nonspecific binding. After drying, serial dilutions of the test sera were added and incubated. The assay was completed using the Vectastain reagents or ^{125}I conjugated goat antihuman IgG to detect specific antibody binding to the protein. To determine the presence or absence of specific antibody binding in the ELISA method, the mean O.D. of the control wells at each dilution of test serum was subtracted from the mean of the O.D. of the antigen wells. The serum was considered positive if this calculated number was greater that that calculated for a standard negative serum control plus a constant derived for each assay. The constant was calculated as the mean variance of O.D.s for all control wells at the 1:16 dilution plus two standard deviations from this mean.

For assays of cellular immunity, VZV proteins were prepared by immunoaffinity chromatography and used to stimulate T-lymphocytes (Arvin et al. 1986a). VZV monoclonal antibodies were coupled to cyanogen bromide activated Sepharose 4B at 20 mg of protein/0.5 g Sepharose. A detergent extract of VZV infected or uninfected melanoma cells was incubated overnight with the antibody coated Sepharose. Bound protein was then eluted with potassium thiocyanate, the Sepharose was removed by centrifugation and the supernatant was concentrated by vacuum dialysis. Tis procedure, in combination with the fact that the VZV protein preparations can be diluted to at least 1:50 before addition to the peripheral blood mononuclear cell cultures, eliminated the problem of lysis of the cells in culture by residual detergents carried from the solubilization of the VZV infected cell monolayer. Peripheral blood mononuclear cells were separated by Ficol-Hypaque gradient and cultured in round bottom microtiter plates; the cells were added at $3x10^5$ cells/well in RPMI and

30% pooled human serum. Whole antigen, made from sonicated VZV infected melanoma cells or immunoaffinity-purified VZV proteins or controls were added at varying dilutions in triplicate to the cell cultures. After incubation at 37°C for five days, the wells were pulsed with tritiated thymidine, cells were harvested and thymidine incorporation was determined by counting in a liquid scintillation counter. The transformation index (T.I.) was calculated as the ratio of the mean cpm for test and control wells at each antigen dilution.

Cellular immunity was also assessed by measuring the production of gamma interferon by T-lymphocytes stimulated with the VZV proteins. Supernatants harvested after the incubation of peripheral blood mononuclear cells for five days were tested for gamma interferon using the radioimmunoassay method (Centocor, Inc., Malvern, PA). The detection of greater than 1.0 U of gamma interferon was considered positive.

The Host Response to VZV Proteins, gp I, gp III, and p 170

Immunity to VZV proteins in Primary and Recurrent Infections Caused by Wild Type Virus. In our initial study of immunity to VZV proteins, IgG antibodies to gp I, gp III and p 170 were measured by radioimmunoassay in sera from otherwise healthy individuals with varicella or herpes zoster (Arvin et al 1986a). Primary VZV infection in healthy subjects was accompanied by the development of IgG antibody against gp I in 63% of subjects who were tested within six weeks after the onset of the varicella rash, to gp III in 59% of subjects and to p 170 in 60% of subjects. The mean titers were 1:200 to gp I, 1:132 to gp III and 1:331 to p 170. The subjects with herpes zoster had much higher IgG antibody titers to each of the VZV proteins. The mean titers in these subjects were 1:29,512 to gp I, 1:15,848 to gp III, and 1:15,848 to p 170. The reactivity of antibodies in sera from patients with recent primary and recurrent VZV infections against the major glycoproteins of the virus, i.e., gp I, gp II and gp III, is also observed by immunoprecipitation (Weigle and Grose, 1984).

IgG antibodies to gp I, gp III and p 170 were detected by radioimmunoassay in most immune subjects tested more than 20 years after primary VZV infection. Using the ELISA method, all naturally immune individuals who were tested had antibodies to gp I and p 170 (Table 2); gp III was not tested (Diaz et al. 1988). The geometric mean titers (GMT) for antibodies to gp I and p 170 in naturally immune individuals were 1:135 and 1:270 (Table 3).

The acquisition of T-lymphocyte proliferation to gp I, gp III and p 170 was demonstrated in healthy subjects in the convalescent phase of

TABLE 2. PERCENTAGES OF CHILDREN AND ADULTS GIVEN LIVE ATTENUATED VARICELLA VACCINE AND OF ADULTS WITH NATURAL IMMUNITY TO VZV WITH DETECTABLE IMMUNITY TO WHOLE VZV ANTIGEN, VZV gpI AND p170

Population	Interval After Immunization	IgG Antibodies > 1:16			T-Lymphocyte Proliferation[a]		
		Whole VZV	gp I	p 170	Whole VZV	gp I	p 170
Children	2 months	21/30 (70%)	8/30 (27%)	13/30 (43%)	19/30 (63%)	10/26 (38%)	7/25 (28%)
	1 year	18/23 (78%)	5/23 (22%)	11/21 (52%)	9/23 (39%)	9/20 (45%)	9/20 (45%)
Adults	2 months	10/13 (77%)	6/11 (55%)	3/11 (27%)	6/10 (60%)	3/11 (27%)	8/11 (73%)
	1 year[b]	10/10 (100%)	9/11 (80%)	5/10 (50%)	10/10 (100%)	3/9 (33%)	6/9 (67%)
Naturally Immune Adults[c]		21/21 (100%)	18/18 (100%)	19/19 (100%)	18/18 (100%)	15/18 (83%)	16/18 (89%)

[a] After second dose
[b] At least 20 years after varicella
[c] T.I. ≥ 3.0 was considered a positive response to whole VZV antigen; T.I. ≥ 2.0 was positive for gpI and p170

From Diaz et al. 1988

TABLE 3. THE GEOMETRIC MEAN IgG TITERS AND THE MEAN TRANSFORMATION INDICES TO WHOLE VZV ANTIGEN, VZV gp I AND p 170 IN CHILDREN AND ADULTS GIVEN LIVE ATTENUATED VARICELLA VACCINE WHEN TESTED AT ONE YEAR AND IN ADULTS WITH NATURAL IMMUNITY TO VZV

	Geometric Mean Titer by ELISA			Mean T-lymphocyte Proliferation[a]		
Population	Whole VZV	gp I	p 170	Whole VZV	gp I	p 170
All vaccines	1:56 (1:100)[b]	1:4 (1:37)	1:6 (1:38)	6.4 ± 1.39	2.3 ± 0.33	3.3 ± 0.81
Children only	1:32 (1:91)	1:3 (1:63)	1:7 (1:38)	4.1 ± 0.96	2.3 ± 0.40	2.9 ± 0.96
Adults only	1:165 (1:165)	1:14 (1:25)	1:6 (1:39)	12.7 ± 0.59	2.2 ± 0.59	4.2 ± 1.55
Naturally Immune Adults	1:468	1:135	1:270	13.0 ± 1.67	3.6 ± 0.57	3.7 ± 0.45

[a] ± Standard error
[b] (Calculated for vaccines with seroconversion only)

From Diaz et al. 1988.

primary VZV infection. Lymphocyte proliferation to gp I, gp III and p 170 persisted in most immune subjects for at least 20 years. Gamma interferon production was also detected in cultures from five of nine naturally immune individuals (56%) of this population after stimulation with p 170 (mean 6.7 U/ml).

In general, healthy subjects with primary VZV infection had somewhat variable patterns of response to specific proteins and did not always develop humoral and cellular immunity against all of the VZV proteins which were tested. Allowing for the possibility that present methods for the detection of immunity to VZV proteins are not optimally sensitive, the successful resolution of primary VZV infection apparently does not require that the host develop or maintain detectable immune responses directed against each major viral protein. Nevertheless, variations in immunity to predominant VZV proteins may imply subtle differences in VZV immunity which could influence the relative severity of primary infection and the susceptibility of the host to the reactivation of the latent virus.

Immunity to VZV Proteins in Vaccine-Induced Immunity

As part of our clinical evaluation of the assays for immunity to VZ proteins, we have investigated the host response to gp I and p 170 among healthy subjects immunized with the live attenuated varicella vaccine Oka-Merck, (Merck, Co., Inc., West Point, PA) (Diaz et al. 1988). Susceptible children received 950 pfu as a single dose while adult vaccine recipients received 2,500 pfu per dose in two injections given at a three month interval. Vaccine-induced immunity to VZV was compared with natural immunity as measured among the healthy adult population with a history of varicella more than 20 years before evaluation.

When the vaccine recipients were tested at two months after immunization, prior to the administration of the second dose to the adult vaccines, IgG antibody was detected to whole VZV and gp I among the adults more frequently but the prevalence of antibody to p 170 was equivalent among children and adults (Table 2; Figure 2). The GMTs to whole VZV antigen and to gp I in children tested at one year were significantly lower than the GMTs for adults at one year after vaccination (whole VZV:p=0.03; gp I:p=0.009, t test) (Table 3). The GMTs to whole VZV antigen, gp I and p 170 in children were also significantly lower than the GMTs of naturally immune individuals (p=0.0001, t test). Although the responses to whole VZV antigen and gp I were more consistent among adults than children, the GMTs to whole VZV antigen, gp I and p 170 in healthy adults with vaccine induced immunity were still significantly lower than

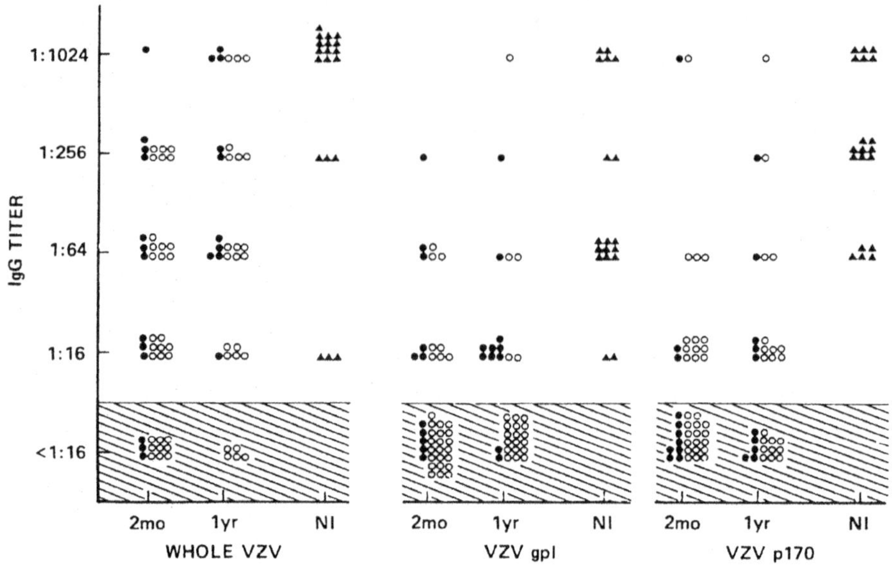

Figure 2. IgG antibody production to whole VZV antigen, VZV gp I and VZV p 170. Antibody titers obtained from vaccinated adults (closed circles) and vaccinated children (open circles) are contrasted with titers obtained from a naturally immune population (NI) represented by closed triangles. Titers for adults are shown at two months after the first dose and one year after a second dose and for children at two months and one year after a single dose. Antibody production is represented by a titer of 1:16 or greater. From Diaz et al. 1988.

those of naturally immune individuals (whole VZV:p=0.017; gp I:p=0.0003; p 170:p=0.0001, t test).

Children and adults given the live attenuated varicella vaccine were also tested for the acquisition of T-lymphocyte responses to the gp I and p 170 proteins of VZV (Figure 3). T-lymphocyte proliferation to gp I did not differ significantly between children and adults with vaccine-induced immunity but the mean T.I. to gp I was significantly lower (p=0.02, t test) when all vaccinees were compared to naturally immune individuals (Table 3). Adults tested at two months and at one year were consistently more likely than children to have T-lymphocyte proliferation to p 170 (p=0.001, x^2) (Table 2). However, when the populations were compared based upon the mean T-lymphocyte proliferation to p 170, the response was equivalent between the two vaccine populations at one year and between each vaccine group and the naturally immune population (Table 3).

Gamma interferon production to gp I was not detected in seven vaccinees tested one to three months after the initial dose of vaccine (0/4 children and 0/3 adults). Two of four children had a gamma interferon response to p 170 whereas none of three adults responded to p 170 after

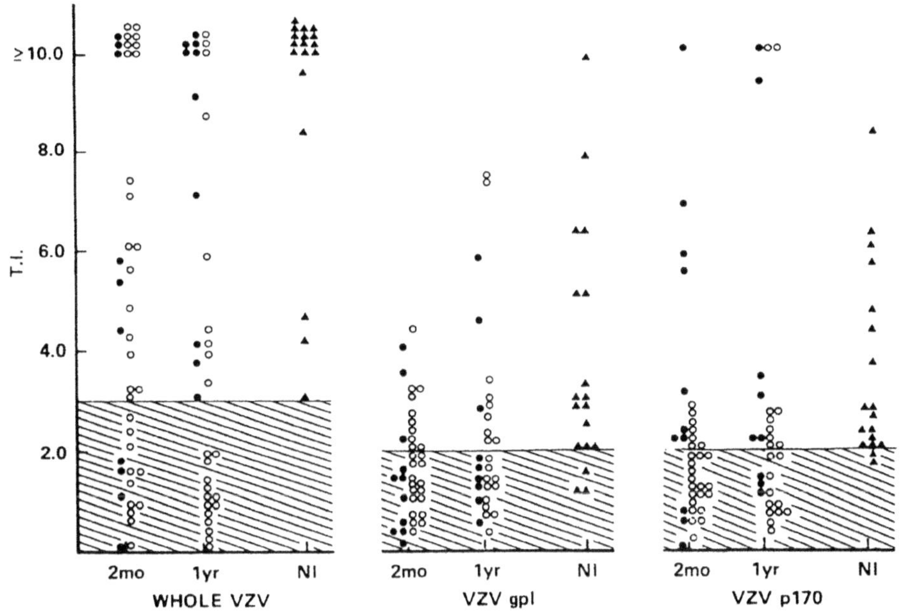

Figure 3. Lymphocyte proliferation to whole VZV antigen, gp I and p 170. Transformation indices from vaccinated adults (closed circles) and vaccinated children (open circles) are contrasted with transformation indices from a naturally immune population (NI) represented by closed triangles. Transformation indices are shown for adults two months after their first dose and one year after a second dose and for children at two months and one year after a single dose. A positive response is represented by a transformation index of ≥3.0 for whole VZV and ≥2.0 for gp I and p 170. From Diaz et al. 1988)

the initial immunization. By 9 to 14 months after the second dose, 38% of eight adults responded to gp I (mean 4.9 U/ml) and 25% responded to p 170 (mean 8.2 U/ml). In contrast, none of four children tested one year after vaccination had gamma interferon production to gp I or p 170 and only one of eleven children tested at two years for gamma interferon production to gp I or p 170 responded to gp I.

The differences in humoral and cellular immunity that were observed between children tested at one year and naturally immune subjects persisted when the children were tested at two years after vaccination. Of seventeen children followed for two years, 31% had antibody to gp I with a GMT of 1/5 and 50% had antibody to p 170 with a GMT of 1:14. T-lymphocyte proliferation to gp I was detected in 41% at two years with a mean T.I. of 2.7±0.7 SE and to p 170 in 29% with mean T.I. of 1.7±0.2 SE.

In summary, when vaccine-induced immunity to the VZV proteins, gp I and p 170, was compared with immunity to these proteins following natural infection, differences could be detected in many vaccinees despite the persistence of antibodies and T-lymphocyte proliferation to whole VZV antigen. Differences in humoral and cellular immunity to VZV that were detected by comparing the naturally immune population with the two populations of vaccine recipients may account for the essentially complete protection against symptomatic reinfection provided by natural immunity, even though immunologic evidence suggests that subclinical reinfection does occur (Arvin et al. 1983). The diminished gamma interferon response following stimulation with VZV proteins in vaccinees compared to naturally immune subjects could affect the clonal expansion of VZV specific T-lymphocytes upon reexposure to the virus. While protective immunity against VZV has not been defined in terms of requirement for a persistent host response to any specific VZV protein or proteins, the breakthrough cases of mild varicella that have been observed after immunization with live attenuated varicella vaccine could be related to differences in immunity identifiable by investigating the immune response to specific VZV proteins.

Immunogenicity of Purified VZV Proteins, gp I and p 170, in the Strain 2 Guinea Pig

As more information is acquired about the immune response to specific VZV proteins, designing a subunit vaccine that does not contain infectious VZV becomes theoretically feasible. To provide some background for such an effort, we have used monoclonal antibodies with specificity for VZV proteins to determine whether immunoaffinity purified preparations of VZV gp I and p 170 are immunogenic in the strain 2 guinea pig (Arvin et al. 1987). The guinea pig model was used because weanling guinea pigs can be infected with VZV (Myers, et al. 1985) and develop antibodies to the major viral proteins that are also targets of the human IgG response as detected by immunoprecipitation (Grose and Freidrichs, 1982). Guinea pigs also acquire cellular immunity after VZV infection as measured by delayed hypersensitivity to whole VZV antigen (Matsunaga et al. 1982).

In our experience, immunity to VZV was assessed following a single inoculation of guinea pig adapted VZV and compared with the response of animals given an initial injection of gp I or p 170, followed by a second dose after one week and a third dose at 6 or 18 weeks (Figures 4 and 5). Four of 8 animals had detectable antibody by 10 days after the first injection of gp I as did 4 of 7 animals given p 170. All of nine animals given infectious VZV remained seronegative at 10 days after inoculation but had seroconverted by 3-6 weeks. VZV antibody titers of >1:16 per-

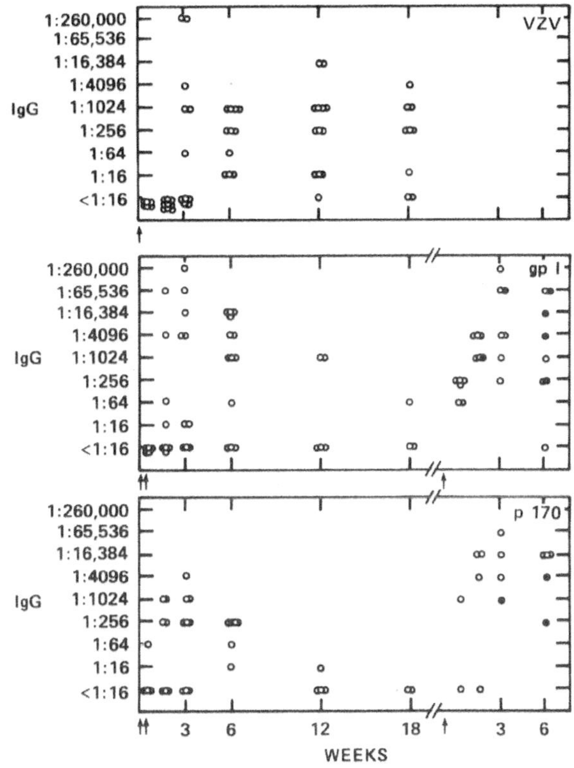

Figure 4. VZV IgG antibody production after the inoculation of strain 2 guinea pigs with infectious VZV, VZV gp I or VZV p 170. This figure shows the antibody titers measured by radioimmunoassay in serum from animals given infectious VZV adapted to guinea pig embryo cells (top), immunoaffinity purified gp I (center) or p 170 (bottom). The titers are presented in relation to the time interval in weeks after the injections which are indicated by the arrows on the horizontal axis. The dark circles indicate the IgG titers of animals who were given the third injection of gp I or p 170 at six weeks rather than 18 weeks after the second dose. From Arvin et al. 1987.

sisted in 18 of 21 samples taken from 12-18 weeks after inoculation with a GMT of 1:100. Seven of 9 animals had IgG titers >1:16, while the GMT in animals given gp I and p 170 declined to <1:16 by 12-18 weeks. When animals in the latter group received a third injection of the respective protein preparation, the antibody titers rose rapidly to high titers, with the GMT at 3 weeks being 1:8,709 for gp I and 1:7,943 for p 170. The sera from the test animals were also analyzed for VZV neutralizing antibodies. Complement-dependent neutralizing antibody against VZV, with titers of 1:16-1:64, was detected as late as 12 weeks in five animals tested after the initial injection of gp I. The neutralizing titers with complement rose to 1:128 in two animals tested 10 days and 3 weeks, respectively, after the third injection of gp I. Five animals given infectious VZV had low titers of complement-independent neutralizing antibodies (1:888-1:16) which were enhanced to 1:16-1:64 by complement. None of six animals given p 170 had VZV neutralizing antibodies at any point after single or multiple injections.

The injection of guinea pigs with VZV gp I and VZV p 170 elicited cellular immunity, measured by T-lymphocyte proliferation (Figure 5).

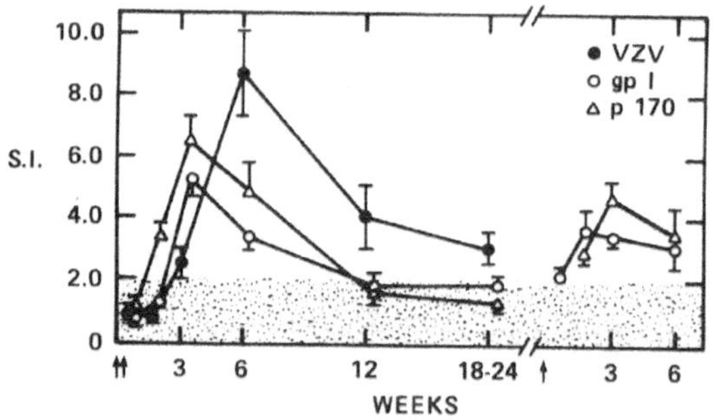

Figure 5. VZV lymphocyte proliferation after inoculation with infectious VZV, VZV gp I or VZV p 170. This figure shows VZV lymphocyte proliferation expressed as the stimulation index (S.I.) for cultures of peripheral blood mononuclear cells from animals given infectious VZV (●), VZV gp I (0), or VZV p 170 (△). The S.I. is the ratio between mean cpm for VZV antigen stimulated cell cultures and the mean cpm for cells incubated with an uninfected cell control preparation, as measured by uptake of 3[H]-thymidine. The S.I. is presented in relation to the time interval in weeks after inoculation, as indicated by arrows on the horizontal axis. From Arvin et al. 1987.

The peak response occurred at three weeks after the initial inoculation with mean peak stimulation index (S.I.) of 5.2 ± 1.03 SE for gp I and 6.6 ± 1.53 SE for p 170. The peak S.I. was not detected until 6 weeks after inoculation of the animals with infectious VZV but it rose to 8.7 + 2.05 SE. The mean lymphocyte proliferation response decreased to less than 2.0 by 12 weeks after immunization with gp I and p 170 while 6 or 8 guinea pigs tested 18-24 weeks after inoculation of infectious VZV maintained S.I. above 2.0. As was observed in the analysis of the humoral immune response the third dose of gp I and p 170 produced an immediate rise in cell-mediated immunity as determined by lymphocyte proliferation (Figure 5).

These studies demonstrated that gp I and p 170 were equally immunogenic in inducing both humoral and cellular immunity in the strain 2 guinea pig. Immunization with the gp I protein prepared by immunoaffinity using a gp I reactive monoclonal antibody also elicited a neutralizing antibody response *in vivo* that was complement-dependent, as was observed by Keller, et al. (1986). Although the non-glycosylated protein, p 170, did not elicit neutralizing antibody *in vivo,* it induced both IgG antibody and T-lymphocyte proliferation in the strain 2 guinea pig which parallels our observation of persistent humoral and cellular immunity to p 170 in most human subjects following VZV infection (Arvin et al. 1986a). However, waning of VZV antibodies and T-lymphocyte proliferation illustrates the general failure of immunization with herpes viral proteins to induce long term immunity. Both peak IgG antibody production an T-lymphocyte proliferation responses were delayed in animals given infectious VZV in comparison with those of animals given gp I and p 170 suggesting that a phase of *in vivo* viral replication occurred. By analogy with the experience with inactivated compared to live viral vaccines, some replication may be a requirement for inducing persistent immunity. Despite the waning of VZV immunity after protein injection, our studies demonstrated a "priming" effect upon the immune response to the third injection of gp I or p 170. The investigation of the immune response to VZV proteins in the guinea pig provides background for testing methods of enhancing the immunogenicity of purified viral proteins (Ho et al. 1986) which can then be compared to the immune response to infectious VZV.

ACKNOWLEDGEMENTS

The work related to HSV was supported by U.S. Public Health Service grants HD-16080 and AI-23592 and by the Valley Foundation, San Jose, CA. The investigations of VZV were supported by grants from the National Institute of Allergy and Infectious Diseases (AI 20459 and AI 22280) and by Merck, Co., Inc. West Point, PA. Adults were immunized as participants in the investigation of varicella vaccine directed by Dr. Anne Gershon, Columbia University (Contract AI 62529). Dr. Arvin is the recipient of a Research Career Development Award, AI 00624.

REFERENCES

Ashley R, J Benedetti, L Corey (1985) Humoral immune response to HSV-1 and HSV-12 viral proteins in patients with primary genital herpes. J Med Virol 17:153-166.

Ashley R, GJ Mertz, L Corey (1987) Detection of asymptomatic herpes simplex virus infections after vaccination. J Virol 61:264-268.

Arvin AM, CM Koropchak, Wittek AE (1983) Immunologic evidence of reinfection with varicella-zoster virus. J Infect Dis 148:200-205.

Arvin AM, E Kinney-Thomas, K Shriver, C Grose, CM Koropchak, E Scranton, AE Wittek, PS Diaz (1986a) Immunity to varicella-zoster viral glycoproteins, gp I (gp90/58) and gp III (gp118) and to a nonglycosylated protein, p 170. J Immunol 137:1346-1351.

Arvin AM, CM Koropchak, BRG Williams, FC Grumet, SKH Foung (1986b) Early immune response in healthy and immunocompromised subjects with primary varicella zoster virus infection. J Infect Dis 154:422-429.
Arvin AM, PA Hensleigh, CG Prober, DS Au, LL Yasukawa, AE Wittek, PE Palumbo, SG Paryani, AS Yeager (1986c) Failure of antepartum maternal cultures to predict the infant's risk of exposure to herpes simplex virus at delivery. N Engl J Med 315:796-800.
Arvin AM, SM Solem, CM Korpchak, E Kinney-Thomas, SG Paryani (1987) Humoral and cellular immunity to varicella-zoster virus glycoprotein gp I and to non-glycosylated protein, p 170, in the strain 2 guinea pig. J Gen Virol 68:2449-2454.
Bernstein DI, MA Lovett, YJ Bryson (1984) Serologic analysis of first-episode non-primary genital herpes simplex virus infection. Presence of type 2 antibody in acute serum. Am J Med 77:1055-1060.
Boucher FB, LL Yasukawa, DS Au, PA Hensleigh, AM Arvin, CG Prober (1988) Longitudinal study of herpes simplex type 2 (HSV-2) seroprevalence and seroconversion among pregnant women. Abstract, 28th Interscience Conference on Antimicrobial Agents and Chemotherapy, Los Angeles.
Brown ZA, LA Vontver, J Benedetti, CW Critchlow, SJ Sells, S Berry, L Corey (1988) Effects on infants of a first episode of genital herpes during pregnancy. N Engl J Med 317:1246-1251.
Brunell PA, C Geiser, Z Shehab, JE Wangh (1982) Administration of live varicella vaccine to children with leukemia. Lancet 2:1069-1072.
Brunell PA, VM Novelli, PM Keller, RW Ellis (1987) Antibodies to the three major glycoproteins of varicella-zoster virus: Search for the relevant host immune response. J Infect Dis 156:430-435.
Corey L, HG Adams, ZA Brown, KK Holmes (1983) Genital herpes simplex virus infections. Clinical manifestations, course, and complications. Ann Intern Med 98:958-972.
Corey L, PG Spear (1986) Infections with herpes simplex viruses. N Engl J Med 314:686-691 and 749-757.
Cremer NE, CK Cossen, GR Shell, L Pereira (1985) Antibody response to cytomegalovirus polypeptides captured by monoclonal antibodies on the solid phase in enzyme immunoassays. J Clin Microbiol 21:517-521.
Davison AJ, CM Edson, RW Ellis, B Forghani, D Gilden, C Grose, PM Keller, A Vafai, Z Wroblewska, K Yamanishi (1986) A new nomenclature for the glycoprotein genes of varicella-zoster virus and their glycosylated producers. J Virol 57:1195-1197.
Davison AJ, JE Scott (1986) The complete DNA sequence of varicella-zoster virus. J Gen Virol 67:1759-1816.
Diaz PS, S Smith, E Hunter, AM Arvin (1988) Immunity to whole varicella-zoster virus antigen, glycoprotein I and p 170 in relation to the immunization regimen with live attenuated varicella vaccine. Submitted.
Friedrichs WE, C Grose (1984) Glycoprotein gp 118 of varicella-zoster virus: purification by serial affinity chromatography. J Virol 49:992-996.
Gershon AA, SP Steinberg, L Gelb, G Galasso, W Borkowsky, P LaRussa, A Ferrara (1984) Live attenuated varicella vaccine: Efficacy for children with leukemia in remission. JAMA 252:355-362.
Gershon AA, SP Steinberg, P LaRussa, A Ferrara, M Hammerschlag, L Gelb, and the NIAID varicella vaccine collaborative study group (1988) Immunization of healthy adults with live attenuated varicella vaccine. J Infect Dis, in press.
Grose C (1987) Varicella-zoster virus: pathogenesis of human disease, the virus and viral replication, and the major viral glycoproteins. In: Hyman RW (ed) Natural History of Varicella Zoster Virus. Boca Raton: CRC Press, p. 1-65.
Grose C, WE Friedrichs (1982) Immunoprecipitable polypeptides specified by varicella-zoster virus. Virology 118:86-95.
Grose C, V Litwin (1988) Immunology of the varicella-zoster virus glycoproteins. J Infec Dis 157:877-881.
Ho RJ, BT Rouse, L Huang (1986) Target sensitive immunoliposomes: preparation and sensitivation. Biochemistry 25:5500-5506.
Hutto C, A Arvin, R Jacobs, R Steel, S Stagno, R Lyrene, L Willett, D Powell, R Andersen, J Werthammer, G Ratcliff, AJ Nahmias, C Christy, R Whitley (1987) Intrauterine herpes virus infections. J Pediatr

110:97-101.
Ito M, T Ihara, C Grose, S Starr (1985) Human leukocytes kill varicella-zoster virus infected fibroblasts in the presence of murine monoclonal antibodies to virus specific glycoproteins. J Virol 54:98.
Keller PM, BJ Neff, RW Ellis (1986) Three major glycoprotein genes of varicella-zoster virus whose products have neutralization epitopes. J Virol 52:293-297.
Lee FK, RM Coleman, L Pereira, PD Bailey, M Tatsuno, AJ Nahmias (1985) Detection of herpes simplex virus type-2 specific antibody glycoprotein G. J Clin Microbiol 22:641-644.
Matsunaga Y, K Yamanishi, M Takahashi (1982) Experimental infection and immune response of guinea pigs with varicella-zoster virus. Infect Immun 37:407-412.
McGeoch DJ, HWM Moss, D McNab, MC Frame (1987) DNA sequence and genetic content of the HindIII 1 region in the short unique component of the herpes simplex virus type 2 genome: identification of the gene encoding glycoprotein G, and evolutionary comparisons. J Gen Virol 68:19-38.
Mertz GJ, RW Coombs, R Ashley, J Jourden, M Remington, C Winter, A Fahnlander, M Guinan, H Ducey, L Corey (1988) Transmission of genital herpes in couples with one symptomatic and one asymptomatic partner; a prospective study. J Infect Dis 157::1169-1177.
Miller A. (1980) Selective decline in cellular immune response to varicella-zoster in the elderly. Neurology 30:582-587.
Myers MG, LR Stanberry, BJ Edmond (1985) Varicella-zoster virus infection of strain 2 guinea pigs. J Infect Dis 151:106-113.
Nahmias AJ, HH Keyserling, G Kerrick (1983) Herpes Simplex. In: Remington JS, JO Klein (eds) Infectious Diseases of the Fetus and Newborn Infant. Philadelphia, WB Saunders, p. 156-190.
Nahmias AJ, H Keyserling, R Bain (1985) Prevalence of herpes simplex virus type-specific antibodies in a USA prepaid medical practice population. Abstract, 6th International Society for STD Research.
Pachl C, RL Burke, LL Stuve, L Sanchez-Pescador, G Van Nest, F Masiarz, D Dina (1987) Expression of cell-associated and secreted forms of herpes simplex virus type 1 glycoprotein gB in mammalian cells. J Virol 61:315-325.
Palumbo PE, AM Arvin, CM Koropchak, AE Wittek (1984) Investigation of varicella-zoster virus infected cell proteins that elicit antibody production during primary varicella using the immune transfer method. J Gen Virol 65:2141.
Preblud SR. (1981) Age-specific risks of varicella complications. Pediatrics 68:14-17.
Prober CG, PA Hensleigh, FD Boucher, LL Yasukawa, DS Au, AM Arvin (1988) Identification of neonates exposed to herpes simplex virus: the value of routine viral cultures at delivery. N Engl J Med, in press.
Rasmussen LR, J Mullenax, M Nelson, TC Merigan (1985) Human cytomegalovirus polypeptides stimulate neutralizing antibody *in vivo* Virology 145:186-190.
Roizman B, B Norrild, C Chan, L Pereira (1984) Identification and preliminary mapping with monoclonal antibodies of herpes simplex virus 2 glycoprotein lacking a known type 1 counterpart. Virology 133:242-247.
Shiraki K, RW Hyman (1987) The immediate early proteins of varicella zoster virus. Virology 156:423-426.
Sullender WM, LL Yasukawa, M Schwartz, L Pereira, PA Hensleigh, CG Prober, AM Arvin (1988) type specific antibodies to herpes simplex virus-2 glycoprotein G in pregnant women, infants exposed to maternal HSV-2 infection at delivery and infants with neonatal herpes. J Infect Dis. 157:164-171.
Townsend AR, J Rothbard, FM Gotch, G Bahadur, D Wraith, AJ McMichael (1986(The epitopes of influenza nucleoprotein recognized by cytotoxic T lymphocytes can be defined with short synthetic peptides. Cell 44:959-968.
Weigle KA, C Grose (1984) Molecular dissection of the humoral immune response to individual varicella-zoster viral proteins during chickenpox, quiescence, reinfection, and reactivation. J Infect Dis 149:741-749.

Whitley RJ, AJ Nahmias, AM Visintine, CL Fleming, CA Alford (1980) The natural history of herpes simplex virus infection of mother and newborn. Pediatrics 66:796-800.
Yeager AS, AM Arvin (1984) Reasons for absence of a history of recurrent genital infections in mothers of neonates infected with herpes simplex viruses. Pediatrics 73:188-193.
Yewdell JW, JR Bennick, M Mackett, L Lefrancois, DS Lyles, B Moss (1986) Recognition of cloned vesicular stomatitis virus internal and external gene products by cytotoxic T lymphocytes. J Exp Med 163:1529-1534.
Zarling JM, PA Moran, LA Lasky, B Moss (1986) Herpes simplex virus specific human T-cell clones recognize HSV glycoprotein D expressed by a recombinant vaccinia virus. J Virol 59:506-509.
Zweerink HJ, L Corey (1982) Virus specific antibodies in patients with genital herpes simplex virus infections. Infect Immun 37:413-421.

DISCUSSION

de la Maza (University of California Irvine Medical Center, Orange, CA):

In one of your slides on the pregnant population, you said that there were 3.4% of the pregnant women that had a past history of herpes. Then the bottom line said that there were 1.7% that were herpes positive, IgG negative.

Arvin A:

No, they gave us a clinical story that they had had genital herpes. They're not viral culture proved.

de la Maza L:

Is this 1.7% of 3.4%?

Arvin A:

No, of the total population.

de la Maza L:

So, based on your definition, 50% of your patients had primary herpes during pregnancy.

Arvin A:

No, I think I've lost something there in trying to get the message across. In fact, only 14% of them had primary infection at the time of delivery, and those ones that were shedding. That slide that you are referring to was just sort of a surveillence study, not done at the time of delivery. But just looking at a total population and trying to see how the history of herpes correlated with whether they had type-2 antibodies or not.

de la Maza L:

If the data is correct, it's either your test is not picking up those patients or you really have . . .

Arvin A:

No, I think those 1.7% probably had type-1 genital herpes.

de la Maza L:

Well, in this country, 10% of the people have genital herpes type-1.

Arvin A:

If you look at primary genital herpes, it's about 10% type-1. If you look at recurrent genital herpes, which is mainly what we're dealing with in these pregnant women, it is much less. It is maybe 3 or 4%. The recurrence rate for type-1 herpes is very different in the genital site than for type-2. So, we realize, of course, that some of the maternal infection at delivery will be type-1. We're not saying that this will solve the whole problem of exposure to babies, because it will not detect the type-1 infection. But the recurrence rate for genital type-1 is very low, and I think is O.K. not to focus on that phenomonen, even though we know it occurs.

de la Maza L:

So, what you are saying basically, is that your test is not picking up a certain number of patients that are herpes-type-2 recurrent, but they are IgG negative.

Arvin A:

No, I not saying that. I think that those people probably had type-1. I cannot prove it, but that would be my guess.

de la Maza L:

If the percentage of herpes type-1 is very low, there is no way you can account for the 50% that you have between your 3.4% and 1.7%.

Arvin A:

Well, what that illustrates to me is the unreliability of the clinical history. Relying only on the history, the person may have had cancroid or something like that. Our whole experience has been that the history, pro or con, is not very helpful. That's what I meant to illustrate there.

Clark J (Humana Hospital, Sunrise, Las Vegas, NV):

We had had a varicella zoster that I confirmed, called the doctor and told him that it was varicella zoster. He wanted to know if there was a possibility of a second infection with varicella zoster. I told him that from what I had read there was no evidence. He assured me that he had several reliable people to tell him that there is.

Arvin A:

You can show in the literature a couple of cases of virologically proved infection that was a second infection that was symptomatic. In our experience, we have never proved one, even though we've tried. I don't think that most of what is called chicken pox and then is followed by true chicken pox was chicken pox to begin with, if you look at all the viruses that can cause vesicular rashes. It would have had to be confirmed as varicella the first time, and it is not easy to look at a skin lesion and say it's a chicken pox lesion or it's not. So, I think what you said was correct.

Trachtenberg A (CA Dept. of Health Services, Office of AIDS, Berkeley, CA):

I was very interested in your slide of anti-gG seroprevalence by age. I had two comments about it. One, you commented off-the-cuff that you would suppose male seroprevalence by age would be similar. I think it's pretty well understood that there is a certain bias for men to have sexual partners with females of a younger age than themselves.

Arvin A:

I don't know anything about that at all. I get your point. It should not be immediately extrapolated, and I'm sure that gay men population studies would be different, in terms of prevalence for type-2 infection as well.

Trachtenberg A:

The second point you were wondering about was the 25-30 year old women who were the highest, then it went down. I think what you may well be seeing there is a kind of cohort effect. Not only will women pick up infection throughout their lives, but you have a cohort that might have gone through a less complex and varied sexual experience than the somewhat older age group, than in the cohort behind them.

Arvin A:

Well, I'm in the somewhat older age group, and I'm not ready to explain it quite in that way, but it's possible.

Participant:

I have a couple of questions that are probably not terribly practical, but are interesting to me. One, when you stated that there is a prolongation of development in antibodies to gG, is that based on a functional correlation with what that protein does?

Arvin A:

Well, that's an interesting question, but I don't think we can answer it because we don't understand functionally what the protein does. It's certainly there on the infected cells, very early on. Why it takes the host so long to respond to it, I don't know.

Participant:

The second part of that question is, you made the comment that in neonates it's perhaps due to an immune disfunction, but my question there would be, is it more of an altered virus or prolonged stage in a certain part of the replication cycle that is doing it, rather than an absent immune response to antibody in that situation.

Arvin A:

Yes, it's certainly possible.

Participant:

A similar question in regards to the response to VZV in that you showed a diminished *in vitro* response to a couple of the proteins. Is that because those proteins aren't expressed in the vaccine strain or produced by the vaccine strain, or is it because you're using immunosuppressed individuals who are immunized.

Arvin A:

Those were absolutely normal individuals who are immunized. Both of the proteins that we've looked at are expressed by the vaccine virus. There is a possibility that the glycoprotein, the gP5 is not made by the vaccine virus, but the other ones are there in.

Participant:

The other part of that is, these proteins have to be seen by the host in context of self, so does this mean that they are seen somehow differently.

Arvin A:

I think that is an interesting thing to speculate about. I think that's possible. Because you're not providing the virus entry in the same way, so how the virus is seen at that phase may differ, whether it's seen with class 1 or class 2, or whether it's seen as what portion of what proteins could really differ, I think. It's possible.

Caulfield C (AIDS Resource Center, Dallas, TX):

You spoke a little bit about the VZ vaccine and its T-cell proliferative function. I wonder what the implications are for T-lymphotrophic viruses in using this vaccine as a treatment.

Arvin A:

Well, I don't think anyone would ever give this live viral vaccine to.....

Caulfield C:

It is a live viral then?

Arvin A:

Yes, it is. It is attenuated, but it is live virus. I wanted to make just one comment if I can about something that Dr. Warford mentioned, and that is the issue about whether we should be doing these cultures at the time of delivery on all women who deliver. I wanted to just make it clear that we do not think that this should become the standard practice. I wanted to be sure that I don't give the impression that, at this time, is indicated. I think she has a wonderful opportunity to tell us whether it's going to be valuable or not. But we, as Dr. Corey and other people work on this problem, do not feel that it is indicated. We have plenty of experience in this problem of herpes during pregnancy that would lead us to follow the recommendation of cursing the darkness rather than lighting the wrong candle. I think that we would be lighting the wrong candle if we said all of a sudden, now everybody should culture every pregnant woman. It would be just as bad as the whole thing with the antipartum culture.

de la Maza L:

Would you recommend culturing everybody that is gG positive?

Arvin A:

I think that that is at least a target population. We have a problem if we do that, though. We still don't know what to do with the exposed babies. That is another complication of identifying those babies. But, we are looking in a research sense at the value of that. Does it make sense to focus on those women? There's a good chance that it will not be the only strategy because of these primary infections that do happen at term.

HEPATITIS DELTA VIRUS: INFECTION AND DISEASE

BONINO FERRUCCIO, BRUNETTO MAURIZIA ROSSANA, ROSINA FLORIANO AND OLIVERI FILIPPO

Division of Gastroenterology, San Giovanni Battista "Molinette" Hospital Corso Bramante 88, 10126, Torino Italy

BIOLOGY OF HEPATITIS DELTA VIRUS

Hepatitis delta virus (HDV) is a unique defective agent whose infection is inextricably linked to that of hepatitis B virus (HBV) (Rizzetto et al. 1987; Verme et al. 1983). The virion is a 36 nm spherical particle containing hepatitis delta antigen (HDAg) and RNA (HDV-RNA); its envelope consists of a composition of proteins similar to that of 22 nm defective particles of HBV (Bergman and Gerin, 1986; Bonino et al. 1981, 1986a). Transmission experiments in chimpanzees suggest that HDV is a distinct infectious agent but its expression requires the helper function of HBV (Purcell et al. 1983). An infectious inoculum induced HBV hepatitis and intrahepatic expression of HDAg in HBsAg negative chimpanzees up to a 1 x 10^{-5} dilution whereas a 1 x 10^{-6} dilution caused only an HBV infection (Purcell et al. 1983). No infection occurred in HBsAg negative chimpanzees using higher dilutions of the same inoculum (>1 x 10^{-7}) while dilutions, up to 1 x 10^{-11} induced hepatitis with intrahepatic expression of HDAg in HBsAg carrier chimpanzees. Experimental transmission of HDV to chimpanzees showed also that immunity against HBV protects against HDV (Purcell et al. 1983; Rizzetto et al. 1987).

The genome of the defective agent, has been reverse transcribed to complementary DNA (cDNA), cloned and sequenced (Denniston et al. 1986; Wang et al. 1986). No significant homology has been observed between HDV-RNA, HBV-DNA, host DNA and nucleic acid obtained from serum and liver of HBV or non-A non-B (NANB) infected animals (Weiner et al. 1987).

HDV-RNA contains 1,679-1,683 nucleotides and five open reading frames (ORF) of more than 100 aminoacids in its genomic and antigenomic strands. The largest ORF, found on the + strand codes for 214 amino acids, was expressed in bacteria and the resultant fusion protein reacted with human antibodies to HDAg (anti-HD) (Wang et al. 1986). *In vitro* translation of RNA derived from HDV infected liver produced two HDAg proteins (24 and 27 Kda) which reacted by immunoblot with polyclonal anti-HDV antibody and with a monoclonal antibody obtained from Epstein-Barr virus-transformed human lymphocytes (Pohl et al. 1987). Serum derived HDAg is composed of two major proteins, with a molecular weight of 24-27 Kda and 27-29 Kda while these proteins are detected in the liver together with the other forms of different molecular weight (22, 14 Kda) (Bonino et al. 1986a).

The lack of a +3'poly-(A) tract on the viral RNA, together with its failure to translate HDAg *in vitro* and the presence of HDAg coding sequences on the antigenomic strand of HDV-RNA, suggests that HDV is a negative strand virus in which the complementary strand of the genome (+) serves as a mRNA (Baroudy et al. 1987). HDV particles contain two RNA species that migrate in gel electrophoresis at 1.75 and 2.0 Kb, respectively; these forms have the same polarity (genomic, +) and it may be that they represent nicked linear and circular forms of the same RNA molecule. The circular RNA molecule has an extensive secondary structure; its self-annealing properties and high GC content (60%) allow the formation of a double stranded rod-like structure resembling that of a number of different viroid-like RNAs found in plants. Different forms of HDV-RNA are found in infected livers including genomic and antigenomic strands, poly-A RNA and double stranded RNA (dsRNA). The dsRNA species might represent replicative forms of HDV-RNA suggesting that HDV's replicative cycle is somewhat similar to that of the satellites of plant RNA viruses (Kaper and Tousignant, 1984). Satellite RNAs of cucumber mosaic virus (CUMOV) replicate in a rolling circle fashion producing large amounts of double stranded forms which accumulate in the cell and hinder the replication of the helper virus RNA. This determines the attenuation of disease symptoms caused by CUMOV. Occasionally, however, this defensive mechanism turns out to be dangerous for the plant; this happens when a CMV carrier plant is superinfected with satellites containing defective necrotizing genes which caused tremendous epidemics of plant necrosis. Satellite/CUMOV interference in plants closely resembles that of HDV and HBV in humans and animals. Furthermore, some sequence homology of HDV-RNA with plant virusoids (Wang et al. 1986) suggests that HDV and these agents may have a common ancestry.

EPIDEMIOLOGY

HDV infection is linked to that of the helper virus. In areas of the world where there is low prevalence of HBsAg in the general population (Western world), HDV infection is virtually confined to intravenous drug abusers and to high-risk populations of HBsAg carriers (recipients of multiple and pooled blood products, hemodialysis patients, mentally retarded persons, etc.) (Gmelin et al. 1987; Ponzetto et al. 1986). In countries with intermediate or high HBV endemicity (Far East Africa, South America and southern Europe in increasing order of prevalence), HDV infection also occurs in normal individuals (Ponzetto et al. 1986). HDV is transmitted after exposure to infectious body fluids during close or intimate contact and in this way spreads efficiently within the household (Bonino et al. 1985a). The likelihood of infection increases where persons are crowded together in unhygienic conditions. This explains why HDV infection is most common in the socially underprivileged communities of tropical areas (Ponzetto et al. 1986).

HDV infection in babies was documented only in those born to HBeAg-positive, HDV-infected mothers (Zanetti et al. 1982). The great majority of chronic carriers of HDV have low levels of HBV replication, and for this reason perinatal transmission of HDV is rare.

The risk of post-transfusion HDV infection appears to be negligible in the case of recipients who are negative for markers of HBV and HDV (Rosina et al. 1985). This is conceivable, since the presence of HBsAg in the serum usually disqualifies the HDV carrier from giving blood. However, the fact that HBsAg synthesis may be inhibited to subminimal levels in asymptomatic HbsAg carriers who are superinfected by HDV points to a potential risk of transmission (Rosina et al. 1985). The risk becomes particularly grave when the recipient is also a carrier of HBsAg and the amount of HDV to be rescued is infinitesimal as a consequence of the carriers persistent and well-established HBV infection. In keeping with the high infectivity of HD in the HBsAg carrier, anti-HD is usually found in the serum of 30% - 100% of HBsAg carrier hemophiliacs who have been treated with commercial clotting factors prepared from large pools of donors (Rosina et al. 1985), whereas this antibody has a lower prevalence or is absent in HBsAg carriers treated with products prepared from small pools of donors. These results indicate that blood-derived material given to HBsAg carriers should be obtained from small pools of the safest donors and it should be carefully checked for all HBV and HDV markers, as well as for HBsAg. Sexual activity is also an important factor in the spread of HDV hepatitis, particularly sexual activity between HBsAg carriers.

HDV DISEASE

HDV hepatitis is not clinically distinguisable from the other forms of viral hepatitis; it occurs between one week (even less in the recipient who is an HBsAg carrier) and eight weeks after infection. The severity and outcome of the disease depend on the mode of infection, that is whether HDV infection coincides with that of HBV (coinfection) or whether it is superimposed on a chronic carrier of HBsAg (superinfection) (Caredda et al. 1985).

When susceptible individuals are exposed simultaneously to both viruses, the duration of HBV infection is the limiting factor of HDV expression. Acute HBV infection, even where it is associated with HDV, rarely evolves to chronicity and HDV cannot persist after the clearance of HBV. Usually, the acute illness secondary to HBV/HDV coinfection is not clinically different from classic type B hepatitis. The serologic response to HDV is often elusive; it can be totally absent or limited to a delayed appearance of anti-HD after 30-60 days from onset (Caredda et al. 1985). Acute self-limited hepatitis is characterized by several serologic profiles which vary from the early appearance of HDAg followed by seroconversion to IgM anti-HD and IgG anti-HD, to an IgM/IgG response in the absence of detectable HD antigenemia, and the sole and short-lived detection of IgM or IgG anti-HD. Titers of IgM and IgG anti-HD remain relatively low ($<10^4$). Progressive HDV hepatitis is accompanied by the rapid rise of IgM and IgG antibodies to high titers ($>10^4$) and their persistence permits diagnosis of chronicity. During the acute phase of HBV/HDV coinfection serum HDAg and HDV-RNA can be detected by sensitive solid-phase immunoassays or slot-hybridization, respectively, providing the specific diagnosis of HDV infection which otherwise might be lost; detection of IgM anti-HD can also be helpful in this context. Occasionally, however, in HBV/HDV coinfection a strong serologic response to HDV develops; this form predominates in drug addicts and may be associated with severe liver damage (Arico' et al. 1985; Craxi et al. 1984; Farci et al. 1983). HDAg appears in the serum during the acute phase of infection with a subsequent rise in titers of IgM and IgG anti-HD. In this case a typical clinical course is biphasic hepatitis in which the acute illness is characterized by two peaks of serum transaminases (ALT) (Caredda et al. 1985; Govindarajan et al. 1986). When a single inoculum or consecutive injections are highly infectious for both HBV and HDV, massive HBV replication supports the rapid rescue of the defective agent, which overwhelms and suppresses the synthesis of HBV gene products and thus determines the first increase of ALT. Subsequently, HBV infection completes its natural course, leading to the second bout of hepatitis. In keeping with this hypothesis, the peak of HD viremia (HDAg) precedes or is associated with the first ALT peak. On the other hand, a strong anti-HD immune response occurs at the time of the second ALT increase when the highest values of HBV replication are usually found. The two necrotic events may pose a higher risk of fulminant hepatitis. However, the outcome of HBV/HDV coinfection is benign in the great majority of cases and 90% of patients recover completely with resultant seroconversion to anti-HBs (Caredda et al. 1985).

The mechanism and the natural course of HDV infection are different in chronic carriers of HBsAg superinfected with HDV (Colombo et al. 1983; De Cock et al. 1986). Persistent and well-established HBV infection promotes the florid replication of the defective virus and predisposes one to severe liver damage. This form accounts for a large proportion (30%-60%) of fulminant HBsAg positive hepatitis (Govindarajan et al. 1985; Hadler et al. 1984). In HDV superinfection, the strong serologic response to HDV is characterized by high titers of IgM and IgG anti-HD. HDV superinfection has the propensity to become chronic, since permanent HBV infection permits the indefinite life of the defective agent. Transition to chronicity of HDV infection simulates progressive type B hepatitis and poses important diagnostic problems in cases where the HBsAg carrier status of the patient is known. The lack of IgM antibody to HBV "core" antigen in the serum or the persistence of low titers of this antibody during the course of acute HBsAg positive hepatitis, should always raise the suspicion of HDV or NANB superinfection in a previously unrecognized carrier of HBsAg (Lavarini et al. 1983). In HBsAg positive carriers a decrease of viral replication usually precedes anti-HBs seroconversion indicating an improvement in the clinical picture, but in some unfortunate cases this event may suggest the inhibition of HBV synthesis exerted by superinfection of HDV heralding a deterioration in the liver disease. Viral DNA and anti-HBs may coexist in the serum of HBsAg carriers coinfected or superinfected HDV this is due to the inhibitory effect exerted by the defective agent on the synthesis of the helper virus and it is documented also in carriers of HBsAg with chronic HDV infection (Bonino et al. 1985b).

Detection of HDV markers is mandatory also in serum of HBsAg negative patients with acute hepatitis since the explosive replication of

HDV during acute superinfection can suppress the synthesis of HBsAg, making its levels undetectable. Since the survival of the defective virus depends on the highly efficient use of the replicative machinery of HBV, HDV may compete strongly with HBV. Obviously, the inhibition exerted by HDV cannot be so intense as to suppress the synthesis of HBV completely, since a certain degree of replication of the helper virus is necessary for the survival of the defective agent (Rizzetto et al. 1985). Occasionally, however, in very fortunate cases supervening HDV infection may induce the termination of the HBsAg carrier state with seroconversion to anti-HBs and clearance of both HBV and HDV. The outcome of acute HDV superinfection may depend on whether the carrier of HBsAg was previously asymptomatic or whether he suffered from HBV related illness (Colombo et al. 1983; De Cock et al. 1986). In patients with prior HBV-induced liver disease, HDV causes further damage, accelerating the course of chronic hepatitis towards cirrhosis. Unfortunately, however, severe and progressive liver disease develops also in healthy carriers of HBsAg who have been superinfected with HDV (Arico' et al. 1985). Only a small proportion experience a self-limited HDV infection while the great majority of patients develop chronic liver disease.

Chronic HDV disease is a multifaced illness, in which clinical, laboratory and histologic features overlap considerably with those of other forms of chronic hepatitis (Colombo et al. 1983; De Cock et al. 1986); nonspecific autoimmune phenomena were observed in 40-60% of cases (Crivelli et al. 1983).

Diagnosing of chronic HDV hepatitis by only testing sera for anti-HD is sometimes difficult since it relies on persistence of high titers of both total and IgM antibodies (Smedile et al. 1983). Specific diagnosis requires detection of HDV-RNA or HDAg in the serum; a 90% correspondence was found between detection of intrahepatic HDAg and serum HDV-RNA and HDAg (Smedile et al. 1986). Current immunoassays do not detect HDAg in the serum of patients with chronic HDV hepatitis where HDAg is masked by high titers of the homologous antibody (anti-HD); HDAg set free after treatment with detergent is immediately captured by the homologous antibody and made unaccessible to the labeled anti-HD. In these sera HDAg can be detected by immunoblotting where HDAg proteins are separated from their antibodies by centrifugation and denaturation before being electrophoresed in SDS-PAGE gels and transferred to nitrocellulose (Bonino et al. 1986a). Intrahepatic HDAg is stained by means of immunohistochemical techniques in frozen or formalin-fixed, paraffin embedded liver specimens using anti-HD labeled with fluorochromes or enzymes. Poor prognosis of HDV chronic hepatitis is usually associated with the persistence of intrahepatic HDAg, however chronic liver disease may persist also through recurrent bouts of hepatitis with quantitative reduction or elimination of HDAg from the liver and subsequent new expression of the viral antigen in the hepatocytes. Intrahepatic HDAg may be undetectable also in the liver of patients who have a long lasting disease and are currently positive for serum HDV-RNA. These observations indicate the limits of a liver biopsy that offers only a single frame of the moving picture which characterizes the evolution of chronic hepatitis and emphasize the diagnostic importance of noninvasive, serological methods for detection of HDV-RNA and HDAg. The association of these two assays allows detection of HD replication in about 80% of chronic HBsAg carriers with high titers of anti-HD in serum providing a specific tool to monitor HDV infection in patients undergoing antiviral treatment.

Chronic HDV hepatitis represents the major cause of cirrhosis in young adults in Italy (Brunetto et al. 1987), it shows a rapid progression in about 15% of cases with the development of clinical symptoms of cirrhosis within one year of the acute illness, and a slower progression in the majority of patients (70%) with development of clinical cirrhosis between 15-25 years after the acute HDV infection, it is benign with remission of liver disease in only 15% of cases (Bonino et al. 1987). The search for predictive markers of unfavourable outcomes of HDV infection was extensive; but the only useful markers appeared to be the persistence of markers of ongoing HDV replication in serum and liver (HDAg and HDV-RNA) and the persistence of florid antibody response to HDAg (both IgG and IgM type), in serum.

PATHOLOGY

Whether HDV is directly cytophatic or liver damage is caused by the immune response is yet unknown. The evidence that liver damage is invariably found in both symptomatic and asymptomatic carriers of HDV and that important cytopathic lesions are present in the HDV infected livers where a single cell may contain up to 300,000 copies of HDV-RNA suggest that a direct liver cell damage might be induced by HDV infection alone. However, transmission experiments in animals showed that expression of HDAg in the liver cell nuclei precedes cell damage and inflammation. After development of the host's immunologic response (serum anti-HDV), immunocomplexes are seen within the hepatocytes, lymphocytes accumulate in portal tracts and sinusoids and liver necrosis takes place (Kojima et al. 1986). Studies in man reported a significant relation between the histologic activity and the number of HDAg positive hepatocytes (Negro et al. 1988). These observations entail that the major cause of liver damage might be the host's immune reaction to viral replication.

The outcome of HDV disease was unrelated to the severity of the initial morphologic lesion and cirrhosis developed in patients displaying piece meal necrosis either with or without bridging necrosis as well as in patients with minimal periportal lesions (Verme et al. 1986). Varying degrees of necrosis, steatosis, cholestasis and inflammation were observed in HDV hepatitis out these features were undistinguishable from those observed in other forms of viral hepatitis. Two histologic features, albeit unspecific, were prevalent in both HDV and NANB hepatitis; namely the eosinophilic degeneration of single cells or clusters of hepatocytes and a microvesicular type of cell degeneration (morula-like). Light microscopic changes occurring during NANB and HDV hepatitis were paralleled by electron micrographic similarities of the ultrastructure of the cells (Dienes et al. 1981). Cytoplasmic membranous alterations and aggregates of nuclear particles first detected in chimpanzees with NANB hepatitis (Dienstag and Alter, 1986) were also detected in animals infected with HDV (Canese et al. 1984). Cytoplasmic changes consisted of convoluted tubules, 150 to 300 nm in diameter, attached to the endoplasmic reticulum, which were composed of double membranes containing an osmiophilic substance. However, neither of these structures was specifically associated with HDV or NANB hepatitis since similar findings were reported in a multitude of other viral and nonviral diseases. Shimizu et al. (1987) found that EBV-transformed lymphocytes from a convalescent NANB-infected chimpanzee synthesized an IgM antibody which identifies the same antigen in the hepatocyte's cytoplasm of NANB and HDV-infected animals. This antibody appeared to be directed against a host specific rather than a virus specific antigen, possibly representing a common product of the infected cells to the infection of different agents. However, these observations, the lack of cross-protection between NANB and HDV hepatitis in cross-challenge transmission experiments and the lack of hybridization between HDV-RNA and nucleic acids from NANB material indicate that the HDV and the putative NANB agents are minimally, if not at all related.

Available data suggest that HDV does not play a direct role in the pathogenesis of hepatocellular carcinoma (Chen et al. 1984), however HDV induces a higher risk of death from cancer for the patient with chronic HBV infection since it causes an early development of cirrhosis in the HBsAg carrier (Bonino et al. 1987, Brunetto et al. 1987).

THERAPY

Treatment of chronic HDV hepatitis has been rather disappointing since neither steroids, nor azathioprine, nor levamisole are helpful (Arrigoni et al. 1983; Rizzetto et al. 1983). We treated 12 patients with alpha 2A recombinant interferon at the dosage of 5 million units per square meter of body surface for 12-weeks (Rizzetto et al. 1986). During therapy four of them had a biochemical and virologic improvement. The clinical and virologic improvement was maintained throughout a one year follow-up only in one patient; in the others the disease relapsed and viremia rebounded to pre-treatment levels after discontinuation of therapy. A temporary antiviral effect of alpha IFN was also reported by Hoofnagle et al. (1987) thus a one year administration is at present under evaluation through a controlled trial involving several Italian centers.

We hope that current and future studies will bear fruit in the search for effective drugs and an anti-HDV vaccine to treat HDV hepatitis and to prevent HDV infection, respectively.

REFERENCES

Arico' S, Aragona M, Rizzetto M, Caredda F, Zanetti A, Marinucci, Diane S,Pastore G, Raimondo G, Craxi A (1985) Clinical significance of antibody to the hepatitis delta virus in symptomless HBsAg carriers. Lancet 2:356-357.

Arrigoni A, Ponzetto A, Actis G, Bonino F (1983) Levamisole and chronic delta hepatitis. Ann Intern Med 98:1024.

Baroudy BM, Smedile A, Korba BE, Bergmann KF, Pohl C, Well FV,Gerin virus. Transcription and replication of hepatitis delta virus (1987) Prog Clin Biol Res 234:89-92.

Bergman KF, Gerin JT (1986) Antigens of hepatitis delta virus in the liver and serum of humans and animals. J Infect Dis 154:702-706.

Bonino F (1986) The importance of hepatitis B viral DNA in serum and liver. J Hepatol 3:136-141.

Bonino F, Hoyer BM, Ford E, Shih J-W K, Purcell RH, Gerin JL (1981) The delta agent: HBsAg particles with delta antigen and RNA in the serum of a HBV carrier. Hepatology 1:127-131.

Bonino F, Caporaso N, Dentico P. Marinucci G, Valerio L, Craxi A, Ascione A, Raimondo G, Piccinino F, Rocca G, Rizzetto M (1985a) Hepatology 1:221-226.

Bonino F, Negro F, Chiaberge E (1985b) Active HBV replication in HBsAg carriers with chronic delta infection. Ital J Gastroentero 17:235.

Bonino F, Heermann KH, Rizzetto M, Gerlich WH (1986a) Hepatitis delta virus: protein composition of delta antigen and its hepatitis B virus derived envelope. J Virol 58:945-950.

Bonino F, Rosina F, Rizzetto M, Rizzi R, Chiaberge E, Tardanico R, Callea F, Verme G (1986b) Chronic hepatitis in HBsAg carriers with serum HBV-DNA and anti-HBe. Gastroenterol 90:1268-1271.

Bonino F, Negro F, Baldi M, Brunetto MR, Chiaberge E, Capalbo M, Maran E Lavarini C, Rocca N. Rocca G (1987) The natural history of chronic delta hepatitis. Prog Clin Biol Res 234:145-152.

Brunetto MR, Baldi M, Bonino F, Goi M, Arico' S, Vanni A, Bianchi FP, Zoli M, Canestrini C, Ascione A, Ciampi R, Isabella L, Craxi A, Pagliaro L, Rizzetto M (1987) Chronic HDV infection:an important cause of HBsAg positive cirrhosis of young adults. Prog Clin Biol Res 234:207-208.

Canese MG, Rizzetto M, Novara R, London WT, Purcell RH (1984) Experimental infection of chimpanzees with the HBsAg associated delta agent: an ultrastructural study. J Med Virol 13:63-72.

Caredda F, Rossi E, D'Arminio Monforte A, Zampini L, Re T, Meroni B, Moroni M (1985) Hepatitis B virus coinfection and superinfection with delta agent: indistinguishable disease with different outcome. J Infect Dis 151:925-928.

Chen DS, Lai MY, Sung JL (1984) Delta agent infection in patients with chronic liver diseases and hepatocellular carcinoma infrequent finding in Taiwan. Hepatology 4:502-503.

Colombo M, Cambieri R, Rumi MG, Ronchi G, Del Ninno E, De Franchis R (1983) Long-term delta superinfection in hepatitis B surface antigen carriers and its relationship to the course of chronic hepatitis.Gastroenterology 85:235-239.

Craxi A, Raimondo G, Longo G, Giannuoli G, De Pasquale R, Caltagirone M, Patti S, Squadrito G, Pagliaro L (1984) Delta agent infection in acute hepatitis and chronic HBsAg carriers with and without liver disease. Gut 25:1288-1290.

Crivelli O, Lavarini C, Chiaberge E, Amoroso A, Farci P, Negro Rizzetto M (1983) Microsomal autoantibodies in chronic infection with the HBsAg associated delta agent. Clin Exper Immunol 54:232-238.

De Cock KM, Govindarajan S, Chin KP, Redeker AG (1986) Delta hepatitis in the Los Angeles area: a report of 126 cases.Ann Inter Med 105:108-114.

Denniston KJ, Hoyer BH, Smedile A, Wells JN, Gerin JL (1986) Cloned fragment of the hepatitis delta virus RNA genome: sequence and diagnostic application. Science 232:873-875.

Dienes HP, Purcell RH, Popper H, Bonino F, Ponzetto A (1981) Simultaneous infection of chimpanzees with more than one hepatitis virus. Hepatology 1:506.

Dienstag J, Alter HJ (1986) Non-A. Non-B hepatitis: evolving epidemiologic and clinical perspective. Seminars in Liver Disease 6:67-81.

Farci P, Smedile A, Lavarini C, Piantino P, Crivelli O,Caporaso N, Toti M, Bonino F, Rizzetto M (1983) Delta hepatitis in inapparent carriers of hepatitis B surface antigen. A disease simulating acute B hepatitis progressive to chronicity. Gastroenterology 5:669-673.

Gmelin K, Roggendorf M, Schlipkoter U, Theilmann L, Bernhardt R, Kommerel R (1987) Delta infection in haemodialyzed patient. J Infect Dis 151:74.

Govindarajan S, De Cock KM, Peters RL (1985) Morphologic and immunohistochemical features of fulminant delta hepatitis. Human Pathol 16:262-267.

Govindarajan S, Valinluck B, Peters RL (1986) Relapse of acute B viral hepatitis - role of delta agent. Gut 27:19-22.

Hadler SC, de Monzon M, Ponzetto A, Anzola E, Rivero D, Mondolfi A,Bracho A, Francis P, Gerber MA, Thung S, Gerin JL, Maynard E,Popper H, Purcell RH (1984) Delta virus infection and severehepatitis. Ann Intern Med 100:339-344.

Hoofnagle J, Mullen K, Peters M. Avigan M, Park Y, Waggoner J, Gerin, Hoyer B, Smedile A (1987) Treatment of chronic delta hepatitis with recombinant human alpha interferon. Prog Clin Biol Res 234:291-298

Kaper JM, Tousignant ME (1984) Viral satellites: parasitic nuclei acids capable of modulating disease expression. Endeavour 8:194-200.

Kojima T, Callea F, Desmyter J (1986) Hepatitis delta antigen (HDAg) in hepatocytes and lymphocyte reaction in immune light and electron microscopic studies: Abstracts of the ASL Meeting, Caracas, Venezuela, September 14-217, p 100.

Lavarini C, Farci P, Chiaberge E, Veglio V, Giacobbi D, Bedarida G,Susani G, Toti M, Almi P, Caporaso N (1983) IgM antibody against hepatitis B core antigen (IgM anti-HBc): diagnostic and prognostic significance in acute HBsAg positive hepatitis. Br Med J 287:1254-1256.

Negro F, Baldi M, Bonino F, Rocca G, De Martini A, Pessarino G, Maran Elavarini C, Rizzetto M, Verme G (1988) Chronic HDV (hepatitis delta virus) Intrahepatic expression of delta antigen, histologic activity and outcome of liver disease. J Hepatol 6:8-14.

Pohl C, Baroudy BM, Bergmann KF, Gerin JL (1987) Identification of viral polypeptides and *in vitro* translation products of the genome of the hepatitis D virus by human monoclonal antioody. London, International Symposium on Viral Hepatitis, abstracts: 36 A.

Ponzetto A, Forzani B, Parravicini PP, Hele C, Zanetti A, Rizzetto M (1986) Epidemiology of hepatitis delta virus infection.Eur Epidemiol 1:257-263.

Purcell RH, Gerin JL, Rizzetto M (1983) Experimental transmission ofthe delta agent in chimpanzees. Prog Clin Biol Res 143:79-90.

Rizzetto M, Canese MG, Arico' S, Crivelli 0, Trepo' C, Bonino F, Verme G(1977) Immunofluorescence detection of a new antigen-antibody system (delta anti-delta) associated to hepatitis B virus in liver and serum of HBsAg carriers. Gut 18:997-1003.

Rizzetto M, Verme G, Recchia S, Bonino F, Farci P, Arico S, Calzia R, Picciotto A, Colombo M, Popper H (1983) Chronic hepatitis in carriers of hepatitis B surface antigen with intrahepatic expression of the Delta antigen. An active and progressive disease unresponsive to immunosuppressive treatment. Ann Intern Med 98:437-41.

Rizzetto M, Ponzetto A, Bonino F, Purcell RH (1985) Superimposed hepatitis and the effect on viral replication in chronic hepatitis B. In the proceedings of the International Symposium on Antiviral Therapy of type B hepatitis. London, November, 11-12.

Rizzetto M, Rosina F, Saracco G, Bellando PC, Actis GC, Smedile A, Trinchero P, Sansalvadore F, Pintus C, Dianzani F, Verme G (1986) Treatment of chronic delta hepatitis with alpha-2 recombinant interferon. J Hepatol 3:2298-2338.

Rizzetto M, Gerin JL, Purcell RH (eds) (1987) The hepatitis delta virus and its infection. Liss AR, New York, USA. pp 167.

Rosina F, Saracco G, Rizzetto M (1985) Risk of post-transfusion infection with the hepatitis delta virus. A multicenter study. N Engl J Med 312:1488-1491.

Sagnelli E, Piccinino F, Pasquale G, Giusti G (1984) delta agent infection: an unfavorable event in HBsAg positive chronic hepatitis. Liver 4:170-176.

Shimizu YK, Purcell RH, Gerin JL, Feinstone SM, Ono Y, Shikata T (1987) Relation between non-A, non-B and delta hepatitis: Studies with monoclonal antibodies. Prog Clin Biol Res 234:51-60.

Smedile A. Dentico P, Zanetti A, Sagnelli E, Nordenfelt E, Actis GC, Rizzetto M (1981) Infection with the delta agent in chronic HVBsAg carriers. Gastroenterology 81:992-997.

Smedile A, Lavarini C, Crivelli 0, Raimondo G, Fassone M, Rizzetto M (1983) Radioimmunoassay detection of IgM antibodies to the HBV associated antigen. Clinical significance in delta infection. J Med Virol 9:131-138.

Smedile A, Rizzetto M, Denniston KJ, Bonino F, Wells F, Verme G, Consolo F, Hoyer B, Purcell R, Gerin J (1986) Type D hepatitis: the clinical significance of HDV-RNA in serum as detected by a hybridization-based assay. Hepatology 6:1297-1302.

Verme G, Rizzetto M, Bonino F (eds) (1983) Viral hepatitis and delta infection. Liss AR, New York, USA.

Verme G, Amoroso P. Lettieri G, Pierri P, David E, Sessa F, Rizzi R,`Bonino F, Recchia S, Rizzetto M (1986) A histological study of hepatitis delta virus liver disease. Hepatology 6:1303-1307.

Wang K-S, Choo Q-L, Weiner JH, Ou JH, Najarian RC, Thayer RM, Mullenbach GT, Denniston KH, Gerin JL, Hougton M (1986) Structure, sequence and expression of the hepatitis delta viral genome. Nature 323:508-514.

Weiner AJ, Wang K-S, Choo Q-L. Gerin JL, Bradley DW, Hougton M (1987) Hepatitis delta cDNA clones: undetectable hybridization to nucleic acids from infectious Non-A, Non-B hepatitis materials and hepatitis B DNA. J Med Virol 21:239-247

Zanetti A, Ferroni P, Magliano E, Pirovano P, Lavarini C, Messaro AL, Gavinelli R, Fabris C, Rizzetto M (1982) Perinatal transmission of the hepatitis B virus and of the HBV-associated delta agent from mothers to offsprings in northern Italy. J Med Virol 9:139-149.

DISCUSSION

Fanning W (Scottsdale Memorial Hospital, Scottsdale, AZ):

Is there any association between non-A and non-B hepatitis and the Delta virus? The second question is, do you find the antigen in gammaglobulin, even though it's heat treated? You mentioned that you could pass the Delta virus through gammaglobulin.

Bonino F:

Initially, we thought there was some relation between the Delta virus and the non-A non-B agent since we found a few similarities by electron microscopy between the two diseases such as nuclear particles and tubular forms seen in the cytoplasm of liver cells. When we had the first clones of Delta RNA to run hybridization experiments using nucleic acid from sera and livers of non-A non-B cases, both from animals and humans, we never found any cross-hybridization. Secondly, there is no cross-protection between non-A non-B and Delta infection in animals. Dr. Shimitsu found an IgM monoclonal antibody reacting with an antigen in both non-A non-B actively infected animals and Delta actively infected animals. Now we know that these epitopes are related to a host protein, may be transformed by the virus infection, but it is not specific of either non-A non-B or Delta. Furthermore, the non-A non-B agent has no homology, at least for the part of genome which is known with the Delta RNA. So, right now, there is no evidence that there is any relation between HDV and non-A non-B viruses. Gammaglobulins are capable of transmitting Delta hepatitis when you use them intravenously in a carrier of hepatitis B. The point is, that when we have to protect somebody for some reason, and we know that he is a carrier of hepatitis B, we should be very careful and literally check the lot we are going to use for anti-Delta antibodies. The infectivity of Delta is tremendously high. It is possible that a minimal amount of infectious material is left in the gammaglobulin preparation. The same preparations have no risk in the normal person.

Goldman M (Neurex Corp, Menlo Park, CA):

I very much enjoyed your talk. I have a question about the genomic similarities betwen the Delta virus and some of these plant pathogens. You mentioned that the size of the genome was roughly equivalent, but I forget, is there any information on genomic homology between the two? Secondly, does anyone know if this similarity in size is important in the encapsulation process?

Bonino F:

I think there are definitely similarities with the sequences for the viroid replication. I think it is one of the most striking findings that this new agent has similarities with plant viruses. Delta replication is quite similar to that of a viroid and very well fits the model. There are similarities in the structure and in the way the multimer is cleaved. We need for that model an RNA dependent RNA polymerase. It appears that not enough genetic material is present in this virus to code for an entire protein like an RNA dependent RNA polimerase. It may be that there is one of those open reading frames coding for a factor that subverts or converts a host polymerase, or a hepatitis B virus polymerase to work as a suitable replication enzyme for Delta. Right now, the only similarity that has been definitely shown is this one between the viroid and Delta RNA at the level of the central conserved region of viroids.

de la Maza L:

Is there any correlation between alcoholic liver hepatitis or alcoholic liver cirrhosis and Delta hepatitis or Delta cirrhosis?

Bonino F:

We have run a few epidemiological studies in Italy. There is actually no evidence that alcohol can enhance or create a risk of getting infected from Delta. Of course, if you already have liver disease related to something else, you have a further damage to your liver. But, definitly, there is no evidence, other than the liver damage to the biological evolution of the infection.

Burnett G (Bernett Clinical Labs, Buena Park, CA):

Dr. Bonino, the studies done in Italy about drug addicts seem to me that they are very striking in the sense that AIDS, for instance, is more prevalent in drug addicts than in male homosexuals. You find there a lot of hepatitis B, and now I learn from your talk, a lot of Delta. I understand that Dr. Purcell has done DNA probes on monocytes or lymphocytes and found that the sexual partners of these drug addicts, many have the HIV incorporated, but without any production of antigens, just dormant. Do you find something similar with the Delta, now that you are using the probes?

Bonino F:

We have done a few studies in humans, but much more work was done in the woodchuck model by Dr. Negro, working at NIH with Dr. Purcell and Dr. Gerin . We know that woodchuck hepatitis virus, like human hepatitis B virus can infect and replicate in lymphocytes. But, as far as we know, with many tissues, mainly spleen, lymphocytes, and other tissues from woodchucks infected with Delta, there is no evidence of Delta replication in tissues other than the liver. Interestingly, we have seen that there is an inhibition of replication of the Delta virus by the woodchuck hepatitis virus; not only in the liver, but also in the spleen. Maybe we do not find Delta because we don't have enough sensitivity or because there is an inhibition due to some factor present in the blood. In Italy, HIV infection is growing among drug addicts. Now we have most drug addicts carrying hepatitis B virus replication, markers of HDV and antibody against HIV. There is still a lot of confusion by the meaning of the interplay between multiple virus infections at the biological and pathological levels because it is very difficult to study these patients.

MOLECULAR EPIDEMIOLOGY OF EPSTEIN-BARR VIRUS OBTAINED FROM PATIENTS WITH THE CHRONIC FATIGUE SYNDROME

JAMES F. JONES, SUSAN BAKER AND JOANNE STREIB

National Jewish Center for Immunology and Respiratory Medicine, Denver, Colorado USA

INTRODUCTION

A chronic illness characterized by fatigue, neurological symptoms (headache), paresthesias, tinnitis, dizziness), pharyngitis, cervical (and other sites) lymph node tenderness/swelling myalgia, arthralgia, dyslogia, depressive symptoms, and sleep disturbances, which has prevented the patient from attending school or work 50% of the time over the last six months, has recently been given the label of Chronic Fatigue Syndrome (CFS) (Holmes et al. 1988). Previous work suggested that this illness may be related to immune responses to the Epstein-Barr virus (EBV) (Dubois et al. 1984; Jones et al. 1985; Straus et al. 1985). Subsequently, considerable interest and skepticism has been expressed in the literature concerning whether these patients have an illness, and which infectious agents (if any) are responsible for this syndrome (Buchwald et al. 1987; Holmes et al. 1987).

The skepticism is based in principle on choice of parameters to support an active disease in general or specifically, an active infection. If one hypothesizes that illness is due to the host response to infection, then some evidence of ongoing primary or reactivated infection must be present. The original observation suggested that patients have elevated antibodies to antigens only expressed during active replication of the virus: the early antigen complex. Subsequent studies have suggested EBV serologies may not be reliable indicators of disease activity (Holmes et al. 1987; Merlin, 1986). While these studies themselves are of questionable merit in disproving a role for EBV, the use of antibody tests alone are not sufficient in establishing a cause and effect relationship between symptomatology and ongoing infection in these patients.

A multifaceted approach using Koch-Evans postulates is underway to address these questions. One postulate is that the host responds to infection in a reproducible manner during infection with a given organism (Evans, 1976). For instance, it is well accepted that peripheral blood leukocytes from patients with active primary EBV infection, or reactivated infection in the case of immunosuppressed individuals, will spontaneously transform in whole mononuclear cell culture preparations (Yao et al. 1985). This phenomenon does not occur, or occurs only very rarely, in asymptomatic seropositive individuals, and is dependent on the presence of EBV and an alteration in specific immune responses. It can be recapitulated in healthy, seropositive individuals by removal of T-cells or the addition of the immunosuppressent agent, cyclosporin A to the culture medium. If such cell lines could be established from the CFS patients it would support the concept that EBV is playing a role in this illness and allow a number of important questions to be addressed.

One such question is whether or not specific viral strains may be present in these individuals. A current approach to this question is the utilization of molecular epidemiology. We have chosen to ask two questions using this approach. The first is to examine EBV DNA with probes derived from EcoR1 digests of EBV which together span over one-third of the gene as recently described by Katz et al. (1988) to describe viral isolates form both immunocompromised and normal hosts. The second uses a probe which differentiates between the linear and episomal forms of the viral genome (Raab-Traub and Flynn, 1986). These latter data would theoretically assist in determining active versus latent infection.

This report describes work in progress, which is examining peripheral blood lymphocytes from patients referred for evaluation of chronic illness and asymptomatic control individuals. Spontaneous transformation occurred yielding lymphoblastoid cell lines in approximately 30% of patients, but in less than 3% of control specimens. Viral DNA patterns reflect those previously seen in infectious mononucleosis (IM) and Burkitt lymphoma (BL) lines. In some instances, spontaneous cultures contained a linear viral genome configuration.

METHODS

Individuals referred for evaluation of chronic illness had heparinized blood and saliva specimens taken for evaluation. Mononuclear lymphocytes were harvested by density sedimentation gradients. Saliva was centrifuged at 1,300 x g and clarified material passed through a 0.2 µm filter. Lymphoid cells were placed in 96 well plates at decreasing concentrations of cells in RPMI 1640, antibiotics, and 10% fetal bovine serum. Saliva was added to cord blood mononuclear cells as described by Moss and Pope (1972).

Cultures were examined weekly and one-half of the media changed at the same interval. Spontaneous cultures were determined to be established when clumps of cells appeared and could be transferred to bottles and be maintained indefinitely, thus establishing lymphoblastoid cell lines (LCL). Type A and B growth patterns were determined as per Rowe et al. (1987).

Genomic DNA was extracted from cells lysed with 0.4 mNaCL, 25 mM Tris pH-7.5, 5 mM EDTA, 0.5% SDS, 200 µg/ml proteinase K, treated overnight at 37°C, extracted once with aqueous buffer saturated phenol, once with phenol/chloroform/isoamyl alcohol (25:24:1) once with chloroform/isoamyl alcohol (24:1) and precipitated with 2 1/2 volumes ethanol. DNA was resuspended and 10 µg digested with BamH1 restriction endonuclease, electrophoresed in 0.6% agarose gels and transferred to Biodyne nylon membrane by using Southern's method (Southern, 1975). The filters were prehybridized with 5X SSC, 5X Denhardt's, 50% formamide, 50 mM $NaPO_4$ pH=6.5, and salmon sperm DNA at 500 µg/ml at 42° for 2 h. They were then hybridized in 5X SSC, 1 X Denhart's, 50% formamide, 10% dextran sulfate, 100 µg/ml salmon sperm DNA and 10^6 CPM ^{32}P labeled probe at 42° for 16 h. The blots were washed with 2X SSC, 0.1% SDS, 3 x 5 min at room temperature with shaking at 100 RPM; then the samples were incubated in 0.2 x SSC, 0.1% SDS, 10 mM $NaPO_4$ pH-6.5 at 65°C, twice for 30 min with shaking. The filters were wrapped in plastic film and autoradiographed with DuPont intensifying screens from overnight to 3 days. Probes used were: BamH1 NJhet, a 9 kb BamH1 fragment, courtesy of Dr. Jeff Sklar, Stanford University (Raab-Traub and Flynn, 1986); EcoR1 C and EcoR1 B, both 25.8 kb EcoR1 fragments, courtesy of Dr. Ben Katz (Fisher et al. 1981). EBV DNA fragments were ^{32}P-dCTP random primer labeled.

RESULTS

Cell cultures

Cells from 75 of 250 patients and 2 of 60 control subjects underwent spontaneous transformation. Cultures were positive within 4 weeks in most instances. Ninety LCL's were derived from the 75 patients and 3 LCL's from the control subjects. Multiple lines were obtained from the same individuals by virtue of several wells in the culture plate becoming positive from the same blood specimen. In addition, serial sampling of the same individual allowed establishment of multiple lines over time.

Saliva from 27 individuals transformed cord blood lymphocytes. Thus, LCL's from the same individual's peripheral blood lymphocytes (PBL) and saliva were available for viral gene analysis. Analysis of the patient's clinical status or antibody titers was not complete at this time; this information is being obtained as part of an ongoing prospective case/control study.

EcoR1 restriction patterns

The primary analysis was made with the EcoR1 C probe. This probe encompasses approximately 20% of the right hand portion of the genome and allows detection of 5-6 BamH1 fragments. Fifty-six of 75 original patient lines (75%) had the same EcoR1 C pattern. This pattern was also seen in the PBL and saliva specimens derived from a patient with acute IM. The patterns of the other LCL's were those commonly seen in BL's or other so-called laboratory lines (Figure 1). One LCL had the deletions

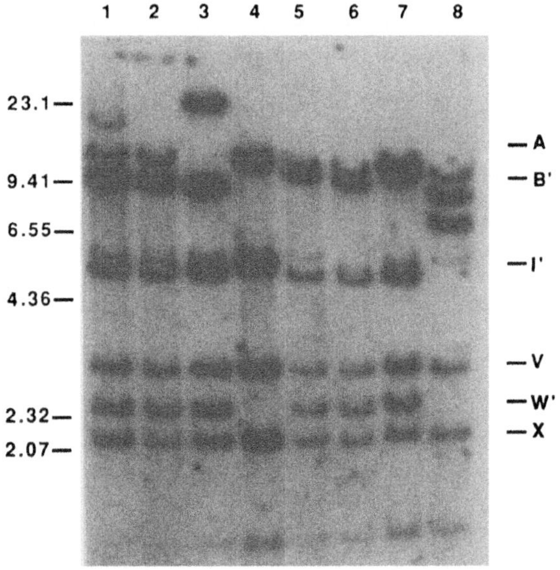

Figure 1. Representative EBV genome patterns isolated from spontaneously transformed cell lines or saliva-induced cord blood cell line and the laboratory strain, B95-8. DNA from the resultant lines were digested with BamH1 and probed with a plasmid-containing FF41 EcoR1 C. The BamH1 and probed with a plasmid-containing FF41 EcoR1 C. The BamH1 fragments contained with the probe are denoted on the right had side of the figure. The numbers on the left denotes the number of kilobase pairs of the fragments. All of the lanes represent lymphoid cell lines from patients except lane 4 which is derived from B95-8 line. Lanes 1 and 2 are from peripheral blood lymphocytes and saliva-infected cord blood cells, respectively from a patient with acute IM. The band in lane 1 of approximately 20 Kb was not seen in subsequent analysis and likely represents a partially digested fragment. Lanes 3 and 8 have the pattern of the Raji and P3HR-1 lines, respectively. The remainder of the specimens have polymorphisms in fragments A', B' and I'.

seen in the marmoset line, B95-8, where BamH1 fragments B', W', and I' are missing. Three lines had the configuration of the Raji BL line. This prototype line is missing a restriction site giving rise to an enlarged BamH1 A fragment. Five lines had the same pattern as the P3HR-1 virus; the prototype nontransforming virus. This line has a fused W' I' fragment. The other lines had restrictions site polymorphisms leading to slight variations in the sizes of BamH1 fragments B', A and I'.

Multiple lines were obtained from the same specimen in several patients. In most instances, the EcoR1 C pattern was the same in each line. In some instances, however, different viral genome patterns were found in a single specimen (Figure 2).

In addition, 16 lines were analyzed with EcoR1 B. This probe allows examination of another 6 BamH1 fragments, including the third internal repeat region, the region that encodes EBNA 1. The majority of the few lines so examined had no rearranged areas. Two lines, which appeared to

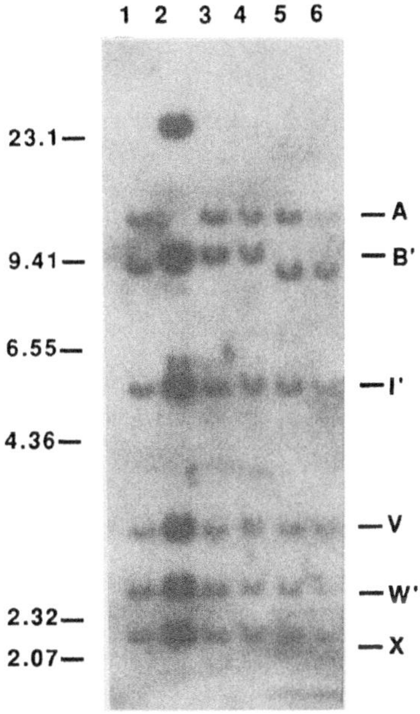

Figure 2. Analysis of EBV DNA in multiple cell lines from the peripheral blood and saliva of the same individuals. DNA specimens were prepared and hybrized as in Figure 1. Lanes 1 and 2 represent cell lines obtained from two different wells of a culture plate established from a single blood sample from one patient. The sample in lane 2 has the same EBV DNA pattern as the Raji line. Lane 2 DNA was obtained from a cord blood cell line established from saliva of the same patient as in lanes 1 and 2, but from a different date. Lane 4 has DNA from a spontaneous lymphoblastoid cell line from peripheral blood of another patient from a third date. Lanes 5 and 6 represent DNA from two lines established from the same individual on the same day. These examples demonstrate that different or the same viral genome patterns may occur from the same patient at single time points, but excretion of the same pattern occurs repeatedly.

be similar to the Raji line (Figure 3) by EcoR1 C analysis, had the expected EcoR1 B rearrangements in the BamH1 Z and E regions (Figure 2) known to be present in Raji virus. One additional cord blood lymphocyte line derived from a patient virus had a polymorphism in K alone. Attempts to use EcoR1 A fragments have been hampered by the instability of fragments used thus far. The sites of potential interest in this region of the viral genome are being examined with other restriction enzyme-derived fragments.

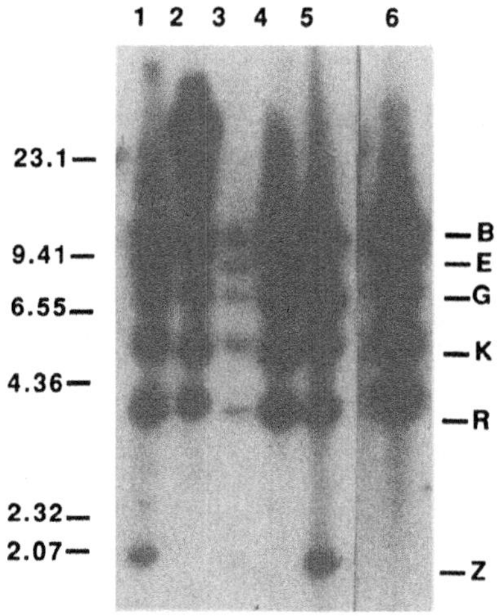

Figure 3. Restrictive fragment patterns after hybridization with EcoR1 B. The fragment sizes recognized by this probe are on the right hand side of the figure. Lanes 1-4 represent DNA obtained from spontaneously transformed lines from patient cells; lane 5 DNA from the LDV-7 line which is a spontaneous lymphoid cell line derived form an adult with mononucleosis, and lane 6 is the BL line, Raji. Lanes 2 and 5 have the same rearrangement as seen in the Raji line, e.g., absence of Band HIZ and a polymorphism in the size of BamH1 E.

Linear vs. episomal

The difference between linear and episomal lines is determined by the configuration of the terminal repeat fragments (Raab-Traub and Flyn, 1986). The NJhet probe binds to the termini of each end of the genome. If the termini are fused, a single (representing a single clone of cells) fragment is seen. Multiple bands of similar size suggest that the cells are from several clones (polyclonal). If multiple small bands are present, of less than 6.6 kb, the termini are not fused and the viral gene is thought to be in linear form. Fifty-one (56%) of the lines had single fused termini, suggesting spontaneous cloning. The remaining episomal lines had multiple bands, indicating polyclonal cultures. An unexpected finding was that lines from serial samples from the same

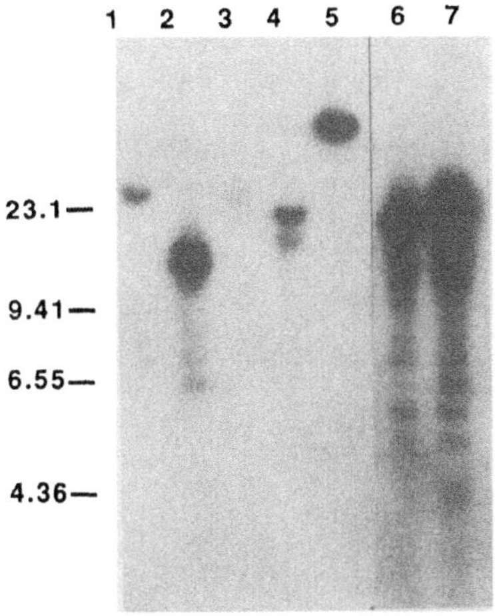

Figure 4. Analysis of EBV DNA from spontaneous-induced and laboratory lines after hybridization with the NJhet fragment. Specimens were prepared and analyzed as in Figure 1 except that hybridization was performed with the NJhet probe. Lanes 1, 3 and 5 represent different monoclonal, episomal forms of the viral genome. Lanes 2, 6 and 7 represent linear forms of the viral genome. Lane 7 is DNA from the prototype B95-8 virus. Lanes 1 and 2 are from the same patient, and were obtained from PBL on different days. These lines have the same EcoR1 C restrictive pattern. Lane 3 is a cord blood line derived from virus-containing supernatants of the line represented in Lane 1; it has a different EcoR1 C pattern than the primary line. Lanes 4 and 5 represent DNA derived from different lines obtained from the same specimen from a single patient. These lines also had different EcoR1C patterns. Lane 4 is an example of a polyclonal episomal line.

patient had the same EcoR1 C pattern, but demonstrated different NJhet patterns. Figure 4 demonstrates several examples of these patterns.

Likewise, multiple lines from a single blood specimen obtained from each of five patients had the same EcoR1 C restriction fragment pattern; but some from each were linear while the remainder were episomal in configuration. Figure 5 demonstrates these findings.

Establishment of Cord Blood Cell Lines

Infection of cord blood cells with supernatants from the spontaneous lines was attempted using viral containing supernatant from 66 individuals. One to six attempts were made to transform cord cells with these virus preparations. Nine cord cell lines were established. In two instances, the resultant cord line contained viral restriction patterns which differed from the spontaneous parent line (Figure 4). A similar phenomenon was seen when saliva samples were incubated with fresh cord blood lymphocytes. The virus restriction pattern from saliva cell line differed from the virus in the spontaneous lymphoid cell line. One spontaneous line that had a linear configuration using the NJhet probe

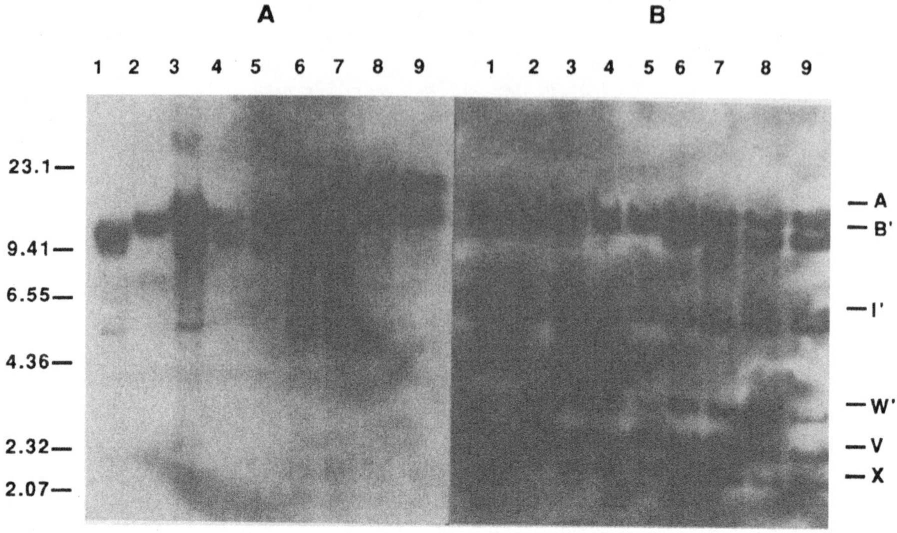

Figure 5. Comparison of multiple lines from single samples from the same individuals using EcoR1 C and NJhet fragment analysis. Lanes 1-5 are derived from one patient, and lanes 6-9 from a second patient. Section A depicts the NJhet hybridization pattern and Section B shows the results of EcoR1 C hybridization. A linear form of the viral genome is present in lanes 1 and 3 from patient one and lanes 6 and 7 from patient 2. Section B demonstrates that the EcoR1 C pattern is the same within the multiple line from the individual patients, but the DNA pattern differs between the patients.

induced four cord cell lines. The primary line and the cord cell lines all had the same EcoR1 C pattern, but the cord lines were episomal in comparison to the primary line.

Distribution of EBNA 2A and EBNA 2B

Type A (EBNA 2A) virus infected cells form irregular clusters with a background "carpet" of lymphoblastoid cells (Rowe et al. 1987). Type B (EBNA 2B) virus infected cells form round clusters and have no background carpet of cells. Twenty-four patient lines have been examined thus far for these characteristics. Sixteen of these lines demonstrate the type A growth characteristics and the remainder fit the criteria for type B virus.

DISCUSSION

The control of spontaneous transformation, as measured by prevention or regression of outgrowth of exogenously infected autologous B lymphocytes is T-cell and possibly NK cell mediated. Cytotoxic and/or suppressor mechanisms are responsible for this control of B cell proliferation (Tosato, 1987).

Individuals in whom these immune systems are not functioning, may have an acquired or a primary immunologic abnormality (Ho et al. 1985). Thus, a pre-existing or acquired immunodeficiency state has been though to be a prerequisite for the spontaneous transformation to occur. The current observations suggest that the processes controlling spontaneous transformation may not be effective in some patients with the CFS. Tosato et al. (1985) demonstrated that the number of virus-containing peripheral blood cells in patients with CFS is no different that the number seen in asymptomatic seropositive individuals. That observation suggests that one explanation for our results, that of an increased number of B cells contain virus, is not correct. It supports the concept that immune surveillance may be inadequate. Experiments to further understand these processes in the current patient population are in progress.

The question of different EBV strains as causative agents of different clinical illnesses has been addressed in several studies over the years. Most of the initial studies have examined laboratory strains which are derived primarily from tumors. Until recently, such studies have not made an association between disease and virus strain. Bornkamm et al. (1980) examined 5 EBV strains originating from BLs (P3HR 1 and CC34-5), nasopharyngeal carcinoma (M-ABA), transfusion mononucleosis (B95-8) and myeloblastic leukemia (QIMR-WIL) by partial denaturation mapping and restriction endonuclease fragment analysis. The first analysis showed, primarily, the presence of a 12,000 kb deletion in the right-hand side of the B95-8 virus. The other findings included differences in the number of internal repeats between M-ABA and B95-8. These workers also observed a deletion in P3HR 1, now known to be in the BamH1 HY region. Their analysis of restriction endonuclease fragments suggested that polymorphisms were present in what is now known as the 3rd and 4th internal repeat regions, encompassing BamH1 fragments K, and B', respectively. Analysis of P3 HR1 lines indicated some variations among them which the authors suggested may have been due to *in vitro* phenomena.

Fischer et al. (1981), approached the question of strain: disease relationship by comparing EcoR1 and BamH1 restriction fragment patterns by examining two different IM isolates propagated in marmoset cells, B95-8 and FF4-1. They found that the deleted portion of B95-8 virus was present in the new strain FF4-1 and that the same sequences were present as those seen in previously considered unique BL lines. They also noted that he number of internal repeat fragments, BamH1 V (W) was less in FF4-1 than in B95-8 virus. These authors concluded that despite changes in the first internal repeat region and in the terminal repeat areas, and perhaps in the fourth internal repeat regions, there were remarkable similarities in overall gene structure between the two original IM isolates. These results support those of Heller et al. (1981) and those cited above by Bornkamm et al. (1980), again suggesting that isolates from different disease entities obtained from widely divergent geographic areas, were basically quite similar.

Bornkamm et al. (1984) asked if viruses associated with malignancy are peculiar to different tumors. They examined restriction fragments of DNA from BL lines and non-malignant cells from the same individuals. Polymorphisms in the BamH1 H, H2 and Y were seen among these lines, but there were no differences between tumor lines and lymphoid cell lines derived from the same patients. They also noted that two viral patterns were seen from the one patient's PBL. One of these patterns was identical to the patient's tumor. Although changes, such as point mutations, which could possibly give rise to biological differences between virus strains, were not detectable in that type of analyses. The authors suggest that particular strains are not associated with tumors.

Evaluations of the P3HR-1 family of viruses (Jijoye and its derivatives) has allowed identification of a 1.8 kb substitution in Jijoye in the EBNA 2 gene region, when compared to the B95-8 or M-ABA viruses. Zimber et al. (1986) used probes specific for M-ABA (type A virus) and Jijoye (type B virus) in the analysis of viral DNA from 38 cell lines of diverse geographic origins. The results showed that type A virus was present in 31 of 38 lines, type B in five, and two lines had neither of the sequences. These authors suggested that since the geographic regions from which type B virus originated, that it was less ubiquitous in the human population.

More recently, Katz et al. (1988) described fragment-linked polymorphisms in virus obtained from immunocompromised and normal individuals. Lines were derived from patients with HIV infection (both ARC and AIDS) and from individuals after liver transplantation. The normal hosts consisted of patients with IM. DNA was incubated with BamH1 enzyme and hybridized with probes containing EcoR1 A, B, and C fragments of the FF4-1 EBV genome, which collectively represent approximately 60% of the virus gene. Their results showed that, in general, restriction patterns differed from patient to patient when patients were epidemiologically unrelated. However, when isolates from parent and child were compared, they were more likely to be similar or the same. It also appeared that saliva and peripheral blood isolates from both normal and immunocompromised individuals were the same for the same individual. In addition, when virus obtained from lymphoid cell lines induced transformation in cord cells the viral genome pattern was the same as the original. The difference in the virus isolations among the subjects were polymorphisms in the BamH1 H and B' regions. Occasionally, loss of restriction sites were also seen resulting in a line with a fusion fragment similar to that seen in P3HR-1 . Their conclusions also suggested that, despite these polymorphisms, the basic structure of the gene was similar, and there were no real differences between immunocompromised individuals and individuals with IM.

Another approach to molecular epidemiology utilized analysis of proteins produced by various lymphoid cell lines which were derived form patients with rheumatoid arthritis, IM, or healthy controls (Sculley et al. 1984). EBV-specific antigens were identified by sera which reacted with both EBNA 1 and EBNA 2 or EBNA 1 alone. The findings of the study were that the proteins in these lines had different molecular weights and some had multiple EBNA antigens. No specific strain of EBV was associated with either rheumatoid arthritis or IM.

The question regarding strain differences associated with the CFS has not been answered by our studies to date. Although one restrictive pattern is prominent, the population from which the results were obtained is still being characterized. More specimens from individuals with other EBV illnesses need to be examined, as well an analysis of well defined groups with CFS.

Despite the need for further work, several interesting observations have been made regarding EBV in cell lines in general derived from these patients. The isolation of viral gene patterns from individuals which were previously thought to be limited to tumor lines, supports the observation by Fischer et al. (1981) that strains are not limited to tumor lines. Secondly, that unlike the observations of Katz et al. (1988), different patterns may be derived from the saliva and PBL from the same patient. Multiple patterns from the same individual, primarily from previously known immunodeficient patients, has been previously reported (Katz et al. 1986). In addition, some of our patients had changes in the viral genome patterns over time.

The fact that some lines had a linear configuration is of considerable interest if one considers the hypothesis that ongoing or reactivated infection may be responsible for production of symptoms. The isolation of both linear and episomal from the same PBL specimen in some patients supports the concept that latent infected cells may undergo spontaneous reactivation. The latter suggestion, however, needs more rigorous proof. This observation is of particular importance if one considers potential use of anti-viral agents, as suggested by Sullivan et al. (1984).

Analysis of the growth characteristics of the spontaneous lines showed a predominance of type A virus, as described by Zimber et al. (1986). Our observations are still too few in number to discuss epidemiologic importance, but they do show that type B virus is not limited geographically, or to tumor cell lines. Moss et al. (1988) have recently suggested that type A virus may be more susceptible to cell mediated cytotoxicity than is type B virus. The suggestion that type A versus type B might be an *in vitro* phenomenon however, raises questions regarding the clinical relevance of the classification of type base or EBNA 2A and EBNA 2B.

One last preliminary observation is that autologous mononuclear cells from one patient (the only one studied to date) from whom linear and episomal lines were established, were able to kill only the autologous linear lines.

Collectively, these data raise interesting questions regarding the possible role of a chronic active or reactivated EBV infection in some patients with CFS. They suggest that the immune response associated with prevention of spontaneous outgrowth may be altered. The strain difference contribution to this alteration remains unanswered. Based on the restriction fragment polymorphism seen in these lines, recurring appearances across several studies, and apparent changes in configuration when virus is added to cord blood cells, one might hypothesize that there is a common EBV "strain" and that alternations in that pattern may be dependent on the cell which is subsequently infected. Evaluation of cloned virus obtained from a variety of "wild" sources will be required to test this hypothesis.

Another analytic tool of potential interest is Western blot analysis of protein production by the different "strain" types along with determination of antibody production by the different patients. This line of investigation is just beginning.

Our preliminary data support EBV as a potential triggering or causative agent in up to 25% of patient with CFS. Further prospective analysis of the molecular epidemiologic characteristics discussed here are necessary in order to establish a causal relationship between EBV and other infectious agents and a chronic illness.

REFERENCES

Bornkamm GW, A Delius, U Zimber, J Hudewentz, MA Epstein (1980) Comparison of Epstein-Barr virus strains of different origin by analysis of the viral DNase. J Virol 35:603-618.

Bornkamm GW, MV Kneble-Boeberitz, GM Lenoir (1984) No evidence for differences in the Epstein-Barr virus genome carried in Burkitt lymphoma cells and non-malignant lymphoblastoid cells from the same patients. Proc Natl Acad Sci USA 81:4930-4934.

Buchwald D, JL Sullivan, AL Komaroff (198 Frequency of "Chronic active Epstein-Barr virus infection" in a general medical practice. JAMA 257:2303-2307.

Dubois RE, JK Seeley JK, I Brus, K Sakamoto, M Ballow, S Harada, TA Bechtold, G Pearson, DT Purtillo (1984) Chronic mononucleosis syndrome. South Med J 77:1376-1382.

Evans AS (1976) Causation and disease: The Henle-Koch postulates revisited. Yale J Biol Med 49:175-195.

Fischer DK, G Miller, L Gradoville, L Heston, MW Weststrate, W Maris, J Wright, J Brandsma, WC Summers (1981) Genome of a mononucleosis Epstein-Barr virus contains DNA fragments previously regarded to be unique to Burkitt's lymphoma lines. Cell 24:543-553.

Heller M, T Dambaugh, E Kieff (1981) Epstein-Barr virus DNA IX. Variation among viral DNA's J Virol 38:632-648.

Ho M, G Miller, RW Atchison, MK Breinig, JS Dummer, W Andiman, TE Starzl, R Eastman, BP Griffith, RC Hardesty, HT Bahson, TR Hakala, JT Rosenthal (1985) Epstein-Barr virus infections and DNA hybridization studies in post-transplantation lymphoma and lymphoproliferative lesions: The role of primary infection. J Infect Dis 152:876-886.

Holmes GP, JE Kaplan, JA Stewart, B Hunt, PF Pinsky, LB Schonberger (1987) A cluster of patients with a chronic mononucleosis syndrome: Is Epstein-Barr virus the cause? JAMA 257:2297-2302.

Holmes GP, JE Kaplan, NM Gantz, Al Komaroff, LB Schonberger, SE Straus, JF Jones, RE DuBois, C Cunningham-Rundles, S Pathwa, G Tosato, LS Zegans, DT Purtilo, N Brown, RT Schooley, I Brus (1988) Chronic fatigue syndrome: A working case definition. Ann Intern Med 108:387-389.

Jones JF, CG Ray, LM Minnich, MJ Hicks, R Kibler, DO Lucas (1985) Evidence for active Epstein-Barr virus infection in patients with persistent, unexplained illnesses: elevated anti-early antigen antibodies. Ann Int Med 102:1-7.

Katz BZ, WA Andiman, R Eastman, K Martin, G Miller (1986) Infection with two genotypes of Epstein-Barr virus in an infant with AIDS and lymphoma of the central nervous system. J Infect Dis 153:601-604.

Katz BZ, JC Niederman, BA Olson, G Miller (1988) Fragment length polymorphisms among independent isolates of Epstein-Barr virus from immunocompromised and normal hosts. J Infect Dis 157:299-308.

Merlin TL (1986) Chronic mononucleosis: pitfalls in the laboratory diagnosis. Human Pathol 17:2-8.

Moss D, IS Miski, SR Burrows, K Burman, R McCarthy, TB Sculley (1988) Cytotoxic T-cell clones discriminate between A- and B-type Epstein-Barr virus transformations. Nature 331:719-721.

Raab-Traub N, K Flynn (1986) The structure of the termini of the EBV as a marker of clonal cellular proliferation. Cell 47:883-889.

Rickinson AB, M Rowe, IF Hart, QY Yao, LE Henderson, H Rabin, MA Epstein (1984) T-cell mediated regression of "spontaneous" and Epstein-Barr virus-induced B-cell transformation *in vitro*: Studies with cyclosporin A. Cell Immunol 87:646-658.

Rowe M, DT Rowe, CD Gregory, CS Young, PJ Farrell, H Rupani, AB Ricksinson (1987) Differences in B cell growth phenotype reflect novel patterns of Epstein-Barr virus latent expression in Burkitt's lymphoma cells. EMBO J 6:2743-2751.

Sculley TB, DJ Moss, RA Hazelton, JH Pope (1987) Detection of Epstein-Barr virus strain variants in lymphoblastoid cell lines "spontaneously" derived from patients with rheumatoid arthritis, infectious mononucleosis and normal controls. J Gen Virol 68:2069-2078.

Southern EM (1975) Detection of specific sequences among DNA fragments separated by gel electrophoresis. J Mol Biol 98:501-517.
Straus SE, G Tosato, G Armstrong, T Lawley, OT Preble, W Henle, R Davey, G Pearson, J Epstein, I Brus, RM Blaese (1985) Persisting illness and fatigue with evidence of Epstein-Barr virus infection. Ann Inter Med 102:7-16.
Sullivan JL, P Medveczky, SJ Foprman, SM Baker, JE Monroe, C Mulder (1984) Epstein-Barr virus induced lymphoproliferation. Implications for antiviral chemotherapy. N Engl J Med 311:1163-1167.
Tosato G (1987) The Epstein-Barr virus and the immune system. Adv Cancer Res 49:75-125.
Tosato G, S Straus, W Hense, SE Pike, RM Blaese (1985) Characteristic T-cell dysfunction in patients with chronic active Epstein-Barr virus infection (chronic infectious mononucleosis). J Immunol 134:3082-3088.
Yao QY, AB Rickinson, MA Epstein (1985) A reexamination of the Epstein-Barr virus carrier state in healthy seropositive individuals. Int J Cancer 35:35-42.
Zimber U, HK Adlkuger, GM Lenoir, M Vuillaume, MV Knebel-Doeberitz, G Laux, C Desgranges, P Wittmann, U-K Freese, U Schneider, GW Bornkamm (1986) Geographical prevalence of two types of Epstein-Barr virus. Virology 154:56-66.

DISCUSSION

Rouse B (University of Tennessee, Knoxville, TN):

That's interesting. Have you ever looked at a situation where you've infected a cell with the virus and then modulated the activity of that cell in some way to see what type of proteins come up?

Jones JF:

We've done it, but not on purpose. We've been asked to look at a number of potential therapeutic agents and one way of doing that is to infect a cell with B958 in the presence or absence of that material. In some of those instances, the viral genome product appears to have changes as far as restriction analysis is concerned. We have not looked at protein products. I think that I can envision some very interesting experiments using biologic agents and immune response modifiers and so on and so forth, but the answer is that we have not deliberately done those experiments.

de la Maza LM (University of California Irvine Medical Center, Orange, CA):

I would like to ask a more general question. Based on your results, what type of follow-up would you recommend for this type of patient? Both from the point of view of the laboratory trying to make a diagnosis and from the point of view of the clinician trying to take care of the patients?

Jones JF:

We're seeing each of these individuals in this study for at least a year and doing these laboratory examinations, five times over that year. The question has come up in the patients in whom we get linear virus line who are ill as to whether or not there should be some attempt at specific therapy. John Sullivan, several years ago, wrote an article with some of his colleagues that examined the linear versus episomal question after whole cell electrophoresis. He made the suggestion that perhaps patients in whom one could identify linear tumors should perhaps be considered for therapy with specific antiviral agents, acyclovir, particularly. We haven't done that yet in the patients with chronic fatique syndrome. We have used that agent long-term, though, in other individuals with more severe disease and have seen some clinical resolution. In one individual with severe hypogammaglobulinemia, we can no longer isolate the virus from her peripheral cells after such therapy. So, with the chronic fatigue syndrome patients, I think we're still just gathering information. I eluded to the PCR studies and those are in progress. I think if we can identify an active replicating virus in a patient, then those patients might be considered for antiviral therapy or other immune-type therapy. I think only under those conditions. So right now, it's still in the information gathering process.

Neuman M (Diagnostic Services, Inc., Naples, FL):

Dr. Jones, I'de like to get your comment on two incidental case findings we've recently observed. One was a 72-year old lady we were asked to see for fever of undetermined origin. She had a 3-month history of recurrent neck masses, pharyngitis, and on CT scan, had a retropharyngeal mass which when needled, offered nothing in regard to diagnostic utility. As a part of her work-up we did EBV titers. She had a VCA of 40,000 and early antigen diffuse of 1:160 and an EBNA of 1:8. From that we suspected that she may have a malignancy. On gross exam, her nasal pharynx appeared normal, but on radionucleotide scan, lit up. She was biopsied, we did impression smears and were able to demonstrate EBNA through IFA complement technique and the histopathology correlated very nicely. Second case is a 46-year old gentlemen, hemophiliac, HIV-positive on Retrovir times 3 months, also had a left axillary mass that decreased while on Retrovir and then recurred. He had a necrotizing process, was biopsied, we again did tissue inprint smears of the node and was found to have EBNA demonstrated with correlating histopathology for a high stage lymphoma. In both cases, the first patient, 90% of the cells demonstrated EBNA, the second case, greater than 60%. My question is, does testing for EBNA in a suspected tumor have any diagnostic relevance?

Jones JF:

It had been the classic way of identifying virus in tumor until the *in situ* procedures were developed. So, yes, it does. Particularly the B cell tumors are EBNA positive. There are tumors, however, in immunosuppressed patients, if one stains for B cell markers and other morphologic criteria, that look for all the world like the typical Burkitt-type cells, but in some recent observations by several groups now, EBV doesn't necessarily have to be in them. So, the conundrum is, if you have a tumor as you described, and it's EBNA positive, than more than likely, it is of EBV origin. The question therefore is whether it's monoclonal or whether it's polyclonal. There seems to be a gradual procession from an oligoclonal or polyclonal proliferation of EBV positive cells that may be tumorous in some individuals, to that of a monoclonal tumor that has a chromosomal rearrangement, which are classical of Burkitt's. Each of those stages with progression have been seen in immunosuppressed individuals with tumors. So, if EBNA is positive, it is very helpful, particularly when you have that high percentage of cells present. The point is, though, you can also have tumors like this that are EBNA negative; so that you need to go further if you have a similar clinical circumstance. I think that the high antibody titers were specifically important, and the lack of anti-EBNA in allowing you to proceed in that diagnostic step. Because, virtually every patient with a severe disease who has developed a tumor has had high anti-EA, high VCA and no anti-EBNA. And no anti-EBNA is very rare in the population at large. In Miller's hands, 95% of us all have anti-EBNA-1 which is what you identify with that screening test. I think that the presence of EBNA in those circumstances is very important, but there are cases where you don't have it but you still may have similar disease.

Neuman M (Diagnostic Services, Inc., Naples, FL):

My second question is, do we have any data at all as to whether high dose acyclovir or gangcyclovir in combination with interferon will serve as a useful adjunct therapeutic modality in regard to radiation therapy or chemotherapy?

Jones JF:

The patients that I described had periodic treatment with acyclovir prior to tumor development, and they did very, very well while on intravenous therapy. Chip Schooley at Harvard has a patient who has this chronic syndrome who he maintains on acyclovir therapy and is doing very well. The apparent prevention of B cell tumors has occurred in immunosuppressed patients in the Pittsburg experience with liver transplants. There is a very high incidence of EBV tumors developing after those transplants, and those tumors have been prevented with the use of acyclovir in those instances. You can't say that they are an adjunct, however, when tumor is already present. That is still an unclear question, whether the virus is actually playing a role in tumor development, or happens to be a bystander. I think it is certainly something deservous of consideration. The unpublished experience is quite varied in using both acyclovir and gangcyclovir in these seriously ill patients. Some patients seem to do better, others go on to have progressive disease. Each patient, though, has been treated as an individual, and there has been great concern about the long-term use, particularly of gangcyclovir, in individuals. So that, I don't think we know enough to answer the question in a standard manner, but I think there is experience is using the agents, some of it is good and some of it is bad.

Sampath A (Harlem Hospital Center, Columbia University, New York, NJ):

Many years ago a comment was made that every case of clinical infectious mononucleosis is an aborted case of Burkitt's lymphoma. On the basis of your findings, would you comment briefly on that?

Jones JF:

Well, it probably isn't. You do get, in most instances, a mono, more than likely, a polyclonal expansion of B lymphocytes which carry the viral genome. Some of those cells which are responding, however, don't carry the virus gene. There are products that Gieovanna Tosota has shown, produced by macrophages which enhance B cell proliferation during this process, so that not all of the cells that are proliferating carry the viral genome. So that probably negates that concept right from the start. Secondly, there appears to be, in Burkitt's lymphoma, a chromosomal anomaly. The mechanism of the chromosomal anomaly is not clear. It seems that, again, in the immunosuppressed patients that you get a polyclonal or oligoclonal B cell response, and then in some of those individuals when it goes on to a monoclonal response, as far as the B cell type that's involved, that there is a chromosomal rearrangement in those cells, but not in the others. So, I think that every case is probably not a Burkitt's lymphoma in the wings. It occurs so rarely that, at least in other populations, that I think that argument is probably not correct.

Gauntt C (University of Texas Health Science Center, San Antonio, TX):

Do the patients with your chronic EB virus infections activate NK cells very well, or at all? Secondly, does the human herpes virus-6 play any role in this disease?

Jones JF:

Fortunately, there was a workshop last week on this syndrome and there were a lot of data presented regarding each of the points you raised. In our prospective study we found no differences in the magnitude of the NK response amongst the six different groups. There have been small outbreaks of a syndrome with an unknown infectious agent that have been studied very well by Ron Herberman and his group, in which they found both. This happened to be in a symphony orchestra, and there were X number of patients with the syndrome and the remainder without. They found just as many individuals in the orchestra without the syndrome who had diminished NK activity as did patients. Then, there was another paper reported from a group in Miami with patients with this syndrome, who likewise, found elevated NK activity. So, right now, part of the problem is probably different methodologies, different concentrations, different ways of reporting the test results, as well as not a very good handle on what one NK activity really is in the population at large, at any one point in time. Although Dr. Herberman's group has shown that it can vary considerably in normal individuals when you measure over time. That's exactly what we saw. The second question is HSV-6. As you all know, that is the newest member of the herpes virus family, which at least in Japan, has been reported to be the agent that causes roseola in children. Seropositivity rates in the United States vary from 25-40% in the population at large. It is highly age dependent. CDC data suggests that there is a higher prevalence of antibodies in young children than in older individuals. Again, we looked at this with some patients in our prospective study and the antibody titers were performed in Houston by Dr. Chang's group at Baylor. They found that again, across the board, in the six groups that we saw, that there was anywhere between 40-45% seropositive rates in patients with this syndrome. Now, there haven't been any prospective studies in patients with this syndrome to see if they were seronegative prior to getting the illness and then followed it out, but I guess that's a possibility. It certainly doesn't appear that it's the long sought after single agent that is, in some peoples' minds, responsible for this new outbreak of this syndrome; that may be in epidemic situations, but certainly in the endemic situations it doesn't appear to be the single agent.

INFECTIOUS COMPLICATIONS OF HUMAN RETROVIRUSES AND HUMAN RETRO-VIRUSES AS INFECTIOUS COMPLICATIONS

JONATHAN W. M. GOLD

Infectious Disease Service, Memorial Sloan-Kettering Cancer Center, 1275 York Avenue, New York, New York 10021

INTRODUCTION

As of June 1, 1988, The World Health Organization had received reports of 96,433 cases of acquired immune deficiency syndrome (AIDS) and had estimated a worldwide cumulative total of over 200,000 cases. In the United States, the Centers for Disease Control reported a cumulative total of 63,726 cases of whom 56% were known to have died. Such grim statistics grossly underestimate the magnitude of the pandemic of human immunodeficiency virus type 1 (HIV-1) infection. Current estimates are of between 5 and 10 million people infected worldwide, with a potential of a 10-fold increase in the next 5 years. Individuals infected with HIV-1 are reservoirs of infection for others and live with what now appears to be a very high probability of developing AIDS (CDC, 1988a; Kreiss et al. 1986; Mann, 1988; Quinn et al. 1986).

AIDS is the end infection with the HIV-1, a human retrovirus (Barre-Sinoussi et al. 1983; Gallo et al. 1984; Gottlieb et al. 1981; CDC, 1987; Masur et al. 1981; Siegal et al. 1981). Mainly because of the AIDS epidemic, there has been a great amount of interest in human retroviruses. Five human retroviruses are presently recognized: human T-lymphotropic viruses (HTLV) I, II and IV, and HIV-1 and HIV-2 (Clavel et al. 1987; Kalyanaraman et al. 1982; Manzari et al. 1987; Poiesz et al. 1980). It seems likely that others will be found. These viruses are characterized by a unique life cycle and an enzyme, reverse transcriptase, which directs the synthesis of DNA from RNA. This newly synthesized DNA must integrate into the DNA of the infected host cell where it remains as a permanent, latent infection or directs the production of new virus particles. Retroviruses are transmitted sexually, perinatally, by transfusion of infected blood and blood products and between drug users who share needles and syringes. To date, AIDS has occurred in homosexual men, intravenous drug users, the recipients of infected blood or blood products, the sexual partners of HIV-1 infected people and infants born to HIV infected mothers.

HIV-1 causes human disease by several mechanisms. First, and most importantly, it causes, as the end result of incompletely understood pathogenetic processes, severe defects in T-cell function, which cause the host to be extraordinarily susceptible to infection and neoplastic disease. These infectious and neoplastic complications (Table 1), occurring in individuals with no known cause for immunosuppression were the first recognized manifestations of AIDS, and led to the original surveillance definition. Subsequently, it was recognized that AIDS patients have a clinically significant defect in B-cell function and an increased susceptibility to infection with microorganisms taking advantage of decreased immunoglobulin levels (Ammann et al. 1984; Lane et al. 1983; Simberkoff et al. 1984). HIV-1 infects macrophages and affects neurocytes through mechanisms which are currently being studied. HIV-1 infected people commonly develop severe neurologic problems including, very commonly, dementia, vacuolar myelopathy and peripheral neuropathy (Snider et al. 1983).

HIV-1 is striking in that it causes illness by permitting other organisms to be virulent in an altered host. As HIV-1 spreads in different populations we are seeing significant alterations in the epidemiologic and clinical patterns of many infectious diseases, including for example syphilis, tuberculosis and measles. It can be expected that as HIV-1 spreads in different countries, it will change the epidemiologic patterns and virulence of infections endemic in those areas. Potential examples include leishmaniasis, trypanosomiasis, malaria and a large variety of other infections (Matheron et al. 1988).

Just as it spotlights and alters infectious diseases, HIV-1 infection is an illness which remorselessly highlights and exposes many inadequacies and inequities in social, political and medical care systems. It poses extraordinary challenges in developed countries: in New York City municipal hospitals, AIDS patients currently occupy one-half of all medical beds. Chronic care facilities for severely debilitated patients are far too few. The cost of medical care for one patient is approximately $150,000 during his lifetime, and this cost, ironically may rise with treatments which are effective in prolonging life, but very expensive such as zidovudine (azidothymidine; AZT) (Scitovsky and Rice,

TABLE 1. COMPLICATIONS OF AIDS

ORGANISMS WHICH TAKE ADVANTAGE OF T-CELL DEFECTS

PARASITES	BACTERIA	FUNGI	VIRUSES
Pneumocystis carinii	Mycobacterium avium-intra-cellulare	Candida albicans	Cytomegalovirus
Toxoplasma gondii	Mycobacterium tuberculosis	Cryptococcus neoformans	Herpes simplex virus
Cryptosporidia	Salmonella	Histoplasma capsulatum	Varicella zoster virus
Isospora belli	Legionella	Coccidioides immitis	JC virus
	Nocardia asteroides		? Adenovirus
	Listeria monocytogenes		

ORGANISMS WHICH TAKE ADVANTAGE OF B-CELL DEFECTS

Streptococcus pneumoniae
Hemophilus influenzae

NEOPLASTIC COMPLICATIONS

Kaposi's sarcoma
B-cell lymphomas

1987). In developing countries, the problems of treating AIDS patients threaten to utterly overwhelm health care systems.

HIV-1 infection also remorselessly exposes human behavior, and forces us to examine it compassionately and unblinkingly. Since most cases are transmitted sexually or through the use of illicit drugs, it emphasizes how little we really know about human sexual behavior in different cultural contexts, and how complex the problems of altering sexual behavior and controlling illicit drug use are.

Despite these frightening statistics, and the impression that AIDS is an illness which is 100% fatal, there have been a number of advances in the treatment of AIDS patients which are considerably increasing the lifespan and quality of life for AIDS patients. Understanding of the life cycle, genetics, biochemistry and molecular biology of HIV-1 and the other human retroviruses is advancing at an explosive pace. It has already resulted in the development of effective, life-prolonging therapy for HIV-1 infected people, and it seems very likely that new, more effective treatments will emerge.

At least three of the five known human retroviruses have been linked with human disease: HTLV-I, HIV-1 and HIV-2.

HTLV-I is associated with adult T-cell leukemia/lymphoma and tropical spastic paraparesis (TSP) or HTLV-I associated myelopathy (HAM) (Gessain et al. 1985; Uchiyama et al. 1977; Yoshida et al. 1982). Seroprevalence studies have emphasized high prevalence rates in Japan, Africa and the Caribbean. In southern Japan, there are striking differences in sero-prevalence between neighboring communities. Transmission is sexual with females of infected partners more likely than males to develop infection. Vertical transmission from mother to child, either transplacentally or by nursing is common. It is possible to prevent transmission to children of infected mothers by substituting bottle feeding for breast feeding (Hino et al. 1987). The incidence of ATL in adults infected with the virus has been estimated at about one per thousand infected persons per year (Kondo et al. 1985; Tajima and Kuroishi, 1985). It is possible that as many as one in 20 to one in 50 infected people will develop ATL. However, attack rates for infected people are imprecise because it is difficult to determine the duration of infection and because the risk seems to increase with age.

Transfusion associated HTLV-I infection has been frequently documented in Japan. Seroconversion has been documented in over 6% of seronegative recipients of seropositive blood (Okochii and Sato, 1985). Although no cases of ATL have been clearly linked to transfusion associated HTLV-I infection, perhaps because of the long incubation period from infection to the development of disease, there are a number of reports of TSP or HAM two to four years following transfusion (Osame et al. 1986).

Recent reports have provided estimates of the prevalence of HTLV-I infection in the United States. In one study, it was found that 6 of 211 multiply transfused leukemia patients were infected as a result of transfusion. It was calculated that this represented a donor prevalence of HTLV-I infection in the donor population at least 2.3 per 10,000 (Minamoto et al. 1988). A similar figure (2.5 per 10,000) was found in a large sero-survey of American blood donors (Williams et al. 1988). Interestingly, the incidence of transfusion associated HTLV-I infection in the population of leukemia patients remained steady between 1978 and 1987, whereas the incidence of transfusion associated HIV-1 infection rose dramatically prior to the screening of blood for HIV-1 antibody, suggesting that HTLV-I was present in the donor population at a relatively constant level while the prevalence of HIV-1 increased dramatically.

The risk of illness from transfusion associated HTLV-I infection appears to be much smaller than from HIV-1 infection. Nevertheless, HTLV-I is a preventable cause of leukemia and neurologic illness. In addition, secondary infection from infected transfusion recipients occurs. Because of these risks, some hospitals in the United States have already begun screening blood donors for HTLV-I on an experimental basis, and universal screening will be introduced when antibody tests are licensed.

HTLV-II was isolated from patients with hairy cell leukemia, although an etiologic relationship is not yet established. Currently used antibody tests (ELISA and Western blot) do not reliably distinguish between HTLV-I and II. HTLV-I/II infections are prevalent among intravenous drug users in certain areas (Robert-Guroff et al. 1986). Because of the difficulties of distinguishing HTLV-I and HTLV-II infection by serologic testing, it is not clear how many infections are due to each virus. Co-infection with HIV-1 and HTLV-I/II is not uncommon in this population. The possibility exists that HTLV-I/II gene products will enhance the replication of HIV-1 in HTLV-I/II infected cells and accelerate the progression to AIDS in such patients (Bartholemew et al. 1987). Conversely, HIV-1 associated immunosuppression has been reported to alter the course of HTLV-I infection leading to the proliferation of cells with a T-suppressor phenotype and a leukemia-like illness (Harper et al. 1986).

HIV-1 is the etiologic agent of AIDS. HIV-2 is a different, but related virus, which causes AIDS. At this time, it is thought to be much more geographically restricted than HIV-1, occurring principally in West Africa. It is important to note that current antibody tests for HIV-1 may not detect antibody to HIV-2 and vice-versa (Clavel et al. 1987; Kalyanaraman et al. 1982; Poiesz et al. 1980).

HTLV-V is a newly described virus isolated from lymphoma patients. A role as an etiologic agent of human disease is not yet established (Manzari et al. 1987).

PATHOGENESIS OF HIV-1-INDUCED IMMUNE DEFICIENCY

Over time, HIV-1 infection leads to an irreversible depletion of T-helper lymphocytes. The mechanisms may include viral lysis of the infected cells and the induction of toxic substances which suppress or destroy T-helper lymphocytes. Loss of T-helper lymphocyte function leads to the extraordinary increased susceptibility to certain neoplasms (Kaposi's sarcoma and B-cell lymphomas) and to opportunistic infections which define AIDS in the absence of other causes of immune deficiency.

The T-helper lymphocyte plays a central role in the regulation of the immune system. Therefore, there are a large number of other defects in immune function present in HIV-1 infected patients some of which are clinically significant and lead to certain kinds of infections. The B-cell defect is quite pronounced and is associated with an increased frequency of *Streptococcus pneumoniae* and *Hemophilus influenzae* infections (Ammann et al. 1984; Lane et al. 1983; Simberkoff et al. 1984). Decreased numbers of circulating neutrophils are common in AIDS patients, and may be clinically significant, but are often surprisingly well tolerated (Whimbey et al. 1986; Witt et al. 1987). Abnormalities of the integument and gastrointestinal tract are common in Kaposi's sarcoma patients and may be associated with bacteremia or cellulitis.

INFECTIONS IN AIDS AND HIV-1 INFECTED PATIENTS

The organisms (Table 1) which take advantage of T-cell defects and infect AIDS are mostly obligate or facultative intracellular parasites which establish latent infections. Reactivation disease may develop many years after the original infection. These infections respond more slowly to treatment in AIDS patients and relapse is very common. Treatment is generally prolonged, often for the life of the patient (Gold, 1985). Many patients at risk for specific infections can be identified by skin testing (*M. tuberculosis*), history (herpes simplex) or serology (*T. gondii*). Knowledge of this susceptibility now leads more and more often to early recognition of disease when treatment is more effective. In addition, prophylaxis is available for some infections and is being developed for others. Increased awareness on the part of physicians and patients, close follow-up of patients with particular attention to the early signs of infection so that treatment can be instituted early, treatment with AZT and, possibly, prophylaxis of infections such as *P. carinii* pneumonia (PCP), has lead to striking improvements in the survival of AIDS patients. Whereas up to 1985, AIDS patients at Memorial Sloan-Kettering Cancer Center died within a median time of 1 year, well over 50% of patients treated since 1986 still survive as of June, 1988.

SPECIFIC INFECTIOUS AND NEOPLASTIC COMPLICATIONS OF AIDS

Numerous reviews of the clinical features of the common infectious, neoplastic and neurologic complications of AIDS and HIV-1 infection have been published. This discussion will highlight some of the newer developments which have occurred in the approach to some of these complications.

Pneumocystis carinii

PCP remains the most commonly diagnosed serious infection and AIDS defining illness. Diagnosis is made by demonstration of the characteristic cysts in pulmonary secretions obtained by bronchoalveolar lavage. Recent experience suggests that examination of expectorated induced sputum specimens may be a useful diagnostic technique although the sensitivity and specificity need to be established (Bigbee et al. 1986). Monoclonal antibodies to identify the organisms in secretions may also simplify the diagnosis (Kovacs et al. 1988). An advantage of bronchoscopy with bronchoalveolar lavage and transbronchial biopsy remains the ability to diagnose other infections, especially cytomegalovirus (CMV) pneumonia, which cannot be diagnosed by the other techniques.

The treatment of PCP with available drugs is very successful. Trimethoprim-sulfamethoxazole and pentamidine are equally effective, although adverse drug reactions occur in more than half of patients. Alternative treatments, especially with Dapsone-trimethoprim, Fansidar or aerosol pentamidine appear effective in certain cases. It is now extremely rare, in some centers, for patients to die of the first episode of PCP, and recovery from successive episodes is becoming the rule. The prophylaxis of PCP has the potential of preventing morbidity and mortality. Trimethoprim-sulfamethoxazole has been proven effective prophylactically (Fischl et al. 1988). Aerosol pentamidine has a satisfying rationale and is widely used, although controlled trials to determine its efficacy have not been completed (Bernard et al. 1988).

Mycobacterium avium-complex

Evidence of infection with this organism is found in about 50% of autopsies by both histopathology and culture. Despite *in vitro* susceptibility to a variety of drugs, treatment of this infection has been extraordinarily difficult. Prophylactic studies are being considered (Hawkins et al. 1986).

Mycobacterium tuberculosis

One of the many disturbing aspects of AIDS and HIV-1 infection has been dramatic changes in the epidemiology and clinical manifestations of tuberculosis. In a reversal of one of the great triumphs of the public health movement, the number of reported cases of tuberculosis in the United States in 1986 showed an increase of 2.6% (CDC, 1988b). This marks the first reversal, unrelated to reporting changes, in the declining incidence of tuberculosis since the beginning of the 20th century. This increase is thought to be due mainly, although not entirely, to cases occurring in HIV-1 infected people in the inner cities of the United States.

Tuberculosis must be considered in HIV-1 infected people with positive tuberculin tests, radiographic evidence of healed granulomatous disease or histories of prior exposure, especially when they are from regions where infection is highly prevalent. Isoniazid prophylaxis should be considered for such individuals. Conversely, HIV-1 testing should be considered in patients with tuberculosis in whom there are risk factors for HIV-1 infection.

Cytomegalovirus infection

CMV disease in AIDS patients includes severe bilateral pneumonia, hepatitis, enterocolitis, chorioretinitis and adrenalitis. CMV chorioretinitis may occur in as many as 20% of AIDS patients. Adrenal insufficiency due to CMV adrenalitis occurs. Specific diagnosis of this infection is difficult and generally requires the demonstration of typical inclusion bodies and tissue reaction in tissue specimens. Gancyclovir (9-(1,3-dihydroxy-2-propoxymethyl)guanine;DHPG) appears to be effective against CMV chorioretinitis and enterocolitis, and possibly against CMV pneumonia. However, maintenance therapy appears necessary for an indefinite period because of the high frequency of relapse when the drug is stopped (Masur et al. 1986; Tapper et al. 1984; Urmacher et al. 1982). Phosphonoformate (Foscarnet) also appears effective against retinitis (Walmsley et al. 1988), but relapses also occur when treatment is stopped.

Cryptococcus neoformans

As with other infections in AIDS patients the diagnosis is sometimes made in patients with the most minimal symptoms. Treatment is with amphotericin B and 5-fluorocytosine. Patients usually require maintenance therapy for the rest of their lives. Fluconazole, an imidazole derivative, is currently undergoing investigation as both initial treatment and maintenance therapy (Kovacs et al. 1985).

Toxoplasma gondii

Toxoplasmosis in AIDS patients is usually a central nervous system disease due to reactivation of a latent infection. Positive serologies for toxoplasma infection (IgG-IFA) tests are present in about 20% of AIDS patients in New York City. In certain European cities, such as Paris and Brussels, the prevalence of latent infection with *T. gondii* is probably 90%. About 30% of seropositive AIDS patients will develop central nervous system toxoplasmosis. Thus, the risk of developing toxoplasmosis in seropositive AIDS patients is extraordinarily high. This easily identified group of patients are obvious candidates for prophylaxis. The development of safe effective prophylaxis for this disease in AIDS patients should obviously be a high priority (Luft and Remington, 1988; Wong et al. 1984).

Subacute encephalopathy of AIDS (AIDS Dementia Complex)

The AIDS dementia complex is thought to be a direct consequence of HIV-1 infection of the brain, not a complicating infection. Neurotoxic substances produced by HIV-1 infected macrophages are being investigated for a pathogenetic role. The exact frequency of abnormalities of the CNS in AIDS patients, and HIV-1 infected patients is still being determined but appears to be extraordinarily high (Snider et al. 1983).

HIV-1 INFECTION AND OTHER INFECTIOUS DISEASES

As HIV-1 infection spreads into different populations, different infections prevalent in that population are modified. The attack rate, clinical course, diagnosis and response to treatment of many illnesses may be altered profoundly when there is co-existing HIV-1 infection. HIV-1 infection has had a profound effect on the epidemiology of tuberculosis (Hawkins et al. 1986). Fatal cases of measles in HIV-1 infected children are reported (CDC, 1988c). HIV-1 appears to alter the clinical course of syphilis. Meningovascular syphilis may be more common in HIV-1 infected patients. Recently, a seronegative case of typical secondary syphilis was reported. It was diagnosed by a biopsy specimen of a skin lesion. Current treatment recommendations may be inadequate. The suggestion has been made that it may be necessary to treat all HIV-1 infected patients with regimens heretofore reserved for neurosyphilis (Johns et al. 1987; Tramont, 1987). Infections with *Campylobacter jejuni* (Perlman et al. 1988) and shigella (Bell et al. 1985) are more severe and prolonged in HIV-1 infected patients, and may relapse despite appropriate treatment.

OTHER INFECTIOUS DISEASE AND THE TRANSMISSION OF HIV-1

It is likely that HIV-1 infection is facilitated by direct contact with infected blood cells. Such contact is likely to occur in the presence of open sores and ulcers. In populations where heterosexual transmission is the principal route of spread of HIV-1, such as in Africa and among partners of intravenous drug users, the role of genital ulcer disease (GUD) in facilitating the spread of infection is becoming apparent. GUD has multiple etiologies including syphilis, chancroid, lymphogranuloma venereum, granuloma inguinale and herpes simplex, all of which are common sexually transmitted diseases (STDs) in various parts of the world. Among African patients in STD clinics and among prostitutes, the presence or history of genital ulcer disease correlates strongly with the development of HIV-1 infection. Thus STDs appear to be important cofactors for the transmission of HIV-1 infection. Studies to confirm the importance of GUD in the transmission of HIV-1 are underway in United States populations.

SUMMARY

The current pandemic of HIV-1 infection is having an ever widening impact on infectious diseases. HIV-1 infection is associated with immune deficiency and neurologic disease. AIDS patients are especially susceptible to infections which take advantage of T and B-cell dysfunction. Many infections represent reactivation of latent organisms. As co-infection with HIV-1 and other organisms occur it is likely that these infections will be more severe, persistent and difficult to treat and will present with altered clinical features. Some of these infections, such as those associated with genital ulcer disease will enhance transmissibility of HIV-1. Others which share the same means of transmission as HIV-1 are being recognized as agents of human disease, especially HIV-2 and HTLV-I. With increasing prevalence of these other retroviruses, mixed infections are becoming more common, and each virus may affect the outcome associated with the other.

REFERENCES

Ammann AJ, Schiffman G, Abrams D (1984) B-cell immunodeficiency in acquired immune deficiency syndrome. JAMA 251:1447-1449.

Barre-Sinoussi F, Chermann JC, Rey F, Nugeyre MT, Chamaret S, Gruest J, Dauguet C, Axler-Blin C, Brun-Vezinet F, Rouzious C, Rozenbaum W, Montagnier L (1983) Isolation of a T-lymphotropic retrovirus from a patient at risk for acquired immune deficiency syndrome (AIDS). Science 220:868-871.

Bartholemew C, Blattner W, Cleghorn F (1987) Progression to AIDS on homosexual men co-infected with HIV and HTLV-I in Trinidad. Lancet 2:1496.

Bell ET, Rotterdam HZ, Tapper ML (1985) Simultaneous shigellosis and cytomegalovirus (CMV) colitis in patients with acquired immune deficiency syndrome (AIDS). In: Abstracts from the International Conference on Acquired Immunodeficiency Syndrome, April 14-17, Atlanta, Georgia, Philadelphia: American College of Physicians, p 49.

Bernard EM, Schmitt HJ, Lifton A, Seltzer M, Dickmeyer MS, Armstrong D (1988) Prevention of Pneumocystis carinii pneumonia with aerosol pentamidine. Abstract 7169. IVth International AIDS Conference, Stockholm, Sweden, June 12-16.

Bigbee TD, Margolskee D, Curtis JL, Michael PF, Sheppard D, Hadley WK, Hopewell PC (1986) The usefulness of induced sputum in the diagnosis of Pneumocystis carinii pneumonia in patients with the acquired immune deficiency syndrome. Am Rev Respir Dis 133:515-518.

CDC (1987) (Revision) Surveillance Case Definition for Acquired Immunodeficiency Syndrome. MMWR (Suppl.) 36:3S-15S.

CDC (1988a) AIDS Weekly Surveillance Report, United States AIDS Program, Center for Infectious Diseases, Centers for Disease Control, May 30.

CDC (1988b) Tuberculosis, final data, United States, 1986. 36:817-820.

CDC (1988c) Measles in HIV-infected Children, United States. MMWR 37:183-186.

Clavel F, Mansinho K, Chamaret S, Guetard D, Favier V, Nina J, Santos-Ferreira M-O, Champalimaud JL, Montagnier L (1987) Human immunodeficiency virus type-2 infection associated with AIDS in West Africa. N Engl J Med 316:1180-1185.

Fischl MA, Dickinson GM, la Voie L (1988) Safety and efficacy of sulfamothoxazole and trimethoprim chemoprophylaxis for *Pneumocystis carinii* pneumonia in AIDS. JAMA 259:1185-1189.

Gallo RC, Salahuddin SZ, Popovic M, Shearer GM, Kaplan M, Haynes BF, Palker TJ, Redfield R, Oleske J, Safai B, White G, Foster P, Markham PD (1984) Frequent detection and isolation of cytopathic retroviruses (HTLV-II-I) from patients with AIDS and at risk for AIDS. Science 224:500-502.

Gessain A, Barin F, Vernant JC, Gout O, Maurs L, Calender A, DeThe G (1985) Antibodies to human T-lymphotropic virus type-1 in patients with tropical spastic paraparesis. Lancet 2:407-410.

Gold JWM (1985) Clinical spectrum of infections in patients with HTLV-III-associated diseases. Canc Res (Suppl) 45:4652S-4654S.

Gottlieb MS, Schroff R, Schanker HM, Weisman JD, Fan PT, Wolf RA, Saxon A (1981) Pneumocystis carinii pneumonia and mucosal candidiasis in previously healthy homosexual men: evidence of a new acquired cellular immunodeficiency. N Engl J Med 305:1425-1431.

Harper ME, Kaplan MH, Marselle LM (1986) Concomitant infection with HTLV-I and HTLV-III in a patient with T8 lymphoproliferative disease. N Engl J Med 315:1073-1078.

Hawkins CC, Gold JWM, Whimbey E, Kiehn TE, Brannon P, Cammarata R, Brown AE, Armstrong D (1986) Treatment of disseminated Mycobacterium avium-intracellulare infections in patients with the acquired immune deficiency syndrome. Ann Intern Med 105:184-188.

Hino S, Doi H, Yoshikuni H, Sugiyama H, Ishimaru T, Yamabe T, Tsuji Y, Miyamoto T (1987) HTLV-I carrier mothers with high titer antibody are at risk as a source of infection. Japan J Cancer 78:1156-1158.

Johns DR, Tierney M, Felsenstein D (1987) Alteration in the natural history of neurosyphilis by concurrent infection with the human immunodeficiency virus. N Engl J Med 316:1569-1572.

Kalyanaraman VS, Sarngadharan MG, Robert-Guroff M, Miyoshi I, Blayney D, Golde D, Gallo RC (1982) A new subtype of human T-cell leukemia virus (HTLV-II) associated with a T-cell variant of hairy cell leukemia. Science 218:571-573.

Kondo T, Nonaka H, Miyamoto N, Yoshida R, Marsue Y, Ohguchi Y, Inouye H, Komoda H, Hinuma Y, Hanaoka M (1985) Incidence of adult T-cell leukemia-lymphoma and its familial clustering. Int J Cancer 35:749-751.

Kovacs JA, Kovacs AA, Polis M, Wright WC, Gill VJ, Tuazon CU, Gelmann EP, Lane HC, Longfield R, Overturf G, Macher AM, Fauci AS, Parrillo JE, Bennett JE, Masur H (1985) Cryptococcosis in the acquired immunodeficiency syndrome. Ann Intern Med 103:533-538.

Kovacs JA, Ng VL, Masur H, Leoung G, Lane HC, Ognibene FP, Shelhamer J, Parillo JE, Gill VJ (1988) Diagnosis of *Pneumocystis carinii* pneumonia: Improved detection in sputum, with use of monoclonal antibodies. N Engl J Med 318:589-593.

Kreiss JK, Koech D, Plummer FA (1986) AIDS virus infection in Nairobi prostitutes. Spread of the epidemic to East Africa. N Engl J Med 314:414-418.

Lane HC, Masur H, Edgar LC, Whalen G, Rook AH, Fauci AS (1983) Abnormalities of B-cell activation and immunoregulation in patients with the acquired immunodeficiency syndrome. N Engl J Med 309:453-458.

Luft BJ, Remington JS (1988) Toxoplasmic encephalitis. J Infect Dis 157:1-6.

Mann J (1988) The Global Picture of AIDS. IVth International Conference on AIDS, Stockholm, Sweden, June 14.

Manzari V, Gismondi A, Barillari G, Morrone S, Modesti A, Albonici L, De Marchis L, Fazio V, Gradilone A, Zani M, Frati L, Santoni A (1987) HTLV-V: A new human retrovirus isolated in a Tac-negative T-cell lymphoma/leukemia. Science 238:1581-1583.

Masur H, Michelis MA, Greene JB, Onorato I, Vande Stouwe RA, Holzman RS, Wormser G, Brettman L, Lange M, Murray HW, Cunningham-Rundles S (1981) An outbreak of community-acquired *Pneumocystis carinii* pneumonia: Initial manifestation of cellular immune dysfunction. N Engl J Med 305:1431-1438.

Masur H, Lane HC, Palestine A, Smith PD, Manischewitz H, Stevens G, Fujikawa L, Macher AM, Nussenblatt R, Baird B, Menill M, Wittek A, Quinnan LV, Parrillo JE, Rookh H, Eron LJ, Poretz DM, Goldenberg RJ, Fauli A, Gelmann EP (1986) Effect of 90(1,3-dihydroxy-2-propoxymethyl)guanine on serious cytomegalovirus disease in eight immunosuppressed homosexual men. Ann Intern Med 104:41-44.

Matheron S, Lepretre A, Ngovan P, et al (1988) Visceral leishmaniasis during HIV infection. Abstract 7213: IVth International AIDS Conference, Stockholm, Sweden, June 12-16.

Minamoto GY, Gold JWM, Scheinberg DA, Hardy WD, Chein N, Zuckerman BS, Reich L, Dietz K, Gee T, Hoffer J, Mayer K, Gabrilove J, Clarkson B, Armstrong D (1988) Infection with human T-cell leukemia virus type-I in patients with leukemia. N Engl J Med 318:219-222.

Okochii K, Sato H (1985) Adult T-cell leukemia virus, blood donors and transfusions: Experience in Japan. In: Dodd RY, Barker LF (eds) Infection, Immunity and Blood Transfusion. New York, Alan R. Liss, Inc., pp 245-256.

Osame M, Izumo S, Igata A, Matsumoto M, Matsumoto T, Sonoda S, Tara M, Shibata Y (1986) Blood transfusion and HTLV-I associated myelopathy. Lancet 2:104-105.

Perlman DM, Ampel NM, Schifman RB, Cohn DL, Patton CM, Aguirre ML, Wang WL, Blaser MJ (1988) Persistent *Campylobacter jejuni* infections in patients infected with human immunodeficiency virus (HIV). Ann Intern Med 108:540-546.

Poiesz BJ, Ruscetti FW, Gazdar AF, Bunn PA, Minna JD, Gallo RC (1980) Detection and isolation of type-C retrovirus particles from fresh and cultured lymphocytes of a patient with cutaneous T-cell lymphoma. Proc Natl Acad Sci USA 77:7415-7419.

Quinn TC, Mann JM, Curran JW (1986) AIDS in Africa: An epidemiologic paradigm. Science 234:955-963.

Robert-Guroff M, Weiss SH, Giron JA, Jennings AM, Ginzburg HM, Margolis IB, Blattner WA, Gallo RC (1986) Prevalence of antibodies to HTLV-I,

-II, -III in intravenous drug abusers from an AIDS endemic region. JAMA 255:3133-3137.
Scitovsky AA, Rice DP (1987) The cost of AIDS. Issues Sci Technol 4:61-66.
Siegal FP, Lopez C, Hammer GS, Brown AE, Kornfeld SJ, Gold J, Hassett J, Hirschman SZ, Cunningham-Rundles C, Adelsberg BR, Parham DM, Siegel M, Cunningham-Rundles S, Armstrong D (1981) Severe acquired immunodeficiency in homosexual males, manifested by chronic perianal ulcerative herpes simplex lesions. N Engl J Med 305:1439-1444.
Simberkoff MS, El-Sadr W, Schiffman G, Rahal JJ Jr (1984) *Streptococcus pneumoniae* infections and bacteremia in patients with acquired immune deficiency syndrome, with report of a pneumococcal vaccine failure. Am Rev Respir Dis 130:1174-1176.
Snider WD, Simpson DM, Nielsen S, Gold JWM, Metroka CE, Posner JB (1983) Neurological complications of acquired immune deficiency syndrome: Analysis of 50 patients.Ann Neurol 14:403-418.
Tajima K, Kuroishi T (1985) Estimation of incidence rates of ATL among ATLV carriers in Kyushu, Japan. Jap J Clin Oncol 15:423-430.
Tapper ML, Rotterdam HZ, Lerner CW, Al'Khafaji K, Seitzman PA (1984) Adrenal necrosis in the acquired immunodeficiency syndrome. Ann Intern Med 100:239-241.
Tramont EC (1987) Editorial: Syphilis in the AIDS era. N Engl J Med 316:1600-1601.
Tuberculosis, Final Data - United States, 1986 (1988) MMWR 36:817-820.
Uchiyama T, Yodoi J, Sagawa K (1977) Adult T-cell leukemia: Clinical and hematologic features of 16 cases. Blood 50:481-492.
Urmacher C, Myskowski P, Ochoa M, Kris M, Safai B (1982) Outbreak of Kaposi's sarcoma with cytomegalovirus infection in young homosexual men. Am J Med 72:569-575.
Walmsley SL, Chew SE, Read H, Vellend H, Salit I, Rachli A, Fanning MM (1988) Treatment of cytomegalovirus retinitis with trisodium phosphonoformate hexahydrate (Foscarnet). J Infect Dis 157:569-572.
Whimbey E, Gold JWM, Polsky B (1986) Bacteremia and fungemia in patients with the acquired immunodeficiency syndrome. Ann Intern Med 104:511-514.
Williams AE, Fang CT, Slamon DJ, Poiesz BJ, Sandler SG, Darr WF, Shulman G, McGouann EI, Douglas DK, Bowman RJ (1988) Seroprevalence and epidemiologic correlates of HTLV-I infection in United States blood donors. Science 240:643-646.
Witt DJ, Craven DE, McCabe W (1987) Bacterial infections in adult patients with the acquired immune deficiency syndrome (AIDS) and AIDS related complex. Am J Med 82:900-906.
Wong B, Gold JWM, Brown AE (1984) Central nervous system toxoplasmosis in homosexual men and parenteral drug abusers. Ann Intern Med 100:36-42.
Yoshida M, Miyoshi I, Hinuma Y, Lange M, Fried R, Grielo M, Mildvan D, Girion J, Tapper ML, Lerner CW, Armstrong D (1982) Isolation and characterization of retrovirus from cell lines of human adult T-cell leukemia and its implication in the disease. Proc Natl Acad Sci USA 79:2031-2035.

DISCUSSION

Fanning W (Scottsdale Memorial Hospital, Scottsdale, AZ):

I took a quick look at one of those slides, I'm not sure I got it right, but it looked like the incidence of HIV in your transfused leukemics had gone down. Did 0.069%, represent chronic leukemics who had gotten transfusions before 1985, perhaps, and that is why the incidence has gone down?

Gold J:

That 0.069% was an overall estimate over a 5-year period, and is an estimate of the average prevalence of antibody to HIV in the donor population for our patients. It is calculated from the prevalence of HIV in the recipients, and the total number of units of blood and blood products transfused. It was our attempt to figure out how many people were infected among the donors who gave blood.

Fanning W:

Can you tell if the incidence is going up or down?

Gold J:

Incidence of HIV-1 in the donors? We couldn't do that analysis because we didn't know when patients received the units which caused infection.

Trachtenberg A (California Dept. of Health Service, Berkeley, CA):

Do you know of any other evidence from around the country similar to what you were showing us about the endemicity of HTLV-1 versus the epidemicity?

Gold J:

The data on the prevalence of HTLV-1 infection in different populations is just being developed. Data that was published five or six years ago suggested that in Berlington, Vermont, 1% of donors were HTLV-1 positive. It showed a surprisingly high prevalence in certain areas (2-3% in Birmingham, Alabama and Houston, Texas), and has actually been retracted. The antibody assays used were not very specific. Except for the American Red Cross Study which is very limited, I don't know of any recently published data on the endemicity in different areas from donor populations, nor do I know of studies showing changes in antibody prevalence over time.

Trachtenberg A:

Have they looked at samples from other points in time, or just the most recent HAINS?

Gold J:

I don't know. The serum banks exist to answer that question from the Transfusion Safety Studies, and these are being looked at, but I don't know the data if it exists.

Lennette E:

Dr. Trachtenberg, you mentioned epidemicity and endemicity. By epidemicity do you mean real epidemics or do you mean clustering of cases.

Trachtenberg A:

By epidemicity for HIV-1, he showed that the rates in his donors was going up from 79 to 85, is that correct? Whereas the prevalence of HTLV-1 in his donors was pretty much the same, from 79-85 which, to my view with that limited amount of data which showed that it's more endemic than HIV which is more epidemic in the donors he was studying.

Gold J:

The point that I was trying to make is that HTLV-1 has been present for a longer time a fairly constant level whereas HIV-1 appeared more recently and spread very rapidly.

Sampath A (Harlem Hospital Center, New York, NY):

With regard to your comment on seronegative secondary syphilis in HIV-1 infected patients, how frequently has this been reported and was this a clinical diagnosis of secondary syphilis or was it confirmed by histopathology?

Gold J:

I'm referring to one case which was reported in the Annals of Internal Medicine. It was very well confirmed. The diagnosis was made by a biopsy of typical secondary skin lesions where treponemas were actually seen within a typical plasmocytic lymphocytic infiltrate. The person subsequently to the diagnosis developed a positive VDRL.

Sampath A:

The anti-treponeamal antibodies persist for a very, very long time, even after successful treatment. Could we assume then that these patients were infected with HIV before their primary lesions appeared?

Gold J:

That's different for the case I just referred to. Most of these people who had the neurosyphilis or the syphilitic meningitis presumably had secondary syphilis and they were presumed to be infected with HIV before they developed syphilis. There are a couple of cases that have been described in England, of people thought to have been adequately treated for syphilis 20 years ago, who presumably developed HIV infection subsequently. They developed gumatous lesions. You can obviously wonder, were they adequately treated during their initial episode or were they subsequently reinfected.

Little, L (BioRad Laboratories, Hercules, CA):

You mentioned the relationship of CMV to the dementia complex in AIDS. Is CMV commonly related to that complex and would you tell me, do we know the cell types that are infected by CMV?

Gold J:

I don't think that it is commonly related. In fact, I think it's generally not related to AIDS dementia complex. The cause of AIDS dementia is really unknown. It's not CMV: I think it's somehow directly related to HIV infection of the central nervous system.

STATISTICAL ISSUES IN ASSESSING THE AIDS EPIDEMIC

VICTOR DE GRUTTOLA AND WILLIAM BENNETT

Harvard School of Public Health and Harvard Medical School
Boston, Massachusetts, USA

INTRODUCTION

Understanding the spread of AIDS and the virus that causes it is much more difficult than is generally recognized. The hope of predicting even the size of the epidemic over the next fifteen years, let alone the population groups that will be affected, is limited by lack of information about the natural history of AIDS and the infectivity of HIV, and by our sketchy knowledge of sexual and needle-sharing behavior.

Current prospective research on large cohorts of homosexually active men (Kingsley et al. 1987) will ultimately yield good estimates of the risk that HIV infection will progress to AIDS (De Gruttola and Mayer, 1987; Padian et al. 1987). But we shall not be able to estimate the infectivity of the virus or to quantify the risk of so-called lower-risk sexual practices on the basis of current studies monitoring the sexual partners of people at risk for AIDS. The rate of transmission beyond the currently identified population of drug users at risk cannot even be guessed from current knowledge about the patterns of sexual and needle-sharing behavior in that group or others (De Gruttola and Mayer, 1988).

Confusion about the epidemic has also been fostered by flaws in the way data have been obtained and by basic errors of inference from the data that are available. The behavior of AIDS has been modeled in ways that only roughly approximate reality; these models have then been overinterpreted. Public health authorities have failed to make data available to statisticians or mathematical modelers. As a result, the development and use of sophisticated methods required to assess the AIDS epidemic has been slowed.

Perhaps the most elementary error of inference has been to draw conclusions about the course of the epidemic from current changes in the doubling time - the interval during which the number of cases increases by a factor of two. Doubling time cannot be used to infer the extent to which the epidemic has run its course in the population, or in any subset of the population. It gives almost no information about the extent to which the disease is spreading from high- to low-risk groups, and it cannot be used to monitor the extent of behavior change in any group (De Gruttola and Lagakos, (1989).

Whenever the future of the epidemic is forecast, some theory of the way it grows must be invoked. All too often, such theories have tacitly entailed assumptions that cannot withstand examination. The simplest procedure has been to draw the curve of past growth, and then project it into the indefinite future with a dotted line that appears to "fit" the first part of the curve (Public Health Service, 1986). Here the naivete is fairly obvious. Most of the existing cases of AIDS result from infection acquired many years ago, when only a tiny proportion of the total number of people at risk had been infected; the rate of growth of the disease must slow as the proportion of susceptible people in the population diminishes.

Any more sophisticated projection requires a construct, or "model," of how the disease is transmitted. For a model to be useful, it must contain values for such parameters as: (1) the rate at which an infectious person transmits the virus by the various possible routes, (2) the rate at which infected people become infectious, (3) the degree of infectiousness at successive stages of the disease, (4) the extent of risky behavior in different populations, and (5) most often neglected, the likelihood that someone selected as a partner by an infected person is himself or herself already infected.

Without these and other pieces of information, the following questions cannot be answered.

Who transmits AIDS and how do they do it?

The answer to this question is far from obvious. AIDS appears not to be highly infectious in the sense that the number of occasions on which it is transmitted appears to be a relatively small fraction of the number of encounters in which it could be transmitted. What is the nature of the variability? Are some people "high transmitters" and others "low"? Or is it that any one person goes through periods of being more and less effective as a transmitter. Is the variability mainly a function of preferred sexual practices? Or do so-called cofactors, such as simultaneous infection with another agent, play a major role?

How much does promiscuity contribute to the growth of AIDS?

The rate at which people acquire new sexual partners influences the growth of the epidemic, but quantifying this factor is not currently possible. The effect of promiscuity on growth of the epidemic is difficult to assess because of the complex way the spread of the virus depends on all of the following: the probability of transmitting the virus during a given sexual encounter (infectivity), the frequency of sexual activity, the rate of acquiring new partners, and the duration of relationships. All of these factors are likely to change over time: Infectivity may increase when clinical symptoms of HIV infection appear; also, people tend to change their behavior as they age or become ill, and as fashions change.

May and Anderson (1987) have shown that if the risk of transmission per partnership is constant regardless of the length of the relationship, then the most promiscuous men have a very large effect on the growth of the epidemic. Under their assumptions, it is much more effective to target the rate of contact with new partners among the most promiscuous as the behavior to alter, rather than the average rate of acquisition of new partners for the population as a whole.

The strength of this effect, however, depends on the nature of infectivity, which is not well characterized. There are two reasons: (1) If the infectious period is short, variability in the rate of acquiring new partners is much smaller then if the infectious period is long. There is much more variation in the number of partners acquired over a 10-year period than over a 2-month period. (2) The assumption that risk of transmission to a partner is constant may be reasonable if there are two kinds of people: a few who usually transmit the virus and a larger number who rarely transmit. On the other hand, if most infected people are somewhat infectious for a long time, growth of the epidemic is much less influenced by the rate at which new partners are acquired. It simply does not matter so much whether the infectious person has many contacts with a few partners or fewer contacts with many partners. Until more is learned about the variability of infectiousness, it will not be possible to compare the relative advantage of reducing the overall average number of new partners as opposed to eliminating very high levels of promiscuity among the small proportion of the population engaging in such behavior.

How risky are anal, vaginal, and oral intercourse?

Very preliminary evidence suggests that the risk of transmission per contact is greatest for anal intercourse and least for oral intercourse (De Gruttola and Mayer, 1987; Kingsley et al. 1987). The evidence is fairly weak, however, and no adequate attempt has been made to answer the important question: Is the risk to an uninfected person still low after practicing one of these behaviors with an infected partner over long periods of time?

Data from a small study of partners of infected men in San Francisco demonstrated that women who had anal intercourse with their partners were about twice as likely to be infected as those who did not (Padian et al. 1987). Because it is not known whether these partners differed in other ways, the data do not demonstrate conclusively that anal sex is riskier than vaginal sex. If, for example, the men who favored anal intercourse with their female partners had been infected earlier in the course of the epidemic than those who did not (as a function, say, of their sexual adventurousness), the apparent risk of anal transmission in this study, relative to vaginal, would be falsely elevated. Even if the observed differences really do result from more efficient anal transmission, there is no reason to infer that unprotected vaginal intercourse is much safer over the long term, simply because repetition can convert any act from low to high risk.

Similarly, although partner studies have failed to demonstrate much of a role for oral intercourse - only one transmission attributable to this route has been documented (De Gruttola and Mayer, 1987), such studies lack the power to detect low rates of transmission. If, for example, the rate of transmission per oral exposure to infectious semen is 10% of that for anal intercourse, it could not now be detected by any ongoing study. Nonetheless, this level of risk could result in a large number of new infections, especially among gay men no longer practicing unprotected anal intercourse.

Any hope of answering questions about the relative risk of insertive and receptive intercourse, vaginal or anal, requires very large partner studies, numbering in the thousands instead of the tens that have been reported on. Estimating the risk of oral exposure to infected semen would require taking sexual histories from tens of thousands of infected men - indeed, from as many as all the (perhaps 20 thousand) gay men who will develop AIDS in 1988.

How many people are infected?

Procedures to estimate the number of infected people in the United States cannot, with existing information, yield accurate results.

One such effort has made use of data on the cohort of gay men in San Francisco - a group originally entered in a study of the natural history of hepatitis B (Hessol et al. 1987). The early appearance of AIDS in this group provided an opportunity to follow the progress of the newer disease as well. Because both the number of men in this cohort who are infected with HIV and the number with AIDS are known, a ratio of those infected to those with disease can be derived. It seems simple enough to count the number of reported AIDS cases in the United States, multiply by the ratio from the San Francisco cohort, and take the result as an estimate of the total number of infected people in the United States. This procedure is, however, invalid because the ratio changes with time. Unfortunately, making additional assumptions to refine the procedure demonstrates any ratio over a very wide range is consistent with existing data (De Gruttola and Lagakos, 1989).

An alternative is to survey rates of HIV carriage from an available population, such as patients seen at clinics for sexually transmitted diseases or for drug rehabilitation. This number can be projected to the entire United States population, provided an estimate of the total number of people at comparably high risk can be obtained - specifically the number engaging in risky sexual practices or needle sharing. Such estimates are fated to imprecision (Public Health Service, 1986).

Current data support estimates ranging from a few hundred thousand to a few million infected people. The commonly cited figure of 1 - 2 million is at the high end of the plausible range; more reasonable is a figure between 500,000 and 1 million (De Gruttola and Lagakos, 1989).

How long after infection does AIDS develop?

Knowing the risk of AIDS after infection would be useful both for predicting the future number of cases and for counseling and treating infected people. Any useful model of the growth of the epidemic requires consideration of the time elapsed between infection and onset of AIDS (latency). An estimate not only of the mean latency, but also of the distribution of latencies is required. Also needed is some information about the way this distribution is affected by the virus's route of entry, the prior health of the host, and, perhaps, other biological characteristics.

The typical delay between infection and the onset of disease may be characterized in either of two ways, as a mean or median value. The mean cannot be directly estimated until a cohort of infected people has been followed long enough so that all who are capable of developing AIDS have been diagnosed. Since the longest possible latency is undoubtedly much longer than the seven years during which the epidemic has been studied, such data will not be available for many more years. Estimating median latency requires only that 50% of infected people in a cohort develop the disease. No cohort has, as yet, reached this point.

Since not even the median, much less the whole distribution of the latency, can be directly established, mathematical models have been used in the attempt to derive it indirectly from available data.

One such effort, for example, has been based on the experience of people who contracted the disease through transfusion (Lui et al. 1986). On its face, this is a sensible approach, if only because a relatively modest investment in data collection provides the date of exposure and the date at which disease appears, at least among the subgroup who develop AIDS relatively rapidly. In reality, the data are not adequate to differentiate between a latency of 5 or 50 years (Lagakos, et al. 1988).

The experience of the San Francisco cohort has also been used to estimate latency. Hessol et al. (1987) have estimated that over 35% of the men in this group would be expected to develop AIDS within 7 years of the date of infection, most of them between years 5 and 7. However, there is an important bias in this analysis.

For the majority of cohort members, who were not infected at entry, the time of infection is known to fall within an interval that varies from several months to a few years. An important minority, however, was already infected when they entered the study. These men, presumably, infected in the late 1970s determine the level of risk calculated for years 5, 6 and 7 after infection. All that is known about them is the date at which seropositivity was first recorded; there is simply no way to infer the time before which these men could not have been infected.

Hessol et al. (1987) analysis treats these two subgroups according to different rules. The time of infection for those who were not infected at entry into the study is assigned to the mid-point of the interval between last negative and first positive test. For those who were already positive, the date of entry is taken as the moment of seroconversion. Such an analysis could create a spurious acceleration in the estimated rate of developing AIDS five years after infection. Appropriate methods for analyzing such "interval-censored" data are available and should be applied immediately to the San Francisco data; if there is a real acceleration in the rate of developing AIDS after the fifth year of infection, the implication for all those infected and for the health-care delivery system as a whole is very grave.

What factors affect the risk of AIDS after infection?

Unfortunately, the effect of such factors as route of exposure, reinfection with HIV or other viruses, or general health status on the risk of AIDS after infection cannot be estimated from currently available data. Age may be an important factor; there is some evidence that children develop AIDS more rapidly after infection than do adults. For example, estimates of median latency from transfusion-related cases of AIDS are 2-1/2 months for children less than four years old at infection, eight years between the ages of 5 and 59, and 5-1/2 years for people over 60 (Medley et al. 1987). Because of the problems described above that are associated with analyzing such data, however, these estimates are not very reliable. The experience of younger patients in a cohort of hemophiliacs in Hershey, Pennsylvania, was very different: patients between the ages of 22 and 64 were more likely to develop AIDS within 6 years than were those between the ages of 2 and 24 (Eyster et al. 1987). This apparent contradiction may result from the fact that the young age group includes both infants, and older children and adolescents who differ in their average latencies or from the fact that age may be associated with chronic antigenic stimulation or impaired antigen processing in people treated with factors produced from concentrated blood. Aside from age, there have been no convincing demonstrations of the effect of any cofactor, although the power to detect such effects from available data is small.

Perhaps the simplest way to identify the effect of cofactors is to look at people infected by different routes and compare the rate at which they develop AIDS. For example, gay men and hemophiliacs could be compared. A simple ratio of the number of hemophiliacs infected to that who have developed AIDS appears to support the notion that they are at lower risk of developing AIDS. A crude estimate of cumulative incidence of AIDS among infected hemophiliacs as of September 1986 is only about 5% (Centers for Disease Control, 1987; Curran et al. 1988). This appears to be much lower than the risk of 36% estimated for gay men in the San Francisco cohort (Hessol et al. 1987). The reason for this apparent difference may simply be that the proportion of hemophiliacs infected for more than the five years usually required to develop the disease is smaller than the proportion among the men in the San Francisco cohort.

The observed difference in rates of infection between the two groups could be a statistical accident even if the average time since infection is the same in both groups. This would be the case if the dates of HIV infection were more concentrated in time for hemophiliacs than they were for the San Francisco men. Many hundreds of hemophiliacs could have been simultaneously treated with the same contaminated blood factor, and thus, have acquired the virus virtually simultaneously; the rate of infection might therefore rise precipitously after some years in which it appeared to be very low.

To be valid, a comparison must be made between cohorts of individuals for whom time of infection either is known or can be estimated. For a cohort of 84 hemophiliacs followed in Hershey, PA, the cumulative incidence of AIDS at six years was 18.07% (+/-7%) observed in the San Francisco cohort (Hessol et al. 1987). Appropriate methods for analyzing these interval-censored data might bring the estimates even closer.

Will there be a heterosexual epidemic?

There is no doubt that people who think of themselves as heterosexual can get AIDS. But, knowing this, we are still a long way from knowing whether an epidemic of AIDS can be sustained purely and indefinitely by transmission from male to female, then female to male, and so forth.

To answer this question from national surveillance and from studies of partners of infected people requires a lot of information about the United States population as a whole - such as the number of non-monogamous heterosexuals, the risk if transmission per sexual contact with an infected partner (a number that is certainly different for infected men and women), the risk of one's partner uses IV drugs. Although this kind of information might at least be guessed at, there are two vital questions about which we really have no idea. (1) What proportion of observed cases of AIDS can be attributed to female-to male transmissions? (2) How long does it take for a man or woman who has been infected to become infectious?

Information on the natural history of infectivity is totally absent. Whether any men have acquired AIDS through heterosexual contact is in doubt, given that only a tiny proportion of men with AIDS would have to lie to account for the roughly 100 cases of AIDS among American-born men attributed to contact with an infected woman as of the end of 1986 (Centers for Disease Control, 1986). On the other hand, if some of the men whose claim that exposure to prostitutes is their only source of risk are telling the truth, the number could be higher than 100, because of the policy of assigning a case of AIDS to heterosexual contact only when all other risk factors appear to have been ruled out.

Has behavior change been adequate to bring the epidemic under control in populations most at risk?

Unfortunately, we cannot assess the efficacy of behavior change by looking at the rate of increase in AIDS cases. Given that the latency of infection is upwards of seven years, the current count of AIDS cases gives us only a blurry, composite picture of what has been happening over the last decade.

Even studying rates of change in HIV infection, which is exceedingly difficult, is not very helpful. Rates of rectal gonorrhea are an obvious choice. However, reporting of these diseases is notoriously unreliable, and it is extremely difficult to know how the the specter of AIDS has altered patterns of care-seeking and reporting. The significance of the newly reported rise in syphilis among heterosexually active non-whites and Hispanics in urban areas could reflect increased sexual activity in this group, a biological alteration of the spirochete, or merely re-entry of the organism into a sexually interacting population from which it had been temporarily eradicated.

INFORMATION AND IDEOLOGY

To construct a realistic model of the AIDS epidemic requires that we know a good deal about the mores, sexual and pharmacological, of various subsets of the American population. And this requires us to ask ourselves whether the world really is made up of people who live their lives as homosexuals or heterosexuals, with a small number of bisexual men as the link.

More complex patterns of behavior not only can be imagined, they probably come closer to reality. From Kinsey's research it seems probably that there are people who go through periods of predominantly homosexual behavior followed by periods of heterosexuality (Kinsey, 1948). There may be yet others whose behavior is predominantly heterosexual but who engage in occasional homosexual acts. These people probably restrict their homosexual behavior to certain acts that they feel do not define them as "gay". Indeed, it may well be that most men who are homosexually active on occasion have no identification with "gay life" at all and would be reluctant even to characterize themselves as "bisexual". Exciting sex could as easily represent a flight from their habitual social role as an expression of it (Humphreys, 1970). Any attempt to count the number of "homosexuals" in the population must somehow deal with the diversity of possible behavior patterns, to the extent that they are relevant to transmission of the disease.

Yet another aspect of sexual behavior that could seriously complicate any effort to predict the course of AIDS is the simple fact that "promiscuous" people are not at all alike. In the first place, people select their partners. The criteria for selection probably include some that are proxies for risk of disease. People who are promiscuous within a relatively small interacting pool of other similar people may have such a low probability of contracting AIDS in the foreseeable future that they really should not be counted in the population at risk. And yet, if they were asked the obvious questions ("How many different partners have you had in the last year? What sexual activities did you engage in?"), they would certainly be counted as "high risk". This is the converse of the situation among gay men in New York, San Francisco, and Los Angeles at the beginning of the epidemic, when introduction of the AIDS virus into limited, highly interactive pools resulted in explosive growth of the epidemic.

Although we cannot yet be certain of the potential for heterosexual behaviors to sustain the AIDS epidemic, it seems quite certain that the brunt of the epidemic in the next 10-20 years will be borne by men who acquire the disease through homosexual contact. To estimate the size of this population requires more sophisticated thinking than a trichotomy of heterosexuals, bisexuals, and homosexuals. To determine the size of the homosexual population, for purposes of tracking the disease, requires essentially that someone decide what the rules of membership are. Homosexuality may refer to a specific transaction between two people of the same sex, to the biological response that permits such a transaction to take place, or to a social and political orientation that is mostly restricted to members of the urbanized white middle class.

Phrased most precisely, the basic question that needs to be answered in order to plan for the near future is, "What is the size of the population at risk of contracting AIDS through homosexual exposure?" This question is impossible to answer because we do not know the distribution of sexual practices within the population - how many people do what with whom. And we also do not know which of the potentially relevant practices can transmit the virus - most especially, we do not know whether oral sexual acts (oral genital or even intense kissing) put people at risk. The behavior of highest risk for gay men is certainly anal intercourse outside closed sexual pools (Kingsley et al. 1987). When speaking of heterosexuals, it may make sense (but probably does not) to think of a dichotomy between monogamy as the main type of closed pool and promiscuity as existing in an open pool. In the case of gay men, it is possible that closed pools of varying size exist - tacit, possibly even unrecognized, menages a trois, quatre, cinq, and so forth. Again, replies to questionnaires from such people could create a misleading impression of their vulnerability to exposure.

This is not a trivial point. Even if we had perfect information on every sexual encounter, if we know the risk of transmission per sexual contact as a function of time since the infectious partner becomes infected, we still could not be sure of constructing a valid model of the transmission of the disease. Simply by altering our assessment of the way people allocate themselves to pools (that is, their probability of selecting a partner with a given risk profile), we could derive a curve of times of infection with arbitrary shape.

This fact should not prevent an effort to model the AIDS epidemic, but it underscores the limits of our ability to predict the future course of the AIDS epidemic. Attempting to find these limits precisely, and to show how studies of different design may extend them, will require unprecedented cooperation among biostatisticians, epidemiologists, social scientists, and mathematical modelers.

ACKNOWLEDGEMENTS

This work was supported by grant No. AI24643 from the National Institutes of Health. Parts of this chapter have been published by Victor De Gruttola and William Bennett in AIDS 1988: AAAS Symposium Papers.

REFERENCES

Centers for Disease Control (1987) Human immunodeficiency virus infection in the United States: A review of current knowledge. MMWR 36:5-6:41.

Centers for Disease Control (1986) Update: acquired immunodeficiency syndrome (AIDS) - United States. MMWR 32:1721.

Curran JW, H Jaffe, A Hardy, WM Morgan, R Selik, T Dondero (1988) Epidemiology of HIV infection and AIDS in the United States. Science 239:553-696.

De Gruttola V, K Mayer (1987) Human immunodeficiency virus and oral intercourse. Ann Intern Med 107:428-429.

De Gruttola V, K Mayer (1988) Assessing and modeling the spread of the human immunodeficiency virus in the U.S. Rev Infect Dis 10:138-150.

De Gruttola V, S Lagakos (1989) The value of AIDS incidence data in assessing the course of the AIDS epidemic. Statistics in Medicine 8: 35-43

Eyster M, M Gail, Ballard, H Al-Mondhiry, JJ Goedert (1987) Natural history of human immunodeficiency virus infection in hemophiliacs: effect of T-cell subsets, platelet counts, and age. Ann Intern Med 107:1-6.

Hessol N, G Rutherford, RM O'Mally, et al. (1987) The natural history of the human immunodeficiency virus infection in a cohort of homosexual and bisexual men: a seven-year study. In: Program and abstracts of the Third International Conference on AIDS. Washington D.C.

Humphreys L, (1970) Tearoom Trade: Impersonal Sex in Public Places. Chicago: Aldine.

Kingsley L, R Kaslow, C Rinaldo (1987) Risk factors for seroconversion to human immunodeficiency virus among homosexuals. Lancet 1:345-351.

Kinsey AC.(1948) Sexual behavior in the human male. Philadelphia: Saunders.
Lagakos S, L Barraj, V De Gruttola (1988) Nonparametric analysis of truncated survival data, with application to AIDS. Biometrika 75:3, 515-523.
Lui KJ, DN Lawrence, WM Morgan, TA Peterman, HW Haverkos, DJ Bregman (1986) A model-based approach for estimating the mean incubation period of transfusion-associated acquired immunodeficiency syndrome. Proc Natl Acad Sci, USA 83:3051-3055.
May R, RM Anderson (1987) Transmission dynamics of HIV Infection. Nature 326:137-142.
Medley GF, RM Anderson, DR Cox, L Billard (1987) Incubation period of AIDS in patients infected via blood transfusion. Nature 328:719-721.
Padian N, L Marquis, D Francis, RF Anderson, GW Rutherford, PM O'Malley, W Winkelstein (1987) Male to female transmission of human immunodeficiency virus. JAMA 258:788-791.
Public Health Service (1986) Public Health Reports 101:341-348.

DISCUSSION

Caulfield C (AIDS Resource Center, Dallas, TX):

Part of my duties is that I supervise a hot line of volunteers that serves a Metropolitan area of about two and a half million people. One question we've been getting recurrently lately, is that members of the cohort study in San Francisco of hepatitis B, fourteen years ago, have just begun to seroconvert, some of them. Some of them have not seroconverted yet. What are the implications for transmission from these individuals, and what are the implications for the blood supply from that data?

De Gruttola V:

As you point out, one of the distributions that is very important in studying AIDS concerns the time from the initial infection or exposure to the virus to seroconversion. I didn't discuss that distribution at all. One reason is that I don't know of any data sources that would permit estimation of that very important distribution. There are a few people who think they know when they became infected, for whom estimated times of seroconversion are available, but the data are too scanty to be able to estimate this distribution. The implications of your question are very important. Unfortunately, from the study of homosexual partners in Boston, there is no way of estimating the risk that someone who has not seroconverted is nonetheless infectious. The participants have too many outside partners to be certain that an individual was infected from a study partner. To estimate the infectiousness of seronegative people one might find individuals who have received contaminated blood transfusions, but have not yet seroconverted, and study their sexual partners.

The implications of infectiousness among seronegatives are obviously very important, both for sexual transmission and for the blood supply. One of the ways that we'll get information about the safety of the blood supply is from cases of AIDS in the future related to blood transfusion that occurred after 1985. This information would make it possible to estimate the number of people donating blood who test negative, but are in fact, infectious. The distribution I call "W" and the times of infectious transfusions should have dropped sharply in 1985 when the screening tests came on-line. If it starts increasing again, then we'll know that in fact, there are individuals who are capable of infecting people who haven't yet seroconverted. But, the problem is that it's going to take quite a bit of time before we can observe that, because we require the occurrence of not only the infectious transfusion, but also, the development of symptoms of HIV infection in the person who receives the transfusion. Individuals who are seronegative, may in fact be infectious, and we don't have a very good handle on how many such people there are. I think we need some good studies to be able to determine that; the existing studies are not adequate.

Trachtenberg A (California Department of Health Services, Berkeley, CA):

In your calculations about infectiousness from the Boston men, first you assume uniform infectiousness, then you try to vary it. Are you making an assumption of uniform susceptibility to infection of their partners when you do that, and if so, how does that affect what you're doing?

De Gruttola V:

What I have modelled is probability of transmission which subsumes infectiousness and susceptibility. Susceptibility to HIV appears to be very high; for example most people who have received contaminated blood transfusions, have evidence of HIV infection. Some estimates are as high as 90% of the people who get contaminated blood transfusions will develop measurable antibody. On the other hand, that isn't necessarily true for sexual contact. If some people are not susceptible through sexual contact, there is as yet no evidence to show it. Obtaining estimates of risk of transmission from a study of sexual partners requires assumptions. These data are really not adequate to study susceptibility (and infectiousness separately). I think some good and larger studies of sexual partners of people infected by blood transfusion, might provide that information.

Trachtenberg A:

I think that we do have some data to suggest that susceptibility is quite varied, both from Africa and recently from Seattle. Particularly, in relation to the presence or absence of genital ulcerative disorders and perhaps, other sexually transmitted diseases.

De Gruttola V:

By susceptibility, I mean the probability of an individual becoming infected given a sufficiently large inoculum of the virus. Conditions like genital lesions, obviously influence the inoculum of virus that someone actually receives. It appears possible that people who have genital lesions would have a greater risk of exposure to the virus through sexual contact. The prevalence of genital lesions would be one of the factors that would affect the probability of transmission which may vary across individuals. An analysis of data from partner studies provides only a rough estimate of risk of transmission. One has to do more investigative work to find out why some individuals are becoming infected and others are not.

Trachtenberg A:

My last questions is, you mentioned once or twice that it would take a very large sample to try to tease apart those differences. Would you care to make a sample size estimate for us?

De Gruttola V:

I think the problem with getting a good estimate of sample size is that you have to have estimates of how many people will seroconvert during the study. In Boston, the number of seroconverters was very small, only five or six, after several years of follow-up of 280 men. That's a very fortunate thing. It means that people seem to be reducing their risk of infection. It also means that if you want to characterize not only infectivity, but also susceptibility, you're going to have to have a very large study to do it. If you wanted precise estimates within five years, it would require thousands of homosexual men, or thousands of heterosexual partners.

HERPESVIRUS TRANS-ACTIVATION OF HUMAN IMMUNODEFICIENCY VIRUS TYPE-1*

JOSEPH D. MOSCA

Henry M. Jackson Foundation; Rockville, MD 20850 and the Department of Immunology and Infectious Diseases; The Johns Hopkins University School of Public Health; Baltimore, MD 21205 (USA)

INTRODUCTION

In humans, HIV-1 infection is characterized by a period of latency followed by progression to acquired immunodeficiency syndrome (AIDS) and AIDS-related complex in some cases (Blattner et al. 1985). Factors that influence HIV-1 latency are poorly understood. Several models have been proposed that might explain how HIV-1, when persisting in the latent form, can be induced by physiochemical stimuli and expressed as infectious virus particles (Folks et al. 1986). These models include transcriptional repression of integrated proviral DNA by DNA-binding proteins, chromatin conformation, and DNA hypermethylation. To understand these processes, permanent cell lines were constructed that contained as a reporter gene the bacterial chloramphenicol acetyltransferase (CAT) gene coding region fused to the HIV-1-LTR (pU3R-III-CAT) or human T leukemia virus (HTLV) type I (HTLV-I)-LTR (pU3R-I-CAT) promoter sequences. In these permanent cell lines HIV-1-LTR but not HTLV-I-LTR expression was suppressed when integrated into the host chromatin (Mosca et al. 1987a). We have provided evidence that the suppression of HIV-1-LTR expression is accompanied by methylation of LTR sequences (Bednarik et al. 1987). Using these permanent cell lines as a model system that appears to mimic the latent integrated HIV-1 provirus, we showed that reactivation of the reporter CAT gene initiated from the HIV-1-LTR can be induced by either treatment with a protein synthesis inhibitor (cycloheximide) or infection with herpesviruses (Mosca et al. 1987a). For herpes simplex virus I

* Originally published in "The Control of Human Retrovirus Gene Expression", Edited by BR Franza, Jr., BR Cullen and F Wong-Staal, Cold Spring Harbor Laboratory Press, 1988.

(HSV-1), we have identified the 5' upstream sequence within the HIV-1-LTR that appears to contain a specific recognition site necessary for the activation by HSV-1 and by the HSV-1 encoded IE110 (ICPO) regulatory protein (Mosca et al. 1987b). However, for reactivation of the HIV-1-LTR by cytomegalovirus (CMV), no specific recognition sequence within the HIV-1-LTR could be found, indicating that different molecular mechanisms exist for HIV-1-LTR activation by HSV-1 and CMV.

LATENT HIV-LTR SEQUENCES IN PERMANENT CELL LINES CAN BE TRANSCRIPTIONALLY REACTIVATED

Modulation by demethylation

In transient CAT expression assays in fibroblasts, the HIV-1-LTR behaves like a constitutive eukaryotic promoter. For example, relative to the simian virus-40 (SV40) promoter, both the HIV-1-LTR and HTLV-I-LTR, yield approximately one-half as much basal CAT activity. However, once

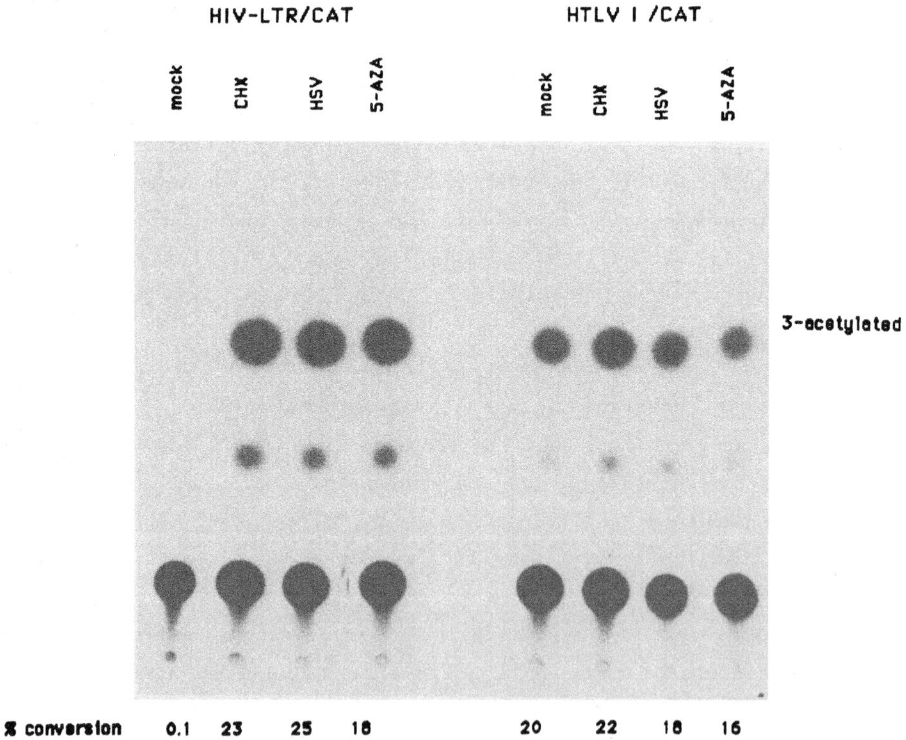

Figure 1. CAT activity from permanent Ltk-cell lines containing either the HIV-1-LTR/CAT or HTLV-1/CAT hybrid gene.

integrated, the HIV-1-LTR but not the HTLV-1-LTR, failed to express CAT constitutively (Figure 1). The control of cellular and viral gene expression is modulated by DNA methylation (Doerfler 1983; Cedar 1984; Doerfler 1984; Keshet et al. 1986) and can be altered by treatment with the nucleoside analog 5-azacytidine (5-azaC), a potent inhibitor of DNA methylation (Jones 1985; Nyce 1986). Therefore, we tested the possibility that methylation of the HIV-1-LTR sequences was involved in HIV-1-LTR suppression in permanent cell lines. Exposure of either the murine LNHIV-CAT Pool (Figure 1) or the simian VNHIV-CAT Pool permanent cell lines (Bednarik, et al. 1987) to 5-azaC resulted in reactivation of the integrated HIV-1-LTR. Similar treatment of either the murine LNHTLV-I-CAT Pool (Figure 1) or the simian VNHTLV-1-CAT Pool permanent cell lines (Bednarik et al. 1987) had no effect on constitutive CAT expression*. Subsequent experiments demonstrated the 5-azaC reactivation of the HIV-1-LTR occurred at the transcriptional level and was related to the degree of methylation within the HIV-LTR enhancer/TAR region as "viewed" by the intrinsic property of the inability of PvuII restriction enzyme to cleave its recognition site if methylated (Bednarik et al. 1987).

Modulation by Cycloheximide and Herpesvirus

In addition to reactivation via demethylation, expression from the HIV-1-LTR, but not the HTLV-1-LTR, can be restored either by treatment with cycloheximide or by infection with herpesviruses (Figure 1). Run-on transcription in isolated nuclei, where only elongation of previously initiated gene transcription occurs, showed no detectable HIV-1-LTR/CAT transcripts in the integrated cell lines. However, after either cycloheximide or HSV infection, CAT transcription was detected suggesting that reactivation occurred at the transcriptional level (Mosca et al. 1987a). Further evidence for transcriptional reactivation was obtained by a concomitant increase in both RNA initiated from the HIV-1-LTR (as measured by S1 nuclease analysis) and CAT enzymatic activity (Figure 2). No constitutive stable CAT RNA was observed in VNHIV-CAT Pool cell line (Figure 2, mock) whereas HSV infection in the absence or presence of phosphonoacetic acid (PAA an inhibitor of HSV replication) gave an increase in HIV-LTR initiated RNA and CAT activity. In addition, the reactivation appeared to be independent of reporter gene and host cell species used, in that human beta-interferon (β-IFN) expression driven by the HIV-1-LTR in MG63 cells (a human osteosarcoma cell line) was also basally silent

---

* Simian VNHIV-CAT Pool and VNHTLV-1-CAT Pool or murine LNHIV-CAT Pool and LNHTLV-1-CAT Pool are G418 selected permanent cell lines containing either HIV-LTR/CAT (pU3R-III-CAT) and HTLV-1-LTR/CAT (pU3R-I-CAT) hybrid gene, respectively.

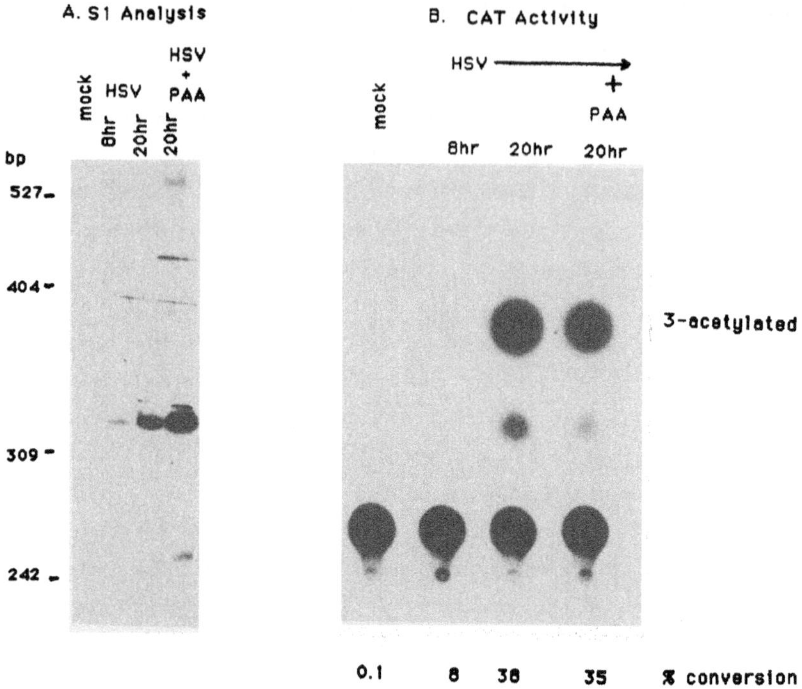

Figure 2. Induction of the HIV-1-LTR by HSV-1 in permanent VERO cell lines.

but was reactivated after cycloheximide treatment or HSV infection (Figure 3A). The intrinsic instability of the human β-IFN mRNA leads to its rapid degradation upon release of the cycloheximide block (Raj and Pitha 1983). Because the RNA in this experiment was hybridized with a genomic human β-IFN RNA riboprobe, the expression of endogenous β-IFN could also be monitored. Whereas cycloheximide treatment showed an increase in endogenous β-IFN-mRNA transcripts (most probably involving stabilization), HSV infection gave little or no increase in endogenous human β-IFN mRNA expression. Unlike the HIV-1-LTR, the integrated HTLV-1-LTR was constitutive and not inducible at the RNA level by either cycloheximide or HSV (Figure 3B). These experiments show that the DNA transfected HIV-1-LTR was inactive and methylated when permanently integrated into the host chromatin and that either 5-azaC or cycloheximide treatment or HSV infection could reactivate expression from the HIV-1 regulatory region.

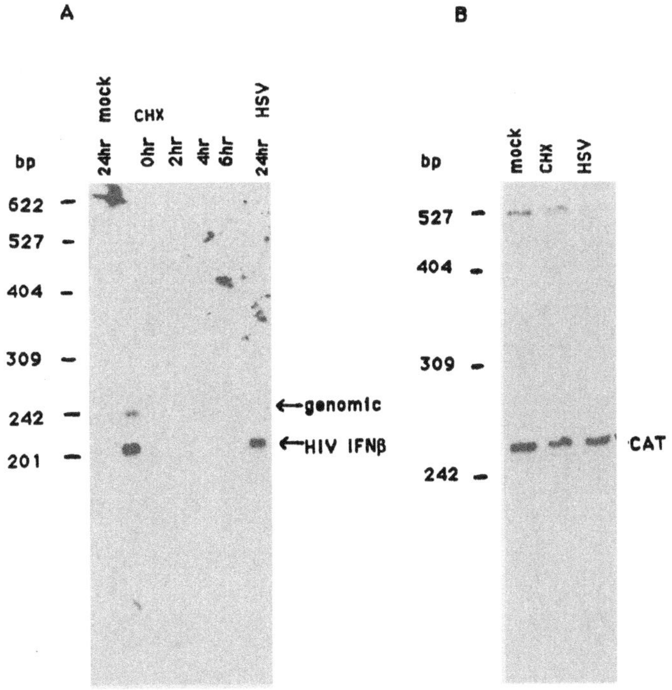

Figure 3. Cycloheximide and HSV-1 induction (A) is independent of reporter gene used to monitor HIV-1-LTR activation and (B) does not activate the HTLV-1-LTR.

IDENTIFICATION OF A REGION WITHIN THE LONG TERMINAL REPEAT THAT RESPONDS TO A TRANS-ACTING FACTOR ENCODED BY HSV-1

HSV-1 Gene Product

As a first approach to identifying the mechanism of HSV-1 reactivation of the HIV-1-LTR, we performed experiments where HSV infection was done in the presence of inhibitors of RNA and protein synthesis (Mosca et al. 1985 and Figure 4). Experiments with the VNHIV-CAT Pool cell line show low constitutive activity, which was increased tenfold in the presence of cycloheximide for 4 hr followed by overnight reversal and 60-fold if the cycloheximide reversal was done in the presence of actinomycin D. For unknown reasons, the cycloheximide activation appeared to be greater in LNHIV-CAT Pool cell lines than in VNHIV-CAT Pool lines. HSV infection in this experiment resulted in a 250-fold activation. When the infection was done in the presence of cycloheximide for 4 hr (to prevent *de novo* viral protein synthesis) followed by release from the cycloheximide block a 310-fold activation was observed. However, when the specificity of the

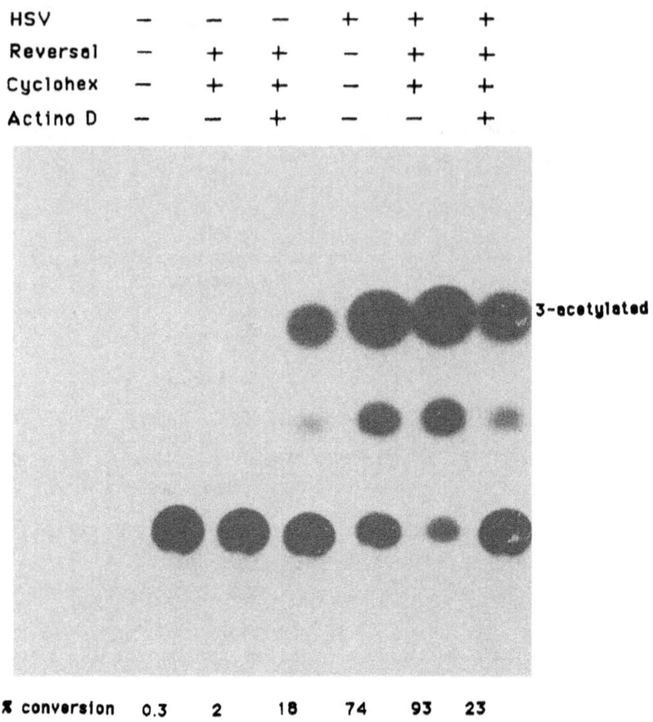

Figure 4. HIV-1-LTR reactivation requires HSV-1 gene expression.

trans-activation was examined by carrying out the infection in the presence of cycloheximide for 4 hr followed by reversal in the presence of actinomycin D to prevent new RNA synthesis the level of CAT activity returned to that observed with cycloheximide and actinomycin D alone in the absence of HSV infection (23% versus 18% conversion). Under this protocol, the only viral proteins synthesized would be immediate-early (IE) gene products encoded by mRNA transcribed in the presence of cycloheximide, however because of the presence of actinomycin D, these IE gene products would not be able to activate the transcription of the integrated HIV-1-LTR gene. Experiments of this type, together with the use of an HSV-1 temperature-sensitive (ts) uncoating mutant, tsB7 (Mosca et al. 1985) showed that (1) Vmw65 (VP16), a late virion structural component that specifically stimulates the expression of the HSV-1 IE genes (Batterson and Roizman 1983; Campbell et al. 1984), was not involved in HIV-1-LTR induction, and (2) HIV-LTR induction required HSV gene expression (Mosca et al. 1987a).

Of the five IE genes expressed after HSV-1 infection, the IE175 gene clearly performs an essential role in HSV-1 delayed-early gene expression and viral replication (Watson et al. 1980). To test whether the

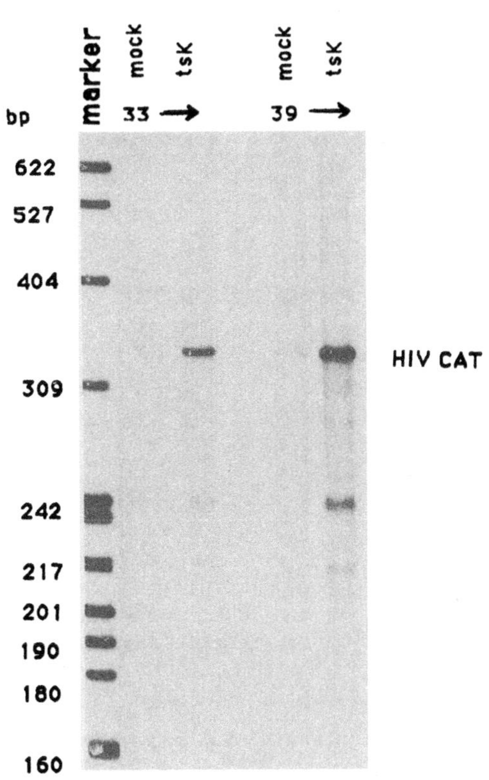

Figure 5. Elimination of the HSV-IE175 gene product involvement in HIV-1-LTR induction using ts mutant in permanant Ltk$^-$ cell lines.

IE175 protein was involved in the reactivation of the HIV-LTR, we took advantage of the availability of the HSV-1 ts mutant tsK, which produces a defective IE175 protein (Preston 1979). Constitutive expression was not observed at either the permissive (33°C) or the non permissive (39°c) temperature in either the VNHIV-CAT Pool (Mosca et al. 1987b) or LNHIV-CAT Pool cell lines (Figure 5). However, infection with the tsK mutant at either temperature reactivated HIV-1-LTR/CAT transcription. These results suggest that the IE175 protein, or at least the particular lesion of the tsK mutant, is not involved in the reactivation of the HIV-LTR by HSV-1 infection.

To identify more definitively which HSV-1 gene product(s) might be involved in HIV-1-LTR reactivation, we tested whether the virion factor (Vmw65 or VP16) or any of the five HSV-IE genes (O'Hare and Hayward 1985) could induce the HIV-1-LTR in transient co-transfection assays (Mosca et al. 1987b; Mosca, unpublished data). By both CAT activity and S1 nuclease analysis, the IE110 and the combination of IE110 and IE175 induced correctly initiated HIV-1-LTR/CAT RNA at least as well as by HSV infec-

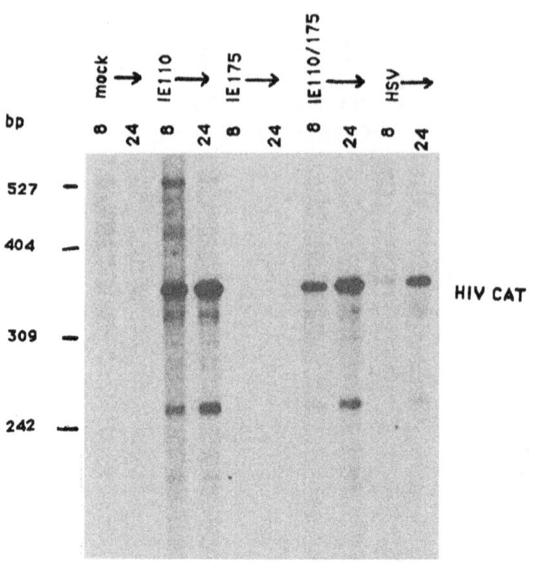

Figure 6. Identification of the HSV-1 gene product required for HIV-1-LTR induction by transient cotransfection assays in Ltk$^-$ cells.

tion in both HeLa (Mosca et al. 1987b) and Ltk cells (Figure 6). Thus, the data from transient co-transfection assays agree with the results obtained for the permanently transfected cells infected with the tsK mutant and indicate that the IE110 gene product may play an important role in the activation of HIV-LTR.

LTR Sequence Requirement

To identify functional regions within the HIV-1-LTR responding to HSV infection, we tested a set of deletion mutants that were previously used to identify the location of cis-acting regulatory sequences in the HIV-1-LTR required for tatIII trans-activation (Rosen et al. 1985). Deletions up to 104 nucleotides 5' from the RNA start site were still inducible by the HSV-IE110 gene product (Figure 7, -104/+80). However, deletions containing only 57 or 45 nucleotides resulted in a dramatic drop in both IE110 and IE110/175 inducibility (Figure 7, -45/+80). In addition, a construction lacking the TAR region (-167/+21) still responded efficiently to IE110 cotransfection (Mosca et al. 1987b). From these results, we conclude that the response elements within the HIV-LTR needed for HSV induction are different from these used for tatIII induction.

To further demonstrate that the tatIII and HSV-1 responsive elements are separate regions within the HIV-1-LTR, we compared the amount of RNA initiated from the HIV-LTR after HSV infection, tatIII transfection or

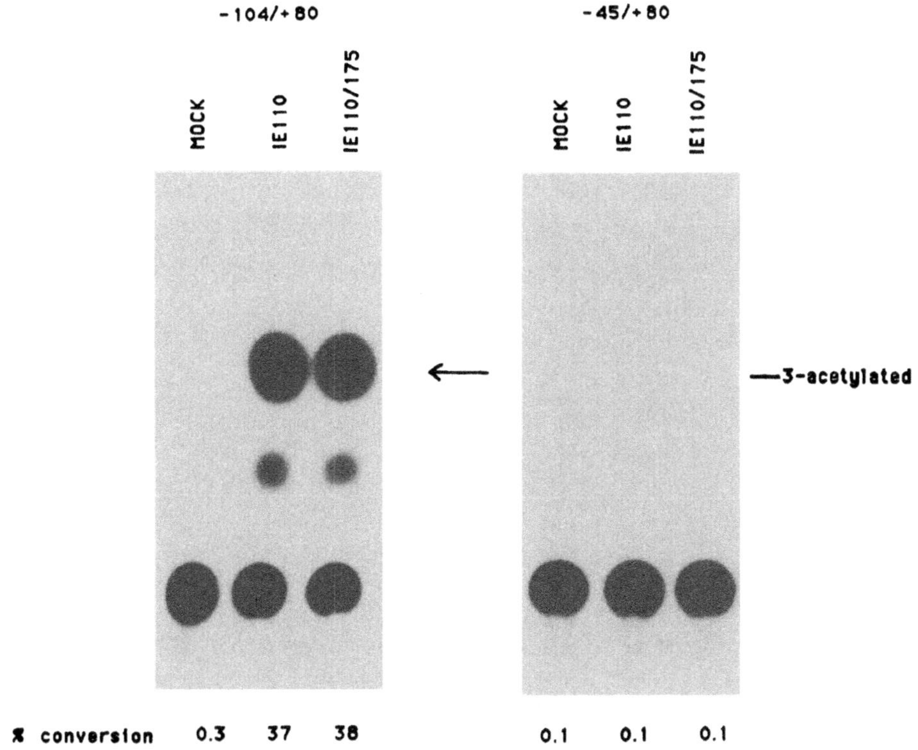

Figure 7. Trans-activation of HIV-1-LTR deletion mutants by HSV-IE genes.

HSV infection after tatIII transfection (Figure 8). In this experiment, we used the mouse β-globin gene as the reporter gene for HIV-1-LTR activation in transient expression assays. Relative to the activation levels obtained with HSV infection or tatIII co-transfection alone, co-transfection of HIV-1-LTR with tatIII followed by subsequent HSV infection resulted in a synergistic response. These data further support the notion of physical separation of the two target recognition sites within the HIV-1-LTR for tatIII and IE110.

The HIV-1-LTR deletion analysis loosely defined the HSV responsive region, however, the deletion analysis was complicated by the fact that they remove the Sp1 binding sites. Since the TATA box and Sp1 sites are usually involved in basal expression of promoter elements, removal of any of these sites by Bal31 nuclease digestion could explain the loss of IE110 trans-activation. Therefore we isolated the HIV-1-LTR sequence from between position -104 and -32 and placed the 73 bp region upstream of a heterologous non-responsive promoter. We chose a deleted mouse

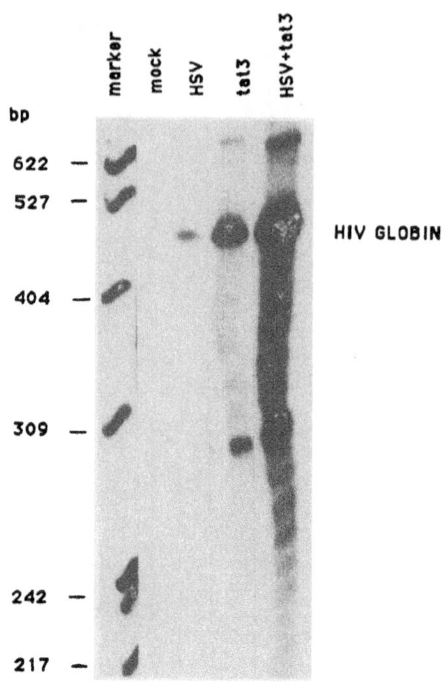

Figure 8. Further demonstration that the tatIII and HSV response elements are separate regions within the HIV-1-LTR.

alpha-4 interferon promoter that contained only 39 bp 5' from the mRNA start site (a TATA box and cap site) and 22 bp of the untranslated region driving CAT as the heterologous target (p-α CAT). We have previously shown that this minimal truncated p-α CAT plasmid was unresponsive to either HSV infection or IE110 trans-activation in transient expression assays in VERO cells (Mosca et al. 1987b). Placement of the 73 bp sequence derived from the HIV-1-LTR 5' to the minimal α-IFN promoter (pHIV/alpha-CAT) conferred both IE110 and IE110/175 responsiveness to the α-IFN promoter after transfection in either HeLa cells (Figure 9A) or VERO cells (Mosca et al. 1987b). However, similar to the IE110 trans-activation of the HIV-1-LTR-transient assays in HeLa cells [but not in either LTK or VERO cells (Mosca et al. 1987b)], no increase in CAT enzymatic activity was observed within 24 hr after glycerol shock (Figure 9B). HeLa cells required the combination of IE110/175 for maximal activation of the pHIV/alpha-CAT plasmid. This result may reflect the intrinsically low level of IE110 expression in HeLa cells compared to other cell types. Our analysis unequivocally located an HSV responsive sequence within the -104 to -57 region of the HIV-LTR.

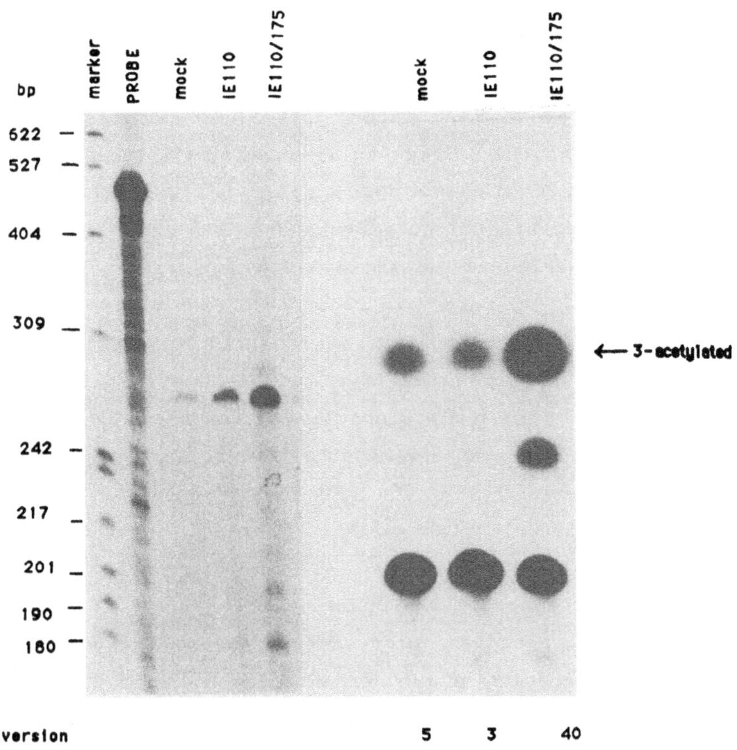

Figure 9. The 73 bp HSV response element within the HIV-1-LTR is constitutive and IE110/175, but not IE110, responsive in transient assays at the protein level in HeLa cells.

CONCLUDING REMARKS

By establishing permanent cell lines containing integrated HIV-1-LTR sequences, a model system has been developed where factors affecting HIV-1 expression can be tested. There has been no evidence for a site specific retroviral integration in infected cells and thus, the integration of the HIV-1-LTR into the cellular chromatin upon infection or DNA transfection will be in both instances a random event. Therefore, we feel that in the present permanent cell lines, the ability to induce HIV-1-LTR expression may be reasonably analogous to reactivation of latent HIV-1 provirus expression in an asymptomatic individual. The data supports the hypothesis that opportunistic herpesvirus infections could trigger the onset of AIDS. By understanding the underlining molecular mechanism involved in provirus reactivation, potential therapeutic agents might be implored that would prevent reactivation.

Further research is needed to define the mechanism of herpesvirus reactivation of latent HIV-LTR expression. The 73 bp sequence responding to HSV trans-activation contains two tandemly repeated core enhancer elements and three Spl binding sites. The HSV specificity towards this region may reside in the particular arrangement of Spl sites (-76 to -45) and/or core enhancer elements (-104 to -80) within the LTR. By using plasmids containing site-directed mutagenesis within the core enhancers (Kappa B mutant, Nable and Baltimore 1987) and natural core enhancer mutations (STLV-III-LTR), both core enhancer elements are required for HSV-IE110 and IE110/175 trans-activation of the HIV-1-LTR (Mosca et al. in preparation). In addition, we have made synthetic oligonucleotides of the two core enhancers, the second core enhancer and the first Spl binding site, and the first two Spl binding site to further identify the particular motif recognized by the HSV trans-acting IE110 gene product. Each of these oligonucleotides will be placed 5' to the minimal alpha-4 IFN promoter CAT construction and HSV responsiveness measured in both transient expression assays and permanent cell lines. Once the recognition sequence is identified, protein binding studies could be performed to aid in our elucidation of the mechanism(s) involved.

In addition to HSV, ultraviolet irradiation (Mosca et al. in preparation) and CMV can reactive the HIV-1-LTR (Mosca et al. 1987b). Although HIV-LTR reactivation is manifested by two viruses from within the same group, the mechanism of their reactivations seems to differ significantly. Unlike HSV, no specific recognition sequence within the HIV-1-LTR has yet been identified for CMV and the mechanism of reactivation by CMV involve post-transcriptional as well as transcriptional events (Mosca et al. 1987b; Figure 4). Obviously, more studies are needed to determine the mechanism of CMV activation of the HIV-1-LTR.

Recent epidemiological studies (Quinn et al. 1987) suggest that herpesvirus infection can increase an individual's susceptibility either to HIV-1 infection or to disease progression. If opportunistic infections should result in an increase in HIV-1-LTR expression and higher virus titers in an asymptomatic HIV-1 carrier, then infections with herpesviruses could be one of the factors contributing to the onset and severity of the AIDS disease. Since HSV replicates in neurones and undifferentiated monocyte and macrophages (Kirchner 1982), it is not unlikely that HSV infection could contribute to HIV-1 activation both in the central nervous system and in cells of lymphoid origin. In fact, it has recently been reported that one can find CMV and HIV-1 within the same cells in brain tissue of AIDS patients. Thus, it is becoming increasingly plausible that herpesviruses may play a role in HIV-1 infection.

ACKNOWLEDGEMENTS

We thank Barbara Schneider for help in preparing the manuscript. These studies were funded by the National Institute of Allergy and Infectious Diseases Grant IR29-AI24489, National Cancer Institute Core Grant 5P30-CA06973 and American Foundation for AIDS Research 000431 (JDM).

REFERENCES

Batterson N, Roizman B (1983) Characterization of the herpes simplex virion-associated factor responsible for the induction of alpha genes. J Virol 46:371-377.

Bednarik DP, Mosca JD, Raj NBK (1987) Methylation as a modulator of expression of human immunodeficiency virus. J Virol 61:1253-1257.

Blattner WA, Biggar RJ, Weiss SW, Melbye M, Goedert JJ (1985) Epidemiology of human T-lymphotropic virus type III and the risk of the acquired immunodeficiency syndrome. Ann Intern Med 103:665-669.

Campbell MEM, Palfreyman JW, Preston C (1984) Identification of herpes simplex virus DNA sequences which encode a trans-acting polypeptide responsible for stimulation of immediate early transcription. J Mol

Cedar H (1984) DNA methylation and gene expression. In: Razin A, Cedar H, Riggs AD (eds.) DNA Methylation: Biochemistry and Biological Significance. Springer-Verlag, New York, pp 147-164.

Doerfler W (1983) DNA methylation and gene activity. Ann Rev Biochem

Doerfler W (1984) DNA methylation and its functional significance: studies on the adenovirus system. Curr Top Microbiol Immunol 108:79-98.

Folks T, Powell DM, Lightfoote MM, Benn S, Martin MA, Fauci AS (1986) Induction of HTLV-III/LAV from a nonvirus-producing T-cell line; implications for latency. Science 231:600-602.

Jones PA (1985) Altering gene expression with 5-azacytidine. Cell 40:485-

Keshet I, Hurwitz-Lieman J, Cedar H (1986) DNA methylation affects the formation of active chromatin. Cell 44:535-543.

Kirchner H (1982) Immunobiology of infections with herpes simplex virus. In: Melnick JL (ed), Monogram Virology Vol 13, pp 1-104.

Mosca JD, Reyes GR, Pitha PM, Hayward GS (1985) Differential activation of hybrid genes containing herpes simplex virus immediate-early or delayed-early promoters after superinfection of stable DNA-transfected cell lines. J Virol 56:867-878.

Mosca JD, Bednarik DP, Raj NBBK, Rosen CA, Sodroski JG, Haseltine WA, Pitha PM (1987a) Herpes simplex virus type-1 can reactivate transcription of latent human immunodeficiency virus. Nature (London) 325:67-70.

Mosca JD, Bednarik DP, Raj NBK, Rosen CA, Sodroski JG, Haseltine WA, Hayward GS, Pitha PM (1987b) Activation of human immunodeficiency virus by herpes virus infection: Identification of a region within the long terminal repeat that responds to a transacting factor encoded by herpes simplex virus 1. Proc Natl Acad Sci USA 84:7408-7412.

Nabel G, Baltimore D (1987) An inducible transcription factor activates expression of human immunodeficiency virus in T-cells. Nature (London) 326:711-713.

Nyce J, Linn L, Jones PA (1986) Variable effects of DNA-synthesis inhibitors upon DNA methylation in mammalian cells. Nucleic Acids Res 14:4353-4367.

O'Hare P, Hayward GS (1984) Expression of recombinant genes containing herpes simplex virus delayed-early and immediate-early regulatory regions and transactivation by herpesvirus infection. J Virol 52:522-531.

Preston CM (1979) Abnormal properties of an immediate-early polypeptide in cells infected with the herpes simplex virus type 1 mutant tsk. J Virol 32:357-369.
Quinn TC, Piot P, McCormick JB, Feinsod FM, Taelman H, Kapita B, Stevens W, Fauci AS (1987) Serologic and immunologic studies in patients with AIDS in North American and African. The potential role of infectious agents as cofactors in human immunodeficiency virus infection. JAMA
Raj NBK, Pitha PM (1983) Two levels of regulation of beta-interferon gene expression in human cells. Proc Natl Acad Sci USA 80:3923-3927.
Rosen CA, Sodroski JB, Haseltine WA (1985) The location of cis-acting regulatory sequences in the human T-cell lymphotropic virus type III (HTLV-III/LAV) long terminal repeat. Cell 41:813-823.
Watson RJ, Clements JB (1980) A herpes simplex virus type 1 function continuously required for early and late virus RNA synthesis. Nature (London) 285:329-330.

DISCUSSION

Pavlakis G (National Cancer Institute, Frederick Cancer Research Facility, Frederick MD):

Have you looked at the kinetics of your reactivation with agents like 5-azacytidine or U.V. irradiation? If it happens early, maybe methylation is not involved.

Mosca J:

The mechanism of demethylation by 5-azacytidine is thought to be through its incorporation into DNA. Whereas the mechanism by ultraviolet light might be envisioned to be through damage and repair resulting in 5-azacytidine incorporation, 5-azacytidine-induced reactivation may not be through DNA incorporation.

Pavlakis, G:

So, how early do you see it?

Mosca J:

Both the ultraviolet and 5-azacytidine reactivation of the HIV-LTR are observed within 8 hr after treatment.

Pullian L (V.A. and University of San Francisco, CA):

That was a really excellent talk. I was interested in this new feeling about LTRs and their differences, and possibly causing HIV virulence, and also, different strains from different sources, for example, brain and macrophage. Have you looked at different strains of HIV isolated from different tissues to see whether or not the LTRs are different? Whether or not they can also be activated?

Mosca J:

Most of our work is done on the initial isolate of Gallo (pHBX2). We have looked at the Simian virus LTR. We used that because it actually has a natural mutation, different from the human LTR region. It only has one core enhancing element. In this case, with just one core enhancer, the activation is not as great. It still activates, but it seems to be that dual core enhancing elements are even better. To answer your questions, one of the problems is that we do not have sequence data on the LTRs of different isolates. Numerous envelope sequences have been determined, and it is thought that differences in the envelope sequences are involved in targeting to macrophages versus T-cell. Two other groups, using two different isolates, also observed HSV reactivation of their HIV-LTR.

McDaniel H (Dallas Fort Worth Medical Center, Grand Prarie, TX):

I, too, was impressed with the intricate specificity and elegance of which you outlined. What I wanted to ask, in terms of trying to relate this to potential effective therapy, do you see any merit in your system to introduce γ-interferon, either from the same species or to treat monocytes and T-cells with an inducer of IFN production, to determine whether the enzyme systems of γ-IFN activate and attack foreign non-native RNA sequences. Would some of these modified genes that you've made be used as effective therapy, in a more physiologic manner?

Mosca J:

Actually, the best way to answer that is to refer you to Dan Bednarik's poster, where exogenous IFN doesn't inhibit virus replication. However, if a cell line is established where IFN synthesis is regulated by the HIV-LTR, HIV infection of this cell line is completely inhibited. The idea is, if we could introduce a chromatin-associated hybrid HIV-LTR/IFN gene in the bone marrow of an infected individual, and the HIV-LTR is suppressed similar to what we observed in our cell lines, then IFN would not be made unless that cell is HIV infected. Expression of the HIV-encoded tatIII gene product would trans-activate the hybrid HIV-LTR/IFN gene to make IFN which, if it behaves like in our transfected cell lines, would completely inhibit HIV replication. In these hybrid constructions, we choose α-IFN, we did not test the effect of either γ- or β-IFN on HIV replication.

REGULATION OF EXPRESSION OF HIV-1 BY VIRAL FACTORS

GEORGE N. PAVLAKIS, BARBARA K. FELBER, CONNIE M. WRIGHT, CHRISTOS CLADARAS, MARGARITA HADZOPOULOU-CLADARAS AND ANGELOS ATHANASSOPOULOS

NCI-Frederick Cancer Research Facility, BRI-Basic Research Program, P.O BOX B, Frederick, Maryland, USA

INTRODUCTION

The expression of the human immunodeficiency virus type 1 (HIV-1) in human cells is regulated by viral factors. This regulation results in the establishment of a steady-state permitting the balanced expression of the viral components. The concerted function of two viral activator proteins tat and rev is necessary for viral expression. Tat and rev are expressed from overlapping reading frames (Figure 1). The action of tat is mediated by a cis-acting element in the R region of the LTR (TAR), while the action of rev is mediated by a cis-acting element in the env region (RRE) (Figure 1). Tat is essential for the accumulation of all viral mRNAs. On the other hand, rev is essential for the accumulation of those viral mRNAs encoding structural proteins.

FUNCTION OF THE TAT PROTEIN

The transactivator or tat is a small basic protein of 86 amino acids encoded by a multiply spliced mRNA (Arya et al. 1985; Siegel et al. 1986; Sodroski et al. 1985, 1986). Tat has been found to be essential for viral replication, because tat$^-$ proviral mutants do not produce any virus after transfection into human cells (Dayton et al. 1986; Fisher et al. 1986a). Tat$^-$ mutants can be complemented in trans by tat expressed from appropriate expression vectors.

Immunofluorescence studies indicate that tat is localized primarily in the nucleus. When tat is over-expressed in monkey COS cells using SV40-based vectors, many cells contain tat in dense bodies in the nucleus resembling nucleoli (Felber et al. unpublished; Hauber et al. 1987). In HeLa cells tat is found in the nucleus, especially around the nucleoli (Felber et al. unpublished). Significant amounts of tat exist also in the cytoplasm of HeLa cells.

Tat is a positive activator of gene expression from the viral LTR promoter (Cullen, 1986; Feinberg et al. 1986; Kao et al. 1987; Muesing et al. 1987; Okamoto and Wong-Staal, 1986; Peterlin et al. 1986; Rosen et al. 1985, 1986; Sodroski et al. 1985; Wright et al. 1986). Tat increases the steady-state levels of LTR-driven mRNAs when the LTR promoter is linked to a variety of indicator genes such as the gene for chloramphenicol acetyltransferase (CAT) (Muesing et al. 1987; Peterlin et al. 1986; Wright et al. 1986), human interleukin-2 (IL-2) (Cullen, 1986), alkaline phosphatase (Berger et al. 1988), tissue plasminogen activator (tpa) and luciferase (Wright and Pavlakis, unpublished). Transcriptional activation by tat has been reported (Hauber et al. 1987; Okamoto and Wong-Staal, 1986; Wright and Pavlakis, unpublished). In addition, an anti-termination mechanism was proposed for tat (Kao et al. 1987). Tat exerts its function via a cis-acting element (TAR) localized within the R

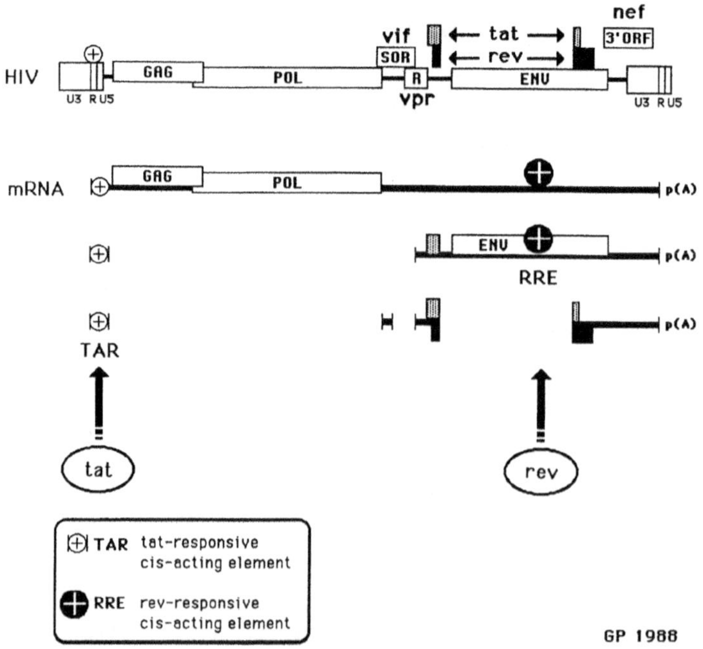

Figure 1. Two positive activators of HIV expression. The genome structure of HIV-1 is shown at the top in its proviral form. The different known open reading frames encoding viral proteins are indicated. A schematic representation of the three classes of viral mRNAs (unspliced, singly-spliced and multiply-spliced) is shown below. The cis-acting elements necessary for the function of tat and rev are indicated. All mRNAs contain the tat-responsive element (TAR) at their 5' end. Only mRNAs encoding viral structural mRNAs contain the rev-responsive element (RRE).

region of the viral LTR (Cullen, 1986; Hauber and Cullen, 1988; Muesing et al. 1987; Peterlin et al. 1986; Rosen et al. 1985; Wright et al. (1986). Therefore, all the viral mRNAs contain this sequence at their 5' end (Figure 1). The position as well as the orientation of TAR within the transcriptional unit is critical for tat function (Muesing et al. 1987; Wright and Pavlakis, unpublished).

An interesting property of tat is that it displays specificity in its function in different cell types (Pavlakis et al. 1988; Wright et al. 1986). In human cells, the extent of activation by tat at the protein level (>500 fold) is much greater than the activation in mouse fibroblasts and monkey kidney cell lines (10-40-fold). This indicates the existence of cell-type specific factors which are important for transactivation. It was also shown that the increase in viral mRNA did not account for the increase in the produced CAT protein, suggesting a complex mechanism of action for tat protein (Berger et al. 1988; Cullen, 1986; Muesing et al. 1987; Wright et al. 1986). The discrepancy between mRNA and CAT protein induction was very prominent in human cells (Felber and Pavlakis, 1988; Wright et al. 1986), where the steady state levels of mRNA increased by 10-to 20-fold in the presence of tat, while the produced CAT protein increased by more than 500-fold. In other cell lines the induction at the levels of mRNA and protein showed a smaller discrepancy (Pavlakis et al. 1988; Wright et al. 1986) or was similar (Gendelman et al. 1986; Peterlin et al. 1986). Other investigators have reported that tat did not affect the levels of viral mRNA (Feinberg et al. 1986; Rosen et al. 1986).

At the present time, the working hypothesis is that tat may act at both transcriptional and post-transcriptional levels to increase all viral mRNAs and the protein produced from the multiply spliced mRNAs. For the production of the structural viral proteins an additional protein, rev, is required.

FUNCTION OF THE REV PROTEIN

Rev is also an essential viral protein because rev$^-$ proviral mutants do not produce any virus after transfections in human cells (Felber et al. 1989a, Feinberg et al. 1986; Hadzopoulou-Cladaras et al. 1989; Sodroski et al. 1986). The presence of rev protein, expressed from appropriate expression vectors fully complements the defects of rev$^-$ mutants. Rev is a small basic protein of 116 amino acids expressed from a multiply-spliced mRNA. Immunofluorescence experiments indicate that rev is a nuclear protein (Felber et al. 1989a). When rev is expressed from the viral LTR promoter in the presence of tat in HeLa cells, rev is localized in the nucleus, especially in the dense bodies resembling the nucleoli.

Several functions have been proposed for rev. Sodroski et al. (1986) reported that, in the absence of rev, an HIV-1 proviral mutant produced the same amounts of mRNA but not viral structural proteins. Feinberg et al. (1986) concluded that the same viral mutant did not produce any unspliced viral mRNA but only spliced mRNA. Therefore, rev was proposed to act on splicing, and the absence of viral structural proteins was attributed to the absence of the corresponding mRNAs. Knight et al. (1987) also reported that while tat was able to increase the levels of an artificial env mRNA expressed from the LTR promoter, both tat and rev were necessary for env protein synthesis. We have examined the mode of action of rev using a collection of proviral mutants (Felber et al., 1989a; Hadzopoulou-Cladaras et al., 1989). Our results demon-strated that all the viral mRNAs are present in the absence of rev, but the ratio of the unspliced to spliced mRNA is greatly decreased. In the absence of rev, the production of gag and env proteins was detectable but low, and no virus was produced. All these defects were complemented in trans by rev protein produced from appropriate expression vectors. The study of the proviral deletion mutants revealed a cis-acting element located within the env region of HIV-1 (rev responsive element, RRE) which is necessary for rev function (Figure 1). Measurements of stability indicated that the stability of the RRE-containing mRNA species is increased in the presence of rev. In addition to the stabilization, rev appears to affect the transport of the unspliced viral mRNA to the cytoplasm. rev exerts its function on the viral mRNAs independent of splicing.

The identified regulatory circuits of HIV-1 are shown in Figure 2. Tat is a positive activator of HIV-1 expression. Rev increases the levels of the unspliced mRNAs and of the structural viral proteins. In addition, rev lowers the levels of the multiply spliced mRNAs that produce tat, rev and nef (3'ORF) proteins. Therefore, rev down regulates its own expression via a negative feed-back mechanism (Felber et al., submitted). Rev also down-regulates the expression of all the other regulatory proteins of HIV-1. The characteristics of rev function indicate that this is a very important regulatory circuit for HIV-1 expression. The only other know HIV-1 product that may down-regulate viral expression is the nef (3'ORF or F) protein. This protein is not essential for the propagation of HIV-1 in tissue culture and the mechan-ism of its action is not known (Fisher et al. 1986b; Terwilliger et

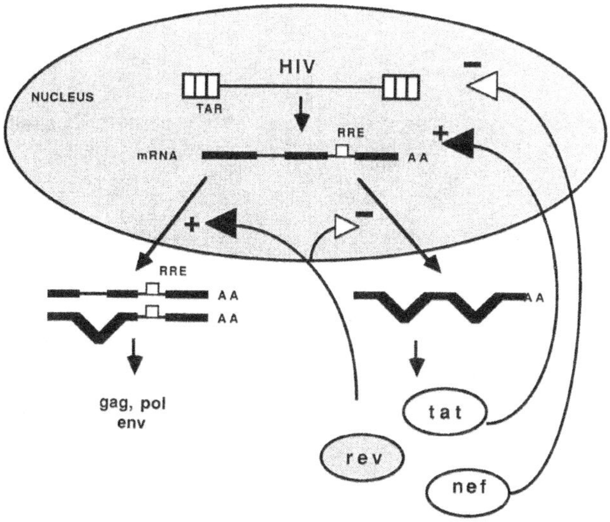

Figure 2. Regulation of HIV-1 expression by tat, rev and nef. Model for the regulation of HIV-1 expression by tat, rev and nef.

al. 1986). Nef is a myristilated protein that displays GTPase, GTP-binding and autophosphorylation activities and was proposed to be a down-regulator of the levels of HIV-1 replication (Ahmad and Venkatesan, 1988; Guy et al. 1987; Luciw et al. 1987).

A QUANTITATIVE BIOASSAY FOR HIV-1 BASED ON TRANS-ACTIVATION

We took advantage of the great activation of the HIV-1 LTR promoter by tat in human cells, to develop a bioassay for HIV-1 based on transactivation. Several human indicator cell lines were generated containing stably integrated copies of the HIV-1 LTR promoter linked to indicator genes such as CAT, which is a convenient reporter gene. Such transcription units are silent and depend on the presence of tat protein for their expression. After infection by HIV-1, tat protein is produced which stimulates both the replication of the virus and the production of CAT enzyme. The enzyme levels reflect the amount of infecting virus. Such human lymphoid cell lines, called indicator cell lines, were used to detect as few as ten HIV-infected lymphoid cells in two days. This bioassay has been shown to be sensitive, fast and specific (Felber and Pavlakis, 1988).

We have generated several indicator cell lines derived from human T cells as well as from human cell lines with monocyte/macrophage characteristics. An important observation from our lab, in collaboration

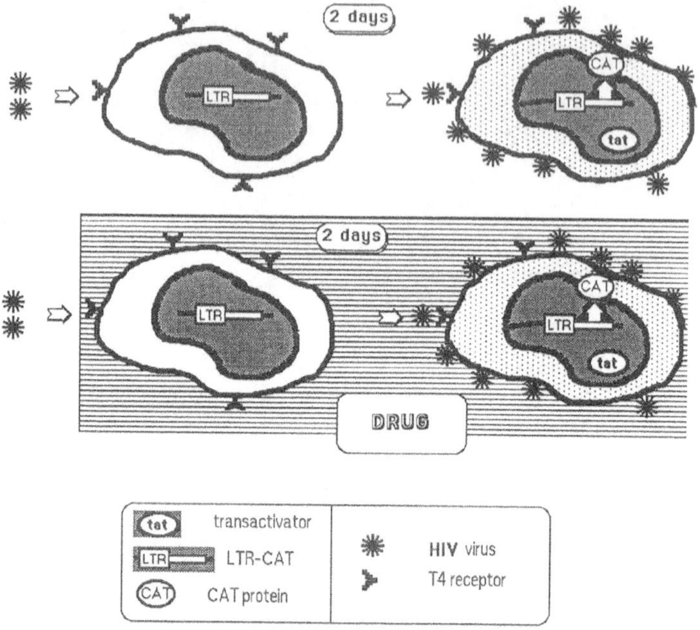

Figure 3. Drug assay. Application of bioassay to screen for anti-HIV-1 agents. Schematic representation of the drug screening protocol using the HIV-1 LTR linked to the CAT gene are infected with HIV-1 either in the absence (top) or the presence of drugs (bottom). Two days later the cells are assayed for CAT production. If a drug affects the infectivity and/or replication of HIV-1, less tat is produced resulting in lower levels of CAT enzyme.

with Eva-Maria Fenjo and Stefan Schwartz from the Karolinska Institute, was that clinical isolates of HIV behave differently in these cell lines. The availability of a variety of indicator cells will allow the quantitation of the differences between the isolates and will aid the characterization of potential HIV-1 subtypes.

An immediate application of this assay has been the specific and quantitative measurement of the effects of drugs on HIV-1 infectivity and replication (Figure 3). Another interesting characteristic of this assay is that it is not dependent on cell killing by HIV-1, therefore, any cell susceptible to infection by the virus can be used for the generation of indicator cell lines. This assay measures the early events of activation by HIV-1 and is fast (2-3 days), specific and quantitative. CAT enzyme can be measured on ELISA plates by a chromogenic assay and inhibition curves have already been derived for several drugs using this method (Pavlakis et al. unpublished). In addition to a model for acute infection, the assay can be used for the establishment of a model of chronic infection by HIV-1. This will allow screening for drugs that affect the expression of the virus in already infected cells. Other indicator genes in addition to CAT have been used successfully for the generation of indicator cell lines. Also, cell lines that depend on the presence of both tat and rev are under construction. These developments will aid the identification of the mechanism of function of the regulatory proteins of HIV-1. In addition, these cell lines can be used for the establishment of assays for drugs affecting the function of specific regulatory HIV-1 products.

ACKNOWLEDGEMENTS

Research sponsored by the National Cancer Institute, DHHS, under contract No. NO1-CO-74101 with Bionetics Research, Inc. The contents of this publication do not necessarily reflect the views or policies of the Department of Health and Human Services, nor does mention of trade names, commercial products or organizations imply endorsement by the U.S. Government.

REFERENCES

Ahmad N, Venkatesan S (1988) Nef protein of HIV-1 is a transcriptional repressor of HIV-1 LTR. Science 241: 1481-1485.

Arya SK, Guo C, Josephs SF, Wong-Staal F (1985) Transactivator gene of human T-lymphotropic virus type III (HTLV-III). Science 229:69-73.

Berger J, Hauber J, Hauber R, Geiger R, Cullen BR (1988) Secreted placental alkaline phosphatase: a powerful new quantitative indicator of gene expression in eucaryotic cells. Gene 66:1-10.

Cullen BR (1986) Trans-activation of human immunodeficiency virus occurs via a bimodal mechanism. Cell 46:973-982.

Dayton AI, Sodroski JG, Rosen CA, Goh WC, Haseltine WA (1986) The trans-activator gene of the human T cell lymphotropic virus type III is required for replication. Cell 44:941-947.

Feinberg MB, Jarrett RF, Aldovini A, Gallo RC, Wong-Staal F (1986) HTLV-III expression and production involve complex regulation at the levels of splicing and translation of viral RNA. Cell 46:807-817.

Felber BK, Pavlakis GN (1988) A quantitative bioassay for HIV-1 based on trans-activation. Science 239:184-187.

Felber BK, Hadzopoulou-Cladaras M, Cladaras C, Copeland T, Pavlakis GN (1989a) The rev (art/trs) protein of HIV-1 affects the stability and sequestering of viral mRNA. Proc. Natl. Acad. Sci. USA 86:1495-1499.

Felber BK, Drysdale CM, Pavlakis GN (1989b). Feedback regulation of HIV-1 expression by rev protein. (Submitted).

Fisher AG, Feinberg MB, Josephs SF, Harper ME, Marselle LM, Reyes G, Gonda MA, Aldovini A, Debouk C, Gallo RC, Wong-Staal F (1986a) The trans-activator gene of HTLV-III is essential for virus replication. Nature 320:367-371.

Fisher AG, Ratner L, Mitsuya H, Marselle LM, Harper ME, Broder S, Gallo RC, Wong-Staal F (1986b) Infectious mutants of HTLV-III with changes in the 3' region and markedly reduced cytopathic effects. Science 233:655-659.

Gendelman HE, Phelps W, Feigenbaum L, Ostrove JM, Adachi A, Howley PM, Khoury G, Ginsberg HS, Martin MA (1986) Transactivation of the human immunodeficiency virus long terminal repeat sequence by DNA tumor viruses. Proc Natl Acad Sci USA 83:9759-9763.

Guy B, Kieny MP, Riviere Y, LePeuch C, Dott K, Girard M, Montagnier L, Lecocq J-P (1987) HIV F/3' ORF encodes a phosphorylated GTP-binding protein resembling an oncogene product. Nature 330:266-269.

Hadzopoulou-Cladaras M, Felber BK, Cladaras C, Athanassopoulos A, Tse A, Pavlakis GN (1989) The rev or art-trs protein of HIV-1 affects viral mRNA and protein expression via a cis-acting sequence in the env region. J. Virol. 63:1265-1274.
Hauber, Perkins EP, Heimer, Cullen BR (1987) Transactivation of human immunodeficiency virus gene expression is mediated by nuclear events. Proc Natl Acad Sci USA 84:6364-6369.
Hauber J, Cullen BR (1988) Mutational analysis of the transactivation-responsive region of the human immunodeficiency virus type I long terminal repeat. J Virol 62:673-679.
Kao SY, Calman AF, Luciw PA, Peterlin BM (1987) Antitermination of transcription within the long terminal repeat of HIV-1 by tat gene product. Nature 330:489-493.
Knight DM, Flomerfelt FA, Ghrayeb J (1987) Expression of the art/trs protein if HIV and study of its role in viral envelope synthesis. Science 236:837-840.
Luciw PA, Cheng-Mayer C, Levy JA (1987) Mutational analysis of the human immunodeficiency virus. The orf-B region down-regulates virus replication. Proc Natl Acad Sci USA 84:1434-1438.
Muesing MA, Smith DH, Capon DJ (1987) Regulation of mRNA accumulation by a human immunodeficiency virus trans-activator protein. Cell 48:691-701.
Okamoto T, Wong-Staal F (1986) Demonstration of virus-specific transcriptional activator(s) in cells infected with HTLV-III by an in vitro cell free system. Cell 47:29-35.
Pavlakis GN, Felber BK, Wright CM (1988) A fusion assay for the detection of HIV infected cells. In Human Retroviruses, Cancer and AIDS Approaches to Prevention and Therapy (D. Bolognesi, ed). Alan R. Liss, Inc. pp. 439-445.
Peterlin BM, Luciw PA, Barr PJ, Walker MD (1986) Elevated levels of mRNA can account for the trans-activation of human immunodeficiency virus. Proc Natl Acad Sci USA 83:9734-9738.
Rosen CA, Sodroski JG, Haseltine WA (1985) The location of cis-acting regulatory sequences in the human T cell lymphotropic virus type III (HTLV-III/LAV) long terminal repeat. Cell 41:813-823.
Rosen CA, Sodroski JG, Chun Goh W, Dayton AI, Lippke J, Haseltine WA (1986) Post-transcriptional regulation accounts for the trans-activation of the human T-lymphotropic virus type III. Nature 319:555-559.
Siegel LJ, Ratner L, Josephs SF, Derse D, Feinberg MB, Reyes GR, O'Brien SJ, Wong-Staal F (1986) Trans-activation induced by human T-lymphotropic virus type III (HTLV-III) maps to a viral sequence encoding 58 amino acids and lacks tissue specificity. Virology 148:226-231.
Sodroski J, Patarca R, Rosen C, Wong-Staal F, Haseltine WA (1985) Location of the trans-activating region on the genome of human T-cell lymphotropic virus type III. Science 229:74-77.
Sodroski J, Chun Goh W, Rosen C, Dayton A, Terwilliger E, Haseltine W (1986) A second post-transcriptional trans-activator gene required for HTLV-III replication. Nature 321:412-417.
Terwilliger E, Sodroski JG, Rosen CA, Haseltine WA (1986) Effects of mutations within the 3' orf open reading frame region of human T-cell lymphotropic virus type III (HTLV-III/LAV) on replication and cytopathogenicity. J Virol 60:754-760.
Wright CM, Felber BK, Paskalis H, Pavlakis GN (1986) Expression and characterization of the trans-activator of HTLV-III/LAV virus. Science 234:988-992.

Mosca J (The Johns Hopkins University, Baltimore, MD):

You don't think the rev protein was involved in the splicing of tat-3 mRNA? Is rev just responsible for the structural proteins, not for the regulatory proteins, is that correct?

Pavlakis G:

The effect of rev on the multiply spliced RNA producing tat, rev and nef is indirect. All viral RNAs are produced from the same promoter. If you reduce the supply of the unspliced precursor, for example, by transporting the unspliced messanger RNA, then there is not enough available to produce high levels of multiply spliced RNA. So, I believe that the effect of rev on the tat mRNA is indirect. We know from all the experiments that we have that the multiply spliced RNAs are not effected by themselves by the rev protein.

MECHANISM(S) OF COXSACKIEVIRUS-INDUCED ACUTE MYOCARDITIS IN THE MOUSE

CHARLES J. GAUNTT, ELMER K. GODENY, C. WILLIAM LUTTON, HELEN M. ARIZPE, NORA M. CRAPMAN, STEVEN M. TRACY, GEORGE E. REVTYAK, ANTHONY J. VALENTE AND MARIUS M. ROZEK

The University of Texas Health Science Center of San Antonio, San Antonio, Texas 78284-7758, (USA)

INTRODUCTION

Acute myocarditis can be induced in several murine strains following inoculation of infant or adolescent mice with a coxsackievirus group B (CVB) (Gauntt, 1987; Herskowitz et al. 1986; Lerner and Reyes, 1985; McManus et al. 1987; Woooruff, 1980). Murine-CVB models have provided significant insight into several mechanisms responsible for this disease process and are particularly attractive because of the pathologic similarities between the induced cardiac lesions and those found in heart tissue from cases of human myocarditis (Lerner and Reyes, 1985; McManus et al. 1987). The CVB have been isolated rarely from heart tissues of these patients. Considerable serologic evidence suggests that serotypes B1-B5 are the major etiologic agents responsible for about half of the clinical cases of myocarditis (Gear and Measroch, 1973; Grist and Bell, 1974; Grist et al. 1978; Lansdown, 1978; Smith, 1970) and coxsackievirus B3 (CVB3) is the serotype most frequently associated with myocardial disease (Lerner and Wilson, 1973).

CVB3-induced myocarditis in the mouse is the most widely studied model of this disease. Data from these studies suggests that cell-mediated immune/delayed hypersensitivity responses involving T lymphocytes play the major role in myocyte destruction, rather than simple virus-induced lysis of cells (Gauntt, 1987; Huber and Job, 1983a; Huber et al. 1980; Job et al. 1986; Paque and Gauntt, 1983; Paque et al. 1978; Woodruff and Woodruff, 1974). Autoimmune mechanisms involving cell mediated immune responses are also thought to play a role in acute CVB-induced myocarditis (Estrin et al. 1986; Huber and Lodge, 1984, 1986;

Huber et al. 1981, 1987). Autoantibodies to cardiac antigens have been found in selected murine strains and may play a role in acute and more likely chronic myocarditis (Beisel et al. 1986; Herskowitz et al. 1985, 1986; Huber and Lodge, 1986; Lodge et al. 1987; Lyden and Huber, 1984; Neu et al. 1987a, Wolfgram et al. 1985, 1986). Indeed, inoculation of mice with murine cardiac myosin induces immunologically-mediated myocarditis only in those strains of mice genetically predisposed to produce high titer autoantibodies against myosin (Neu et al. 1987b). Thus, the genetic background of the host has a major influence on determining the mechanism(s) mediating CVB-induced myocarditis as well as the severity of the disease.

FACTORS INFLUENCING CVB3 MURINE MYOCARDITIS

Inoculation of mice by any of the several usual routes induces myocarditis. However, the severity of disease resulting from a CVB3 infection depends on several variables (Table 1). The genotype of the challenge virus is a major factor in determining the severity of the induced myocarditis. Laboratory-derived (Gauntt et al. 1984; Lutton et al. 1985b; Roesing et al. 1979; Trousdale et al. 1979) and naturally occurring CVB3 variants (Tracy and Gauntt, 1987) have been described.

TABLE 1. FACTORS INFLUENCING CVB3 INDUCED MURINE MYOCARDITIS

Variable	Comments
Virus Genotype	Laboratory and naturally-occurring myocarditic and amyocarditic variants have been described.
Murine Strain	Genetic background in some strains determines mechanism(s) of induction and severity of disease.
Gender	Males generally more susceptible; virus variant plays a role in gender susceptibility.
Age at Inoculation	Mice younger than two weeks of age generally die; the younger a mouse is at infection generally correlates with increased severity in myocarditis.
Immunologic Status	Immunosuppressive drugs administered near time of inoculation exacerbate disease.
Stress	Infected mice forced to swim daily for brief periods develop more severe disease than non-stressed counterparts.

Although the molecular basis for CVB virulence has not been determined, inferences can probably safely be made from studies on virulence in the very similar polioviruses (Evans et al. 1985; Kohara et al. 1985; Omata et al. 1986). By analogy to the poliovirus data, discrete single nucleotide changes in the 5'-noncoding region of the CVB3 genome (Pallansch, Tracy, Chapman and Gauntt, unpublished data, 1988) and possible changes in the viral capsid proteins may be in part responsible for myocarditic virulence in mice.

Mouse strain is also a major variable influencing the disease. Some strains of mice inoculated with a myocarditic CVB3 variant develop an acute fatal disease. Other strains may develop acute disease which resolves whereas some strains infected with CVB progress from acute to chronic disease, with autoimmune responses directed against heart tissues (Herskowitz et al. 1986). In other strains, a humoral immune response may be stimulated from the outset and contribute significantly to the subsequent myocarditis (Huber and Lodge, 1986).

Many studies have found males to be more susceptible than females to CVB3-induced myocarditis (Wong et al. 1977; Woodruff 1980). Sex hormones are reported to influence the extent of CVB3-induced myocarditis (Huber et al. 1982), possibly through increasing the infectivity of the virus for neonatal cardiac endothelial cells and myocytes (Lyden et al. 1987). In addition, female mice produce significantly fewer virus-induced cytotoxic T lymphocytes (CTL) with specificity for virus-infected neonatal myocyte or fibroblast target cells (Huber et al. 1982; Wong et al. 1977). In some mouse strains gender is not a factor, whereas in certain others, females appear to develop more severe myocarditis (Gauntt et al. 1984). The influence of gender on the severity of induced myocarditis is probably closely related to the genotype of virus under study. In many strains, mice younger than four weeks either succumb to CVB3 infection or develop more severe myocarditis han older siblings. This is likely a result of immaturity of their nonspecific and immune defenses (McManus et al. 1987; Woodruff, 1980).

Classic studies (Woodruff, 1980) showed that depletion of T lymphocytes in mice prior to CVB3 challenge largely prevented induction of myocarditis. However, CVB3 infection of athymic mice demonstrate that monocyte/macrophages can, in absence of mature T lymphocytes, produce inflammation in the heart. Studies of potential immunotherapy in murine models of acute CVB3-induced myocarditis showed that administration of corticosteroids, the non-steroidal anti-inflammatory drug ibuprofen, cyclophosphamide, cyclosporine or levamisole early in infection exacerbated myocarditis (McManus et al. 1987). Administration of β-interferon

(IFN) within 24 hr of infection reduced the number of lesions whereas, paradoxically administration of anti-β-IFN antiserum at 3 days post-inoculation of CVB3 also reduced the number of myocarditic lesions (Lutton and Gauntt, 1984). Since IFN did not reduce virus titers in heart tissues but did reduce the number of lesions, these opposing effects of IFN on modulating severity of myocarditis are probably mediated through lymphocyte and macrophage activities. BIOLF-70, a substituted guanine derivative, reduced lesion number if administered around the time of infection; this drug increased natural killer cell activity and decreased reactivity of CTL for virus-infected targets (Gauntt et al. 1985). Thus, these divergent studies show that alterations induced in the immunologic status of the host may have either severe consequences or provide some protection to the CVB3-infected mouse.

Finally, a group of young 2-3 week-old mice infected with CVB3 and forced to swim in warm water 30 minutes twice a day had significantly increased mortality due to congestive heart failure compared with a group of CVB3-infected mice not forced to swim; survivors in the former group exhibited more severe myocarditis (Gatmaitan et al. 1970; Reyes et al. 1984). In an attempt to explain this result, Reyes and colleagues (1984) inoculated both groups of mice with ^{51}Cr-labeled T lymphocytes from CVB3 infected mice. Heart tissues from mice forced to swim had a 4-fold greater abundance of T lymphocytes than did heart tissues of CVB3-infected mice not subjected to physical stress (Reyes et al. 1984). Stressed animals appear to have higher titers of virus in their heart tissues. In addition, exercise alone can enhance release of T lymphocytes from the thymus (Reyes et al. 1981a). The combination of these two factors may result in a higher probability of T cell exposure to virus-induced antigens in heart tissues with resulting clonal expansion and the possibility for more severe myocarditis.

ACUTE CVB3-INDUCED MYOCARDITIS IN THE ADOLESCENT MALE CD-1 MOUSE

Histologic examination of heart tissues from CVB3-inoculated adolescent male CD-1 mice reveals focal myocarditic lesions containing necrotic myocytes, mononuclear and polymorphonuclear cells, lymphocytes and fibroblasts (Figure 1). On occasion, calcium deposits are found in lesions in some heart tissues. Severe dystrophic calcification is observed in myocarditic lesions in heart tissues of mice made deficient in natural killer (NK) cells by immunodepletion (Godeny and Gauntt, 1987a). The earliest studies on CVB3-induced myocarditis observed that

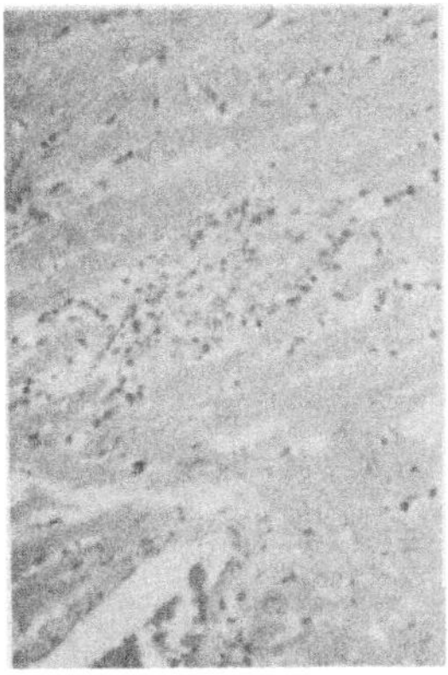

Figure 1. Myocarditic lesions in a coronal section of mouse heart tissue taken 8 days p.i. of CVB3. Hematoxylin and eosin stained. 55X.

virus titers in heart tissues were maximum and then progressively decreased several days before lesions were observed (Figure 2). Virus titers begin to diminish in heart tissues around day 4 post inoculation (p.i.) and nascent lesions become detectable at the light microscope level. Lesion size and numbers increase through day 7 p.i. It was these kinds of data that led to suggestions that most myocyte destruction is likely due to host immune mechanisms rather than lysis of cells by virus. Single and small foci of necrotic myocytes can be detected around 3 days p.i. Over the next 5 days, necrotic myocytes are subsequently found in lesions in progressively smaller numbers relative to the number of infiltrating leukocytes. The extent of myocyte damage becomes most evident between 9-12 days p.i. when the lesions resolve into scar tissue and leukocytes have emigrated from the lesion site.

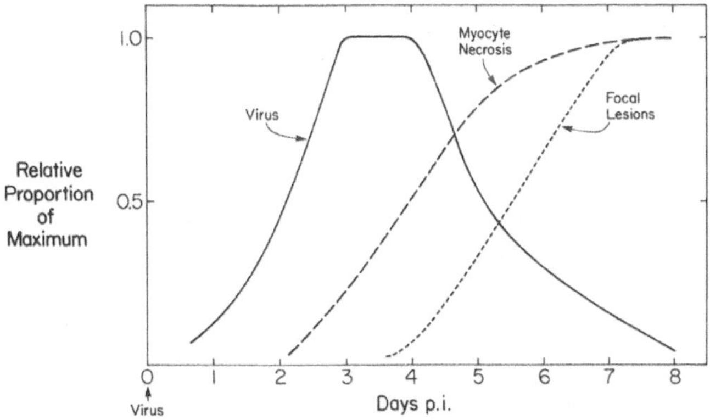

Figure 2. Schematic representation of the temporal courses of virus content, myocyte necrosis and focal lesion formation in the myocardium of an adolescent male CD-1 mouse inoculated with a myocarditic variant of CVB3.

A schematic depicting the temporal sequence of nonspecific and immune defense mechanisms mobilized against infection by a myocarditic CVB3 variant is presented in Figure 3. Within hours of inoculation of the virus into the peritoneum, a viremia develops which persists for the next two days. During the first 24 hrs, virus replicates in unidentified cells and is released into the blood. The cells most likely involved in virus replication are endothelial cells. Virus does not replicate in splenic leukocytes although it appears to be transported by these cells during days 2-4 p.i. Virus replication stimulates the production and release of both β- and γ-IFNs (Gauntt et al. 1988). The interferons probably activate NK cells (Godeny and Gauntt, 1986). In this schema, NK cell activity in the spleen is transient and typical of this type of non-specific defense against an infection. Activated NK cells can lyse CVB3-infected murine fibroblasts *in vitro* and are among the earliest cells found in the developing lesion (Godeny and Gauntt, 1987a). Anti-CVB3 neutralizing antibody of the IgM class is detected on 4 day p.i. and stable high titers are present by day 5 p.i. This antibody must contribute to the decline in infectious virus in the blood (Figure 1) and most likely contributes to the decline of virus titers in heart tissues (Figure 2). Splenic CTL which can lyse a variety of CVB3-infected primary cultures of murine cells (Huber et al. 1987; Woodruff, 1980) are detected by day 5 p.i. Maximum CTL activity is found around 6-7 days p.i. then subsequently declines. Data from several laboratories over the past decade have led to the conclusion that CTL play little or no role in viral clearance but are a major contributor to pathologic lesions in the heart (Gauntt, 1987; Woodruff, 1980). There are three different subsets

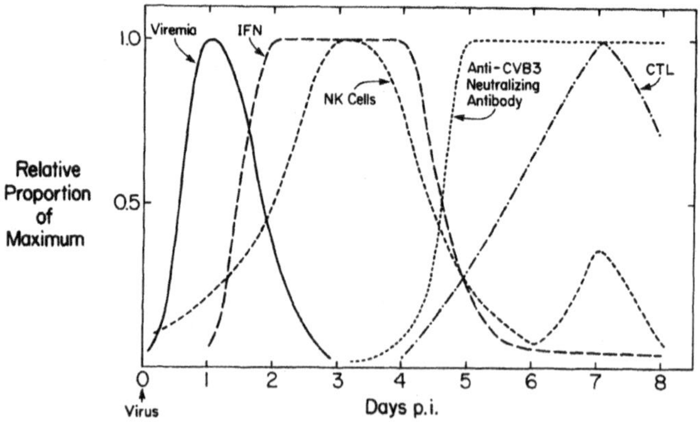

Figure 3. Schematic representation of nonspecific and immune defenses mobilized as a result of infection of adolescent male CD-1 mouse heart tissues with a myocardic CVB3 variant. Virus, IFN and anti-CVB3 neutralizing antibody titers were measured in blood; NK cell and CTL activity were measured in splenic lymphocytes.

of virus-induced splenic CTL ($thy1^+$ asialo GM_1^-) that have been detected in mice (Huber et al. 1987). One subset ($L3T4^+$) lyses virus-infected target cells. The other two subsets ($Lyt2^+$; surface antigen(s) on the other subset unidentified) lyse only uninfected target cells and may be involved in autoimmune pathology.

IDENTIFICATION OF NONCARDIAC CELLS IN HEART TISSUES OF CVB3-INFECTED MICE

The sequence of entry and identity of leukocytes infiltrating infected heart tissues was investigated by *in situ* autoradiography using an ^{35}S-labeled double antibody technique (Godeny and Gauntt, 1987b). Examples of cells identified using this technique are shown in Figure 4 and a summary of the data is presented in Table 2. Lesions became histologically identifiable in the myocardium on day 4 p.i. Increases in lesion size and number continued through day 6 p.i. T lymphocytes (Figure 4C) and NK cells (Figure 4E) were diffusely distributed throughout all

TABLE 2. TEMPORAL SEQUENCE OF LEUKOCYTES INFILTRATING CVB3-INFECTED HEART TISSUES[a]

Cell Type	Presence of Specific Cell (|---|) Day p.i.							
	1	2	3	4	5	6	7	8
Virus-replicating	—	—	—	—	—			—
Macrophages				—	—	—	—	
T lymphocytes ($Thy1^+$)				—	—	—	—	
Suppressor/cytotoxic ($Lyt2^+$)				—	—	—	—	—
Helper ($L3T4^+$)					—	—	—	—
Natural killer				—	—	—	—	—

Lesions:					
small(s), medium(m), large(l)	s	m	l	l	l
number/section (range)	1-5	3-20	22-102	88-168	36-156

[a] Mice were challenged with 10^5 pfu/cell of CVB3. Five mice were sacrificed each time point and hearts fixed in formalin. Two coronal sections approximately one-third from the apical tip were sequentially reacted with a specific primary antibody, ^{35}S-labeled affinity-purified anti-immunoglobulin secondary antibody and then autoradiography was performed. Two additional adjacent sections were stained with hemotoxylin/eosin for detection and counting of lesions. Data modified from Godeny and Gauntt (1987b).

lesions observed. T lymphocytes of the $Lyt2^+$ (suppressor/cytotoxic) subset were detected focally in areas within lesions (data not shown) in a manner similar to that shown for macrophages (Mac+; Figure 4D). Macrophages were found only in lesions from days 4-7 p.i. Helper T cells ($L3T4^+$) were not detected until day 5 p.i. In summary, macrophages, T lymphocytes and NK cells were among the earliest cells detected in developing lesions. Macrophages emigrated from lesions prior to day 8 p.i.

DETECTION OF VIRUS-REPLICATING CELLS IN THE INFECTED MYOCARDIUM

Virus-replicating cells were detected in infected heart tissues by two *in situ* techniques. An *in situ* ^{35}S-labeled antibody technique was used to detect cells producing virus capsid antigens. Photomicrographs showing foci of viral antigen positive cells are shown in Figure 5. Foci

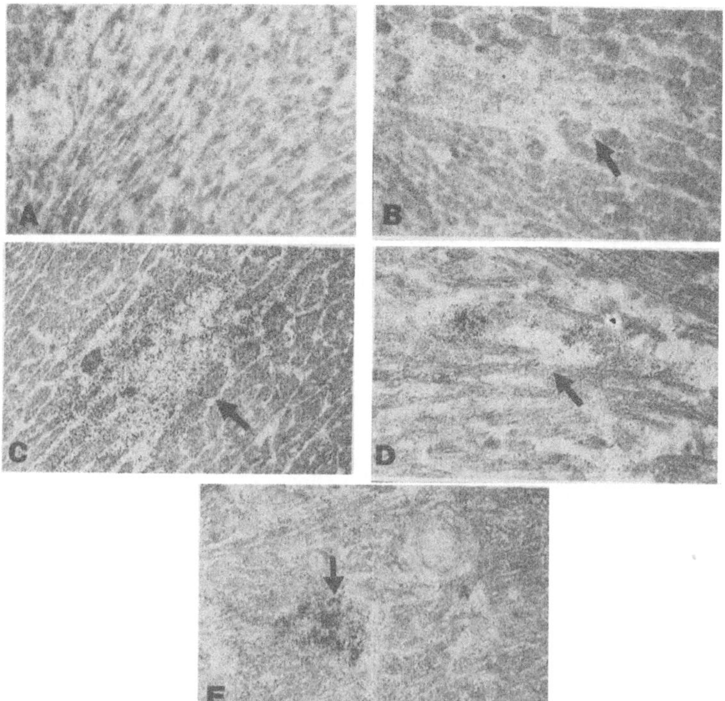

Figure 4. Inflammmatory cells identified in heart tissues of $CVB3_m$ inoculated mice by *in-situ* ^{35}S-labeled antibody autoradiographic detection of cell-specific antigenic surface markers. Normal (A) or CVB3-infected (B) sections on 8 days p.i. without primary antibody but with ^{35}S-labeled secondary anti-immunoglobulin antibody show minimal nonspecific binding of the radiolabeled antibody probe. Arrows identify myocarditic lesions. Specific monoclonal antibody identify the presence of T lymphocytes ($Thy1^+$ cells; C) and macrophages on day 6 p.i. (Mac 1; D). Rabbit anti-asialo GM_1 antibodies identify NK cells on day 6 p.i. (E) H&E at 400X.

of grains were larger on day 3 p.i. compared to foci found on day 8 p.i.These foci were present in infected heart tissues from days 1-5 p.i., a manner similar to that shown for macrophages (Mac+; Figure 4D). Macrophages were found only in lesions from days 4-7 p.i. Helper T cells ($L3T4^{+}$) were not detected until day 5 p.i. In summary, macrophages, T lymphocytes and NK cells were among the earliest cells detected in developing lesions. Macrophages emigrated from lesions prior to day 8 p.i. but surprisingly, NK cells remained up to 10 days p.i. with a maximum number being found on day 3 p.i. (Godeny and Gauntt, 1987b), the time at which CVB3 titers are generally maximum in heart tissues. Foci of virus-replicating cells were never found in nascent or developing lesions even on days 8 or 9 p.i. when sections contained a large number of lesions and a rare focus of virus-infected cells.

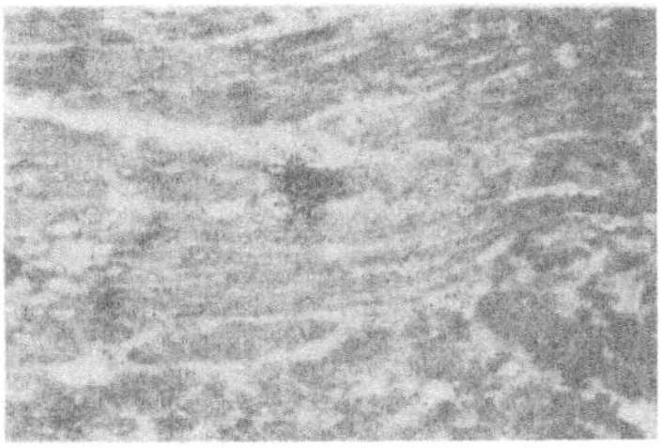

Figure 5. *In situ* ^{35}S-labeled antibody autoradiographic detection of virus-replicating cells in infected heart tissues using antibody against purified CVB3. Foci of grains over cells in heart tissues taken from CVB3-inoculated mice on day 8 p.i. H&E at 400X.

The presence of virus-replicating cells in heart tissues was confirmed by *in situ* hybridization of tissues with a negative strand RNA probe complementary to the capsid region. Tracy and colleagues (1985a, 1985b) molecularly cloned the CVB3 genome. The plasmid pCBIII29, derived from the viral capsid coding region, was digested with Pstl and the resulting 2.1 kb fragment inserted into the pGemini-1 transcription plasmid (Promega Biotec). After digesting the plasmid with Hind III to cleave within the viral insert, the RNA polymerase of bacteriophage T7 was used to transcribe 1.2 kb of tritium-labeled negative strand RNA. This RNA was alkali-hydrolyzed to fragments of 100 to 200 base length and then used as a probe. As a negative control, the pGemini-1 plasmid without the CVB3 capsid coding region insert was transcribed and the product used in identical reactions. The results of probing previously frozen (liquid N_2)-paraformaldehyde-fixed heart sections with the negative strand RNA probe are presented in Figure 6. Very low background grains were found in both negative control assays. Foci of grains indicating hybridization of the negative strand viral RNA probe to nascent CVB3 genomic positive strand RNA were observed over heart sections taken at 3 or 8 days p.i. Foci of grains over heart sections taken at three days p.i., the time of maximum virus titers in heart tissues, were larger than those observed in heart sections taken at 8 days p.i. Again, these foci of grains were never found over lesions in heart tissues. Taken together the two *in situ* assays suggest that virus infection of unidentified cells in the heart induces an inflammatory response which, between days 4 and 5 p.i., quickly eliminates virus-replicating cells.

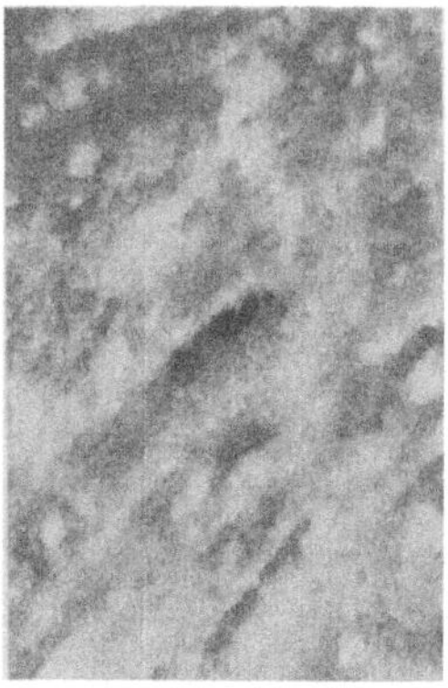

Figure 6. Autoradiographic detection of CVB3 genome synthesis in infected heart tissues by *in situ* hybridization of a negative strand ^{3}H-uridine-labeled RNA probe complimentary to the capsid region of the CVB3 genome. Foci of grains resulting from hybridization of the virus-specific negative strand RNA probe with heart tissues taken from mice on day 8 p.i. H & E stained section at 490X.

VIRUS-INDUCED PRO-INFLAMMATORY EVENTS

CVB3 is a small nonenveloped positive-strand RNA genome virus, a member of the picornaviridae. Recently, we presented the first evidence that this virus could alter the surface of infected cells (Lutton and Gauntt, 1986). During non-lytic replication of CVB3 in primary cultures of neonatal skin, infant heart (10-14 days old) or adolescent heart (4 to 6 weeks old) fibroblasts, double fluorescent staining was used to show that the surface of virus-replicating cells (12-24% of all cells) exhibited an alteration. This virus-induced surface change was detected by the binding of the lectins *Ulex europaeus* agglutinin I (UEAI), specific for α-L-fucosyl residues, and soybean lectin, specific for α/B-N-acetyl-galactosoaminyl residues. Binding of fluorescent-labeled preparations of UEAI was greater than soybean lectin and thus UEAI was used in studies. UEAI binding to CVB3-infected fibroblasts was minimal at 24 hr p.i. Treatment of fibroblasts with β-IFN prior to infection with CVB3 reduced both binding of UEAI and virus yields. Binding of UEAI to CVB3-infected fibroblasts also reduced lysis by CTL from CVB3-infected mice (Lutton, 1986). To determine if this virus-induced cell surface alteration was present in infected heart tissues, the *in situ* method using ^{35}S-labeled antibody was applied (Godeny and Gauntt 1987b). UEAI binding to cells in infected heart tissue is shown in Figure 7. Foci of grains were detected only on day 3 p.i., the time of maximal virus titers in the heart. UEAI did not bind to uninfected heart tissues.

More recently we have begun investigating the production and release of soluble pro-inflammatory mediators by CVB3-infected fibroblasts and by heart tissue from CVB3-infected mice. The results of several *in vitro* studies are summarized in Table 3. Primary cultures of neonatal skin or adolescent heart fibroblasts were challenged with CVB3, washed and incubated in MEM containing 1% bovine serum albumin and antibiotics (MEM-BSA). Supernatant fluids from normal or CVB3-challenged fibroblasts were harvested at various times and assayed for chemotactic activity for non-elicited murine peritoneal macrophages using Boyden (blind well) chambers and 5μm pore size polycarbonate filters using the technique described by Valente et al. (1984). Cultures of CVB3-infected fibroblasts produced and released a factor which was detected at 6 hr p.i. and produced by cultures of uninfected fibroblasts. Normal fibroblasts produce low levels of this factor constituitively, possibly reflecting an immune surveillance mechanism. Production of prostaglandins was also investigated by incubating mock- and CVB3-infected cultures of fibroblasts in MEM-BSA. At various times p.i., culture fluids were removed and directly assayed for PGE_2 or 6-keto $PGF_{1\alpha}$ by RIA as previously described (Campbell et al. 1980, 1982). Infected fibroblasts produced and released approximately 1.7-fold more PGE_2 than uninfected cells over the period 24 to 72 hr p.i. Virus-induced stimulation of PGE_2 was detectable at 6 hr p.i.

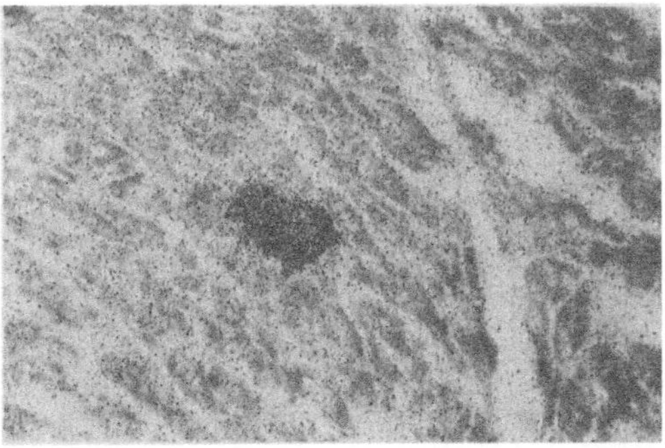

Figure 7. *In situ* ^{35}S-labeled antibody autoradiographic detection of UEAI binding to heart tissues taken from mice at 3 days p.i. Binding of UEAI was followed by addition of rabbit anti-UEAI antibody and ^{35}S-labeled donkey anti-rabbit antibody. H & E stained section at 400X.

TABLE 3. SOLUBLE PRO-INFLAMMATORY MODULATORS PRODUCED AND RELEASED BY CVB3-INFECTED FIBROBLASTS[a]

Mediator	Fold-Increase in CVB3-Infected Fibroblasts/ Normal Fibroblasts[a]
chemotactic factor(s) for macrophages	1.7 - 2.0
PGE_2	1.7
6-keto $PGF_{1\alpha}$	none

[a] Cultures of neonatal skin or adolescent heart fibroblasts were challenged with 100-200 pfu/cell of $CVB3_m$. After 1 hr for adsorption of virus, cultures were washed three times and incubated in MEM containing 1% bovine serum albumin and antibiotics.

[b] Statistically significant (Student's t test) increases in values from infected (24-72 hr p.i.) over uninfected normal cells ($p<0.05$). In the chemotactic activity assay, 12-24 macrophages per high powered field were generally found attracted to media from normal cells over the 24-72 hr p.i. period compared to 24 to 42 macrophages responding to chemotactic activity in media from infected cells cultured over this period. Baseline levels in uninfected cells (means for 6 separate cultures) for PGF_2 and 6-keto $PGF_{1\alpha}$ were 1.3 - 1.9 and 0.4 - 0.9 μg/ml between 24 and 72 hr p.i. for PGE_2 and 6-keto $PGF_{1\alpha}$, respectively.

and continued through 72 hr p.i. Virus infection did not stimulate production of 6-keto $PGF_{1\alpha}$, the stable form of prostacyclin. This prostaglandin is normally produced by endothelial cells and thus served as a negative control in these experiments.

The effects of a CVB3 infection on cardiac arachidonic acid metabolism were examined by isolating and quantifying specific eicosanoids produced by heart tissues taken from normal or CVB3-inoculated mice at 3 or 8 days p.i. Minced heart tissues were incubated with ^{14}C arachidonic acid for 30 min. Radiolabeled metabolites were extracted from culture fluids and separated by high performance liquid chromatography (HPLC) using selected elution program systems for prostaglandins or hydroxyeicosatetraenoic acids (HETEs) (Revtyak et al. 1988). Specific metabolites were identified by comigration of radioactive peaks with standard compounds. Prostaglandins were quantitated by RIA and HETEs by thin layer chromatography. The data from several experiments are summarized in Table 4. *In vivo*, production of PGE_2 was not increased in infected compared to normal heart tissues. However, heart tissues taken at 3 days p.i. consistently produced significantly more 6-keto $PGF_{1\alpha}$ than

TABLE 4. SOLUBLE ARACHIDONIC ACID METABOLITES PRODUCED AND RELEASED BY MINCED HEART TISSUES FROM CVB3-INFECTED MICE[a]

Metabolite	Fold-Increase in Product (Infected/Normal)[b] Normal	3 day p.i. mice	8 day p.i. mice
PGE_2	1.0	1.0	1.0
6-keto $PGF_{1\alpha}$	1.0	1.5[c]	1.0
12- and 15-HETE	1.0	1.0	1.7[c]

[a] Mice were inoculated with 10^5 pfu of CVB3. On days of sacrifice, cervical dislocation was performed and hearts quickly removed, minced and incubated with 1µCi ^{14}C arachidonic acid (1 mM) for 30 min. Products were identified by HPLC and quantitated by RIA or thin layer chromatography (12- and 15 HETEs).

[b] Baseline levels in normal hearts were (mean pmol/mg wet weight heart tissue ± S.E.M. on 6 separate assays): PGE_2 (12 ± 2), 6-keto $PGF_{1\alpha}$ (8 ± 1), 12- and 15 HETEs (38 ± 1).

[c] $p < 0.05$ (Student's t test).

either normal heart tissues or infected heart tissues taken at 8 days p.i. Significantly increased quantities of 12 and-15-HETEs were produced only by heart tissues taken at 8 days p.i. These eicosanoids are not stored by cells and are thought to be produced and act locally as mediators of inflammation and regional blood flow (Baenziger et al. 1979; Marcus, 1984). These small changes in soluble mediator levels induced by a CVB3 infection may have important consequences for the host: the induction of local inflammatory responses to infected foci of cells in heart tissues which results in myocyte destruction and subsequently in scar tissue formation.

CONCLUSIONS

Major events which follow inoculation of a mouse with CVB3 to result in induction of myocarditis are presented as a hypothesis in Figure 8. We propose that interaction of CVB3 with endothelial cells is the initial event. At this time, it is not known if virus replication occurs in this type of cell or if a nonproductive virus interaction is sufficient to alter production of soluble mediators and effect local homeostatic conditions. Attachment of CVB3 to nonpermissive bovine coronary artery endothelial cells results within 6 hr in the synthesis and release

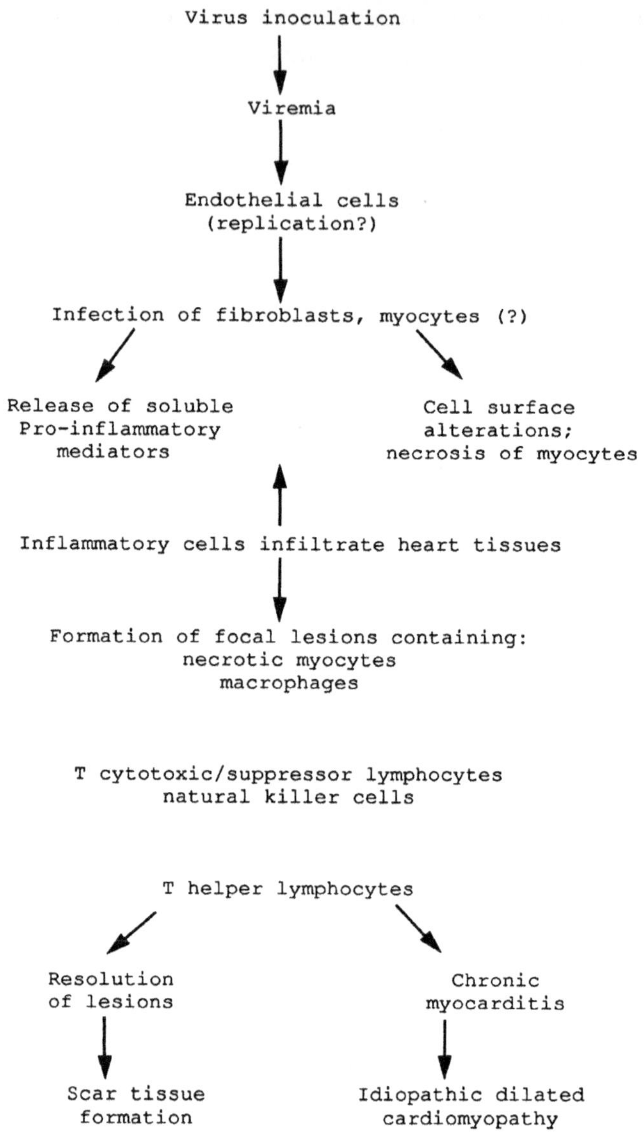

Figure 8. Hypothesis on mechanism of induction of acute murine myocarditis by CVB3 and potential sequelae.

of 6-keto $PGF_{1\alpha}$ at levels 12-to-20 higher than those released by unidentified cells and levels are maintained for a 24 hr period (Revtyak and Gauntt, unpublished data, 1988). If replication does not occur in cardiac endothelial cells, does synthesis of additional prostacyclin (6-keto $PGF_{1\alpha}$ is the stable form) sufficiently alter conditions locally to permit entrance of virus through tight junctions to subjacent fibroblasts? Obviously, virus gains access to heart tissues and replicates in an unidentified cell type(s). At present myocyte undergoing necrosis can be identified histologically at 2 and 3 days p.i. and prior to lesion formation. Fibroblasts cultured from adolescent heart tissues are permissive for CVB3 replication. Thus, these two cell types are candidates as hosts for replication of CVB3 *in vivo*.

We have identified only a few of the many soluble mediators likely to be produced and released by infected fibroblasts and which may contribute to inflammation. Fibroblasts are suspected of producing IFN because addition of anti-ß IFN antibody to infected fibroblast cultures increases virus yields 6- to 600-fold, yet IFN is not detected in supernatant fluids from these cultures (Lutton and Gauntt, 1985a). IFN alone can potentiate synthesis of PGE_2 in several human fibroblast cell lines (Fuse et al. 1982; Yaron et al. 1977) and $PGF_{2\alpha}$ in one fibroblast cell line (Fuse et al. 1982). IFNs ß and γ are produced in CVB3-infected mice and are thus present to activate synthesis and release of arachidonic acid metabolites. Prostaglandins can activate macrophages for cytotoxicity (Gordon, 1986; Gordon et al. 1986) and suppress leukocyte responses to viral antigens *in vitro* (Baker and Thomas, 1985).

Leukocytes are generally believed to be the major cells capable of synthesizing leukotrienes (Lewis and Austen, 1984; Samuelsson, 1983; Stjernschantz, 1984). Presently we are investigating synthesis of leukotrienes in the CVB3-murine system. Leukoproteins are chemotactic for macrophages (Gordon, 1986). Leukoprotein B4 is chemotactic for fibroblasts (Mensing and Czarnetzki, 1984) and is reported to be toxic for human myocytes in culture (Maisch B, Dienesch C, Wendl I (1986) International Symposium on Inflammatory Heart Disease, Wurzurg Germany, Abstract P15).

We have used two different approaches over the past decade to demonstrate that CVB3 infections generate potential pro-inflammatory changes on the cell surface. Initially, infected heart tissues at 8 days p.i. were shown to contain hypertonic salt-extractable proteins which could elicit production of a cell migration-inhibition-factor (MIF) by peritoneal T cells from CVB3-inoculated mice (Gauntt et al. 1979; Lutton et al. 1985b; Paque et al. 1978, 1979). This relatively insensitive biological assay did not detect production of MIF at earlier times nor did it permit quantitation of levels. The α-L-fucosyl surface alteration, detected initially on infected neonatal skin and adolescent heart fibroblasts, also was detected in infected heart tissues at the time maximum virus quantities are present and prior to detection of nascent lesions. Activated macrophages possess a receptor for mannosyl-fucosyl residues (Stahl and Gordon, 1982; Stahl et al. 1985). Macrophages can be activated in part by IFN and are among the earliest cell type identified in nascent lesions. Thus, we suggest that this surface alteration may be one signal to macrophages to initiate an inflammatory responses.

Activated NK cells initiate lysis of infected fibroblasts within 4 hr p.i. (Godeny and Gauntt, 1987a), a time well within the eclipse period for CVB3 in these cells. This early NK cell response to a virus-induced cell surface change or to a released soluble chemical signal could explain the initial appearance of NK cells in nascent lesions. However, the persistence of NK cells in lesions devoid of virus-replicating cells, i.e., lesions detected on days 6-10 p.i., cannot be explained at the moment.

What causes dispersion of the inflammatory cells? Additional chemical mediators produced by cells in the lesion are probably required. Such mediators might be general leukotoxic or chemokinetic agents or they could be specifically directed against one or more cell(s) forming the lesion as is apparently the case for macrophages departing from lesions between days 7 and 8 p.i. (Table 2). Chemical signals of this type would have obvious therapeutic potential. Their detection and the understanding of their mechanism(s) of action are part of our long term goal which is to understand how CVB3 induces myocarditis in the mouse and apply the knowledge we acquire to treatment of the disease in man.

ACKNOWLEDGEMENTS

We thank Mary Devadoss for excellent assistance in preparing this manuscript. This work was supported by Public Health Service Grant HL-21047 from the National Heart, Lung and Blood Institute, the ERACE Foundation, Los Angeles, CA and the American Heart Association, Texas Affiliate.

REFERENCES

Baenziger NL, Becherer PR, Majerus PW (1979) Characterization of prostacyclin synthesis in cultured human arterial smooth muscle cells, venous endothelial cells and skin fibroblasts. Cell 16:967-974.

Baker DA, Thomas J (1985) The effect of prostagalndin E_2 on the initial immune response to herpes simplex virus infection. Am J Obstet Gynecol 151:586-590.

Beisel KW, Wolfgram LJ, Herskowitz A, Rose NR (1986) Variation in the susceptibility of congenic inbred mice to coxsackievirus B3-induced myocarditis among different strains of mice. J Immunol 136:1846-1852.

Campbell WB, Gomez-Sanchez CE, Adams BV (1980) Role of prostaglandins in angiotension-induced steroidogenesis: Absence of an effect by prostaglandin E2. Hypertension 2:472-476.

Campbell WB, Holland OB, Adams BV, Gomez-Sanchez CE (1982) Urinary excretion of prostaglandin E_2, prostaglandin F_2, and thromboxane B_2 in

normotensive and hypertensive subjects on varying sodium intakes. Hypertension 4:735-741.
Estrin M, Huber SA (1987) Coxackievirus B3-induced myocarditis. Autoimmunity is $L3T4^+$ T helper cell and IL-2 independent in BALB/mice. Am J Pathol 137:335-341.
Estrin M, Smith C, Huber S (1986) Coxsackievirus B3-myocarditis: T-cell autoimmunity to heart antigens is resistant to cyclosporin A treatment. Am Pathol 125:244-251.
Evans D, Dunn G, Minor P, Schild G, Cann A, Standway G, Almong J, Currey K, Maizel J, Jr (1985) Increased neurovirulence associated with a single nucleotide change in a non-coding region of the Sabin type 3 poliovaccine genome. Nature (London) 314:548-550.
Fuse A, Mahmud I, Kuwata T (1982) Mechanism of stimulation by human interferon of prostaglandin synthesis in human cell lines. Cancer Res 42:3209-3214.
Gatmaitan BG, Chason JL, Lerner AM (1970) Augmentation of the virulence of murine myocarditis B3 myocardiopathy by exercise. J Exp Med 131:1121-1136.
Gauntt CJ (1987) Cellular and humoral immune responses in coxsackievirus myocarditis. In Kawai C, Abelmann WH, Matsumori A (ed.) Cardiomyopathy Update 1. Pathogenesis of Myocarditis and Cardiomyopathy. Recent Experimental and Clinical Studies. Tokyo: University of Tokyo Press, p 49-61.
Gauntt CJ, Trousdale MD, LaBadie DRL, Paque RE, Nealon T (1979) Properties of coxsackievirus B3 variants which are amyocarditic or myocarditic for mice. J Med Virol 3:207-220.
Gauntt CJ, Gomez PT, Duffey PS, Grant JA, Trent DW, Witherspoon SM, Paque RE (1984) Characterization and myocarditic capabilities of coxsackievirus B3 variants in selected murine strains. J Virol 52:598-605.
Gauntt CJ, Arizpe HM, Kung JT, Ogilvie KK, Cheriyan U (1985) Antimyocarditic activity of the guanine derivative BIOLF-70 in a coxsackievirus B3 murine model. Antimicrob Agents Chemother 27:181-191.
Gauntt CJ, Godeney EK, Lutton CW (1988) Host factors regulating viral clearance. *In*: Huber SA (ed.) Myocarditis: Etiology, Pathogenesis and treatment. Pathol Immunopathol Res (In Press)
Gear JHS amd Measroch V (1973) Coxsackievirus infections of the newborn. Prog Med Virol 15:42-62.
Godeny EK, Gauntt CJ (1986) Involvement of natural killer cells in coxsackievirus B3-induced murine myocarditis. J Immunol 137:1695-1702.
Godeny, EK, Gauntt CJ (1987a) Murine natural killer cells limit coxsackievirus B3 replication. J Immunol 139:913-918.
Godeny EK, Gauntt CJ (1987b) *In situ* immune autoradiographic identification of cells in heart tissues of mice with coxsackievirus B3-induced myocarditis. Am J Pathol 129:267-276
Gordon S (1986) Biology of the macrophage. J Cell Sci Suppl 4:267-286.
Gordon S, Crocker P, Morris L, Lee S-H, Perry VH, Hume D (1986) Localization and function of tissue macrophages. *In*: Biochemistry of Macrophages. Ciba Fdn Sym 118, Pitman: London, pp. 54-67.
Grist NR, Bell EJ (1974) A six year study of coxsackievirus B infections in heart disease. J Hyg (London) 73:165-172.
Grist NR, Bell EJ, Assaad F (1978) Enteroviruses in human disease. Prog Med Virol 24:114-157.
Herskowitz A, Beisel KW, Wolfgram LJ, Rose NR (1985) Coxsackievirus B3 murine myocarditis: wide pathologic spectrum in genetically defined inbred strains. Human Pathol 16:671-673.
Herskowitz A, Traystman MD, Beisel KW (1986) Murine viral myocarditis - new insights into mechanisms of disease. Heart Failure 2:86-91.
Huber SA, Joo LP (1983a) Differences in cytolytic T cell response of BALB/c mice infected with myocarditic and non-myocarditic strains of coxsackievirus group B, type 3. Infect Immun 39:1419-1427.
Huber SA, Job LP (1983b) Cellular immune mechanisms in coxsackievirus group B, type 3 induced myocarditis in Balb/c mice. *In*: Spitzer JJ (ed) Myocardial Injury, Plenum Publishing Co., New York, pp. 491-508.
Huber SA, Job LP, Woodruff JF (1980) Lysis of infected myofibers by coxsackievirus B3-immune T lymphocytes. Am J Pathol 98:681-694.

Huber SA, Job LP, Woodruff JF (1981a) Sex-related differences in the pattern of coxsackievirus B3-induced immune spleen cell cytotoxicity against virus infected myofibers. Infect Immun 32:68-73.
Huber SA, Job LP, Auld KR, Woodruff JF (1981b) Sex-related differences in the reproduction of cytotoxic spleen cells active against uninfected myofibers during coxsackievirus B3 infection. J Immunol 126:1336-1340.
Huber SA, Job LP, Auld KP (1982) Influence of sex hormones on coxsackie B3 virus infection in Balb/c mice. Cell Immunol 67:173-189.
Huber SA, Lodge PA (1984) Coxsackievirus B3 myocarditis in BALB/c mice. Evidence for autoimmunity to myocyte antigens. Am J Pathol 116:21-29.
Huber SA, Lodge PA (1986) Coxsackievirus B3 myocarditis. Identification of different pathogenic mechanisms in DBA/2 and BALB/c mice. Am J Pathol 122:284-291.
Huber SA, Lodge PA, Herzum M, Estrin M, Olszewski J (1987) The role of T lymphocytes in the pathogenesis of coxsackievirus B3 myocarditis. *In*: Kawai C, Abelmann WH, Matsumori A (ed.) Cardiomyopathy Update 1. Pathogenesis of Myocarditis and Cardiomyopathy. Recent Experimental and Clinical Studies. Tokyo, University of Tokyo Press, pp. 9-21.
Job LP, Lyden DC, Huber SA (1986) Demonstration of suppressor cells in coxsackievirus group B, type 3 infected female Balb/c which prevent myocarditis. Cell Immunol 98:104-113.
Kohara M, Omata T, Kameda A, Semler B, Itoh H, Wimmer E, Nomoto A (1985) *In vitro* phenotypic markers of a poliovirus recombinants constructed from infectious cDNA clones of the neurovirulent Mahony strain and the attenuated Sabin 1 strain. J Virol 53:786-792.
Lansdown ABG (1978) Viral infections and diseases of the heart. Prog Med Virol 24:70-113.
Lerner AM, Reyes MP (1985) Coxsackievirus myocarditis with special reference to acute and chronic effects. Prog Cardiovasc Dis 27:373-394.
Lerner AM, Wilson FM (1973) Virus myocardiopathy. Prog Med Virol 15:63-91.
Lewis RA, Austen KF (1984) The biologically active leukotrienes. Biosynthesis, metabolism, receptors, functions and pharmacology. J Clin Invest 73:889-897.
Lodge PA, Herzum M, Olszewski J, Huber SA (1987) Coxsackievirus B3 myocarditis. Acute and chronic forms of disease caused by different immunopathogenic mechanisms. Am J Pathol 128:455-463.
Lutton CW (1986) Studies of pathogenic mechanisms of coxsackievirus B3 in a murine model of virus-induced myocarditis. Ph.D. Dissertation, The University of Texas Health Science Center at San Antonio, Texas.
Lutton CW, Gauntt CJ (1985) The ameliorating effect of IFN-ß and anti-IFN-ß on coxsackievirus B3-induced myocarditis in mice. J Interferon Res 5:137-146.
Lutton CW, Gauntt CJ (1986) Coxsackievirus B3 infection alters plasmal membrane of neonatal skin fibroblasts. J Virol 60:294-296.
Lutton CW, Gudvangen RJ, Nealon TJ, Plaque RE, Gauntt CJ (1985) Cellular immune responses in mice challenged with an amyocarditic variant of coxackievirus B3. J Med Virol 17:345-357.
Lyden DC, Huber SA (1984) Aggravation of coxsackievirus, group B, type 3-induced myocarditis and increase in cellular immunity to myocarditis and increase in cellular immunity to myocyte antigens in pregnant Balb/c mice and animals treated with progesterone. Cell Immunol 87:462-472.
Lyden DC, Olszewski J, Feran M, Job LP, Huber SA (1987) Coxsackievirus B3-induced myocarditis. Effect of sex steroids on viremia and infectivity of cardiocytes. Am J Pathol 126:432-438.
Marcus AJ (1984) The eicosanoids in biology and medicine. J Lipid Res 25:1511-1516
McManus BM, Gauntt CJ, Cassling RS (1987) Immunopathologic basis of myocardial injury. Cardiovasc Clin 18:163-184.
Mensing H, Czarnetzki BM (1984) Leukotriene B4 induces *in vitro* fibroblast chemotaxis. J Invest Dermatol 82:9-12.
Neu N, Beisel KW, Traystman MD, Rose NR, Craig SW (1987a) Autoantibodies specific for the cardiac myosin isoform are found in mice susceptible to coxsackievirus B3-induced myocarditis. J Immunol 138:2488-2492.

Neu N, Soe NR, Beisel KW, Herkowitz A, Gurri-glass G, Craig SW (1987b) Cardiac myosin induces myocarditis in genetically predisposed mice. J Immunol 139:3630-3636.

Omata T, Kohara M, Kuge S, Komatsu T, Abe S, Sembler B, Kaneda A, Itoh H, Arita M, Wimmer E, Nomoto A (1986) Genetic analysis of the attenuation phenotype of poliovirus type 1. J Virol 58:348-358.

Paque RE, Gauntt CJ (1983) Cellular and humoral immune responses in the induction and expression of viral-induced myocarditis. *In:* Robinson JF, O'Connell J (ed.) Myocarditis: A Precursor to Cardiomyopathy, Collamore Press, D.C. Heath and Co: Lexington, MA, pp. 47-69.

Paque RE, Gauntt CJ, Nealon TJ, Trousdale MD (1978) Assessment of cell mediated hypersensitivity against coxsackievirus B3 viral-induced myocarditis utilizing hypertonic salt extracts of cardiac tissue. J. Immunol 120:1672-1678.

Paque RE, Straus DC, Nealon TJ, Gauntt CJ (1979) Fractionation and immunologic assessment of KCl-extracted cardiac antigens in coxsackievirus B3 viral-induced myocarditis. J Immunol 123:358-364.

Revtyak GE, Johnson AR, Campbell WB (1988) Cultured coronary artery endothelial cells synthesize HETEs and prostacyclin. Am J Physiol 245:C8-C19.

Reyes MP, Lerner AM, Ho KL (1981a) Diminution in the size of the thymus in mice during forced swimming. J Infect Dis 143:292.

Reyes MP, Ho KL, Smith F, Lerner AM (1981b) A mouse model of dilated-type cardiomyopathy due to coxsackievirus B3. J Infect Dis 144:232-236.

Reyes MP, Smith FE, Lerner AM (1984) An enterovirus-induced murine model of an acute dilated-type cardiomyopathy. Intervirology 22:146-155.

Roesing TG, Landau BJ, Crowell RL (1979) Limited persistence of viral antigen in coxsackievirus B3-induced heart disease in mice. Proc Soc Exp Biol Med 160:382-386.

Samuelsson B (1983) Leukotrienes: mediators of immediate hypersensitivity reactions and inflammation. Science 220:568-575.

Smith WG (1970) Coxsackie B myopericarditis in adults. Am Heart J 80:34-46.

Stahl PD, Gordon S (1982) Expression of mannosyl-fucosyl receptor for endocytosis on cultured primary macrophages and their hybrids. J Cell Biol 93:49-56

Stahl PD, Wileman TE, Shepherd VL (1985) The mannose receptor of macrophages: A current perspective. *In*: van Furth R (ed.) Mononuclear Phagocytes. Characteristics, Physiology and Function, Martinus Nijhoff: Dordrecht, The Netherlands, pp. 59-65.

Stjernschantz J (1984) The leukotrienes. Med Biol 62:215-230.

Tracy S, Chapman N, Liu H (1985a) Molecular cloning and partial characterization of the coxsackievirus B3 genome. Arch Virol 85:157-163.

Tracy S, Liu H-L, Chapman NM (1985b) Coxsackievirus B3: Primary structure of the ^{51}non-coding and capsid protein coding regions of the genome. Virus Res 3:263-270.

Tracy SM, Gauntt CJ (1987) Phenotypic and genotypic differences among naturally occurring coxsackievirus B3 variants. Eur Heart J (Suppl) 8:445-448.

Trousdale ME, Paque RE, Nealon T, Gauntt CJ (1979) Assessment of coxsackievirus B3 ts mutants for induction of myocarditis in a murine model. Infect Immun 23:486-495.

Valente AJ, Fowler SR, Sprague EA, Kelley JL, Suenram CA, Schwartz CJ (1984) Initial characterization of a peripheral blood mononuclear cell chemoattractant derived from cultured arterial smooth muscle cells. Am J Pathol 117:407-417.

Wolfgram LH, Beisel KW, Rose NR (1985) Heart-specific autoantibodies following murine coxsackievirus B3 myocarditis. J Exp Med 161:112-1121.

Wolfgram LJ, Beisel KW, Herskowitz A, Rose NR (1986) Variations in the susceptibility to coxsackievirus B3-induced myocarditis among different strains of mice. J Immunol 136:1846-1852.

Wong CY, Woodruff JJ, Woodruff JF (1977) Generation of cytotoxic T lymphocytes during coxsackievirus B3 infection. III. Role of sex. J Immunol 119:591-597.

Woodruff JF (1980) Viral myocarditis. A review. Am J Pathol 101:425-483.

Woodruff JF, Woodruff JJ (1974) Involvement of T lymphocytes in the pathogenesis of coxsackievirus B3 heart disease. J Immunol 113:1726-1734.
Yaron M, Yaron D, Gurari-Rotman M, Revel M, Linder D, Zori U (1977) Stimulation of prostaglandin E production in cultured human fibroblasts by poly (I) poly (C) and human interferon. Nature (London) 267:457-459.

DISCUSSION

Pullian L (Veterans Administration Hospital, San Francisco, CA):

I had done some work on human cells with coxsackie B, but I was very interested in your mouse model. Have you done double labels of myosin heavy chain staining or vimenten to be sure of what cell(s) is infected in the hearts of mice inoculated with CVB3 in your mouse model?

Gauntt C:

No, I've not done that.

Pullian L:

Or, more importantly, after an acute infection, have you determined whether a cell in the myocardium might be persistently infected?

Gauntt C:

No. I've not done that and I am not aware that anyone else has either. There are a number of people who are convinced that it is the myocyte and no other cell which is infected, but they've not taken that discriminatory step you've suggested. In terms of persistence, coxsackieviruses group B apparently do persist in the nude mouse and in some murine strains.

Chang R (University of California, Davis, CA):

Can you tell us just how frequently the coxsackieviruses invade the human myocardium? Is there anything you know about it?

Gauntt C:

There's absolutely no information on the frequency of coxsackievirus group B infections of the myocardium in humans who get these infections.

Jones J (National Jewish Center for Immunology and Respiratory Medicine, Denver, CO):

Charlie, does the virus persist in any other tissues besides the myocardium?

Gauntt C:

Virus persists in the pancreas for as long as in the myocardium. The virus is also found in the liver and in the spleen, but it gets cleared very rapidly from these two organs.

Jones J:

In one of your first slides, you suggested that stress, and I guess I don't know what you mean by stress, but is stress equated in that system with exercise?

Gauntt C:

Stress in this animal model is equated with exercise. Nothing is known about the effect of mental stress. Physical stress in humans is suspected of effecting myocarditis and the stress may include shoveling snow, marathon running, or heroic deeds during times of accidents. Those kinds of stress for sustained periods of hours are the kinds of sress that some people believe exacerbate myocarditis during the latent phase of the acute disease.

Schnurr D (State of California, Dept. of Health Services, Berkeley, CA):

I wanted to make one comment on the persistance of virus. In nude mice, we have isolated coxsackie B3 for up to three months from the myocardium after infection. We couldn't isolate virus from the spleen or the liver or pancreas after about three weeks. I will ask you: if the fibroblasts is an important target cell, would you expect to see inflammatory disease in a number of other organs, since fibroblasts are such universal cells?

Gauntt C:

Yes, I would. We've thought about that a number of years. Why don't we see the sort of destruction seen in the myocardium in other tissues? I think the answer is that the major cell of other organs can divide.

Indeed, fibroblasts are in other organs. They probably become infected, contribute to immunopathology, but, except for the heart and the brain, the major cells in the other organs can undergo division to replace necrotic cells. But, myocytes do not divide in the adolescent so when one is lost adjacent cells may hypertrophy or the necrotic cells are replace by scar tissue. However, if you had myositis you can replace muscle cells.

Schnurr D:

Have you ever examined mice sacrificed at seven days post infection to see if there was inflammatory disease in other organs?

Gauntt C:

I've not done that Dave, but a number of people have and there's definite inflammation of the liver.

Participant:

Do you have any idea on the mechanism for altering the surface of infected fibroblasts?

Gauntt C:

No, I don't. This virus, being a picornavirus contains only protein and RNA. There are no lipid and no glycoproteins, so that the cell surface change has to be of cellular origin. I can't tell you that the surface change is a glycoprotein, because it could very well be something like a glycosphyngolipid, which is on the surface of lymphocytes, or it could be a glycolipid. We don't know what's connected to the sugar that is binding the lectin, but whatever it is, it's definitely of cell origin since the virus doesn't code for glycoproteins and has no envelope.

ADENOVIRUS GENE PRODUCTS THAT AFFECT HOST IMMUNE MECHANISMS

PHYLLIS R. FLOMENBERG[1] AND MARSHALL S. HORWITZ[2]

[1]Department of Medicine, Veterans Administration Medical Center, 1st Avenue at E. 24th Street, New York, NY; [2]Departments of Microbiology-Immunology, Cell Biology and Pediatrics, Albert Einstein College of Medicine, 1300 Morris Park Avenue, Bronx, New York, USA

INTRODUCTION

The clinical and pathological manifestations of viral diseases are the net result of virus destruction or transformation of cells and the host immune responses to these infections. The role of specific cell-mediated and humoral immune responses, as well as nonspecific immune responses, in viral pathogenesis is not well understood (Sissons and Oldstone, 1985). Viruses present unique challenges to the immune system because all are intracellular pathogens; some directly infect cells of the immune system and some can establish persistent and/or latent infections. Viral infections may be unusually severe or persistent in immunocompromised hosts, especially those with T cell defects. In contrast, in the attempt to eradicate virus, the normal immune response in itself may cause pathology.

Adenoviruses (Ads) represent a useful DNA virus model for study of mechanisms of viral pathogenesis. Ads cause a variety of common respiratory syndromes as well as occasional gastrointestinal, urinary, and ophthalmologic disease (Horwitz, 1985a). In addition, Ads have the ability to establish persistent and latent infections. Ads have been more extensively characterized at the molecular level than any other human DNA viral pathogen. However, present animal models of Ad pathogenesis are suboptimal due to poor replication of human Ads in animals.

Ads code for proteins that directly influence the immune response. The genes for these viral proteins are mostly contained in an early transcription region, E3, which will be the focus of this review. Study of these Ad proteins has provided insights into the role of immune responses to viral infections. One E3 gene product is a glycoprotein, designated E3-19k, that inhibits cell surface expression of major histocompatibility complex (MHC) class I antigens (Burgert and Kvist, 1985). Another protein, coded by the E3 14.7K open reading frame (ORF), permits infected cells to resist lysis by tumor necrosis factor (TNF) (Gooding et al. 1988). Additional viral genes, such as EIA and EIB, may play a role in the interaction between Ads and the immune system, and these data will be summarized. Although Ads have transforming properties, the emphasis will be on the manifestations of acute infectious diseases and viral persistence, rather than on tumor formation by this family of viruses.

EPIDEMIOLOGY

Adenoviruses, which are widely disseminated in humans and animals, cause a variety of disease syndromes. There are 42 serotypes that infect humans (Horwitz, 1985a; 1985b, in press). They were initially grouped based on hemagglutination of rat and rhesus red blood cells. Currently, they are classified into 6 major groups (A-F) by molecular criteria such as relatedness of their nucleic acid sequences by DNA-DNA hybridization (Table 1; Wadell, 1984). These properties correspond to many of the bio-

TABLE 1. CLASSIFICATION SCHEMES FOR HUMAN ADENOVIRUSES (MAST-ADENOVIRUS H)

Sub-Group	Hemagglutination Groups		Serotypes	Tumors in Animals	Transformation in tissue Culture	Percent of G+C in DNA
A	IV	(little or no agglutination)	12,18,31	High	+	48-49
B	I	(complete agglutination of monkey erythrocytes)	3,7,11,14 16,21,34,35	Moderate	+	50-52
C	III	(partial agglutination of rat erythrocytes)	1,2,5,6	Low or none	+	57-59
D	II	(complete agglutination of rat erythrocytes)	8,9,10,13 15,17,19, 20,22,-30 32,33,36-39	Low or none	+	57-61
E	III		4	Low or none	+	57-59
F	III		40,41	Unknown		

logic manifestations of Ad infection such as the patterns of clinical disease in humans (Table 2) or the ability of some of these viruses to transform rodent cells in a model system for oncogenesis. For example, group C Ads (types 1,2,5,6) cause upper respiratory symptoms, tonsillitis and mild conjunctivitis, primarily in young children (Brandt et al. 1969). Some group B Ads (Ads 3,7) cause lower respiratory tract disease such as pneumonia which can be fatal or result in severe chronic pulmonary pathology (Similia et al. 1971). Other members of group B (Ads 11,21) have a predilection for the urinary tract where they can cause hemorrhagic cystitis (Mufson et al, 1973; Numazaki et al. 1973). Severe Ad conjunctivitis has been a property of some of the group D agents such as Ads 8,19,37 (Kemp et al. 1983). Although many Ads can be isolated from the intestine and some persist for long periods (Ads 12, 18, 31), only group F (Ads 40, 41) have been correlated with gastrointestinal manifestations of disease (deJong J et al. 1983).

Ads can cause serious morbidity and mortality in immunocompromised hosts, although not as common a problem as various herpesviruses. Ad infection should be considered in the differential diagnosis of interstitial pneumonia, hepatitis, gastroenteritis, hemorrhagic cystitis, and meningoencephalitis in the immunosuppressed host (Kelsey, 1978; Landry et

TABLE 2. ILLNESSES ASSOCIATED WITH ADENOVIRUS INFECTIONS

Disease	Individuals Most at Risk	Principal Serotypes
Acute febrile pharyngitis	Infants, young children	1,2,3,5,6,7
Pharyngoconjunctival fever	School-age children	3,7,14
Acute respiratory disease	Military recruits	3,4,7,14,21
Pneumonia	Infants, young children	1,2,3,7
Pneumonia	Military recruits	4,7
Epidemic keratoconjunctivitis	Any age group	8,11,19,37
Pertusis-like syndrome	Infants, young children	5
Acute hemorrhagic cystitis	Infants, young children	11,21
Gastroenteritis	Infants, young children	40,41
Acquired immunodeficiency syndrome (AIDS) and other immunosuppression	Homosexuals, intravenous drug abusers, renal and bone marrow transplant recipients.	34,35

al. 1987; Shields et al. 1985; Zahradnik et al. 1980). Most often, Ads of the common lower serotypes have been incriminated in these infections. Ad5, in particular, has been grown from the livers of hepatic transplant patients (Koneru et al. 1987). However, we have observed a relatively high incidence of the uncommon group B Ad34 and the closely related Ad35 in AIDS patients (deJong PJ et al. 1983; Horwitz et al. 1984). Ad35 isolates have also been obtained from other immunodeficient individuals such as bone marrow transplant patients, renal homograft recipients, and patients with severe combined immunodeficiency disease (Shields et al. 1985; Stadler et al. 1977). In our survey of patients with AIDS or AIDS-related complex, 12.6% (14/111) had Ad34 or Ad35 in their urine cultures. Although there have been rare cases of fatal Ad34 or Ad35 disseminated infection reported, most patients had asymptomatic viruria. We have been unable to find the reservoir for Ad34/35 in other patients such as those surveyed with a variety of febrile illnesses or in normals. Ad35-infected AIDS patients did not make Ad35-specific antibodies and pooled human gammaglobulin had little antibody to these agents (Flomenberg et al. 1987a). Therefore, it is unclear whether the frequent isolation of Ad35 in AIDS patients is due to reactivation of latent virus or represents primary infection. The unique association of Ad35 with immunocompromised patients was the initial impetus that led us to examine the broader issues of viral-host immune system interactions.

ADENOVIRUS STRUCTURE

Ads are nonenveloped, regular icosahedra which contain a double stranded (ds) DNA genome of 35 kilobases (kb) (Figure 1). The protein capsid is composed of 3 different types of viral polypeptides designated hexon, penton base and fiber; antigenic determinants on these proteins are important in the serologic classification of Ads (Wigand et al. 1982). The fiber protein, which projects from each of the 12 vertices and mediates attachment to cells, is probably an important determinant of tissue tropism. The core structure of the virion is composed of ds DNA with four additional proteins. One of the proteins (terminal protein) is covalently linked to each 5' end of the viral DNA and is important for initiating DNA replication (Lichy et al. 1982). There are additional virus coded gene products that are not assembled into the virion. These include other proteins involved in DNA replication and proteins which regulate the expression of both viral and cellular gene products (Horwitz, 1985b). Some of these will be described below.

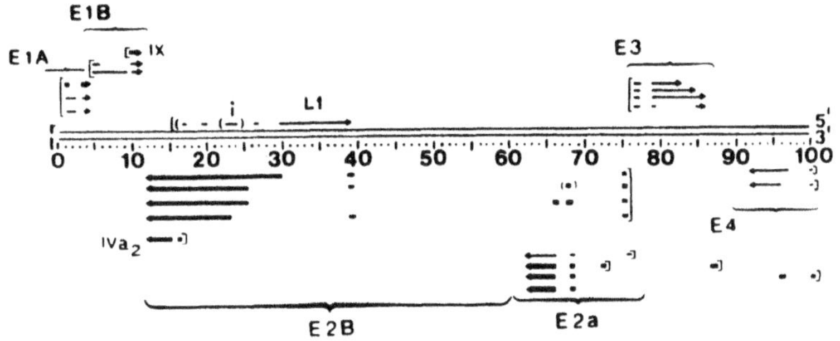

Figure 1. Genomic map of Ad2 early regions (E1-4) are designated and subdivisions within E1 and E2 shown. The thickness of the arrow corresponds with the abundance of that particular mRNA species. The early regions of interest in this manuscript are primarily E1A, E1B and E3 which are transcribed to the right using the top DNA strand as template. The L1, IVa2 and IX regions code for late proteins but synthesis of their mRNAs begins during the early phase of the infection. The E2a region has multiple promoters as shown. The L1 transcript which uses the major late promoter, contains the tripartite leader and, in some cases, a fourth segment designated "i". Modified from Persson and Philipson (1982).

ADENOVIRUS INFECTIOUS CYCLE

Recent data has extended our understanding of the early events of Ad-cell interaction (Pastan et al. 1987). Attachment to the cell is probably mediated by binding between fiber and a receptor on the cell surface. Purified fiber can inhibit virus attachment and attachment can be prevented by antibodies against fiber (Philipson et al. 1968). There is limited data about the specific viral receptor (Hennache and Boulanger, 1977; Svensson et al. 1981). The virus enters the cell via receptor-mediated endocytosis (Fitzgerald et al. 1983), a pathway utilized by many growth factors, hormones, and toxins. Soon after entry, the virus disrupts the endocytic vesicle and is released into the cytoplasm in a partially uncoated form (Chardonnet and Dales, 1970). In a process that probably involves microtubules (Dales and Chardonnet, 1973), this viral core is transported to the nucleus where the viral DNA can be replicated.

The Ad replicative cycle is divided into early and late phases, with the latter commencing with the onset of viral DNA replication (Figure 1). A variety of early (E) proteins are expressed in a regulated process from four complex transcription units, designated E1 (with components E1A and E1B), E2, E3, and E4 (Figure 1). EIA is a potent transactivating region that controls the expression of the E1B, E2, E3 and E4 regions (Lewis and Matthews, 1980) and affects host gene expression as well (Nevins, 1982).

The E2 region codes for three proteins needed for DNA replication; one of these is an Ad DNA polymerase (Lichy et al. 1983). The E3 region will be described in detail below. The E4 region codes for a number of polypeptides, including a 34K protein that interacts with the E1B-55K protein. This complex is required for completion of the Ad lytic cycle (Cutt et al. 1987).

After synthesis of the early Ad proteins, Ad DNA replication occurs, followed by the expression of an excess amount of late structural proteins such as hexon, penton base and fiber. During the late phase, which begins at 7-9 hours post infection, host DNA and protein synthesis are reduced by greater than 90% (Ginsberg et al. 1967; Horwitz, 1971). After synthesis of viral proteins and DNA, virions are assembled in the nucleus and released upon cell lysis. The infected cell develops specific morphologic changes that help the pathologist diagnose Ad infection on biopsy or autopsy tissue (Chang et al. 1958). Infected cells have enlarged nuclei containing amphophilic or basophilic inclusion bodies surrounded by a rim of cytoplasm. Some of these cells are referred to as "smudge" cells when the inclusion fills the entire nucleus and obliterates the nuclear membrane (Horwitz, 1985a). The inclusions represent sites of viral assembly which contain virions and unassembled viral protein and DNA. Ads do not cause syncytia or multinucleated giant cells as occur during herpes virus infection. As in tissue culture, the epithelial cells are the primary target for Ad cytopathology *in vivo*.

LATENCY AND MALIGNANCY RESULTING FROM ADENOVIRUS INFECTION

Ads were initially discovered during attempts to establish tissue culture lines from tonsils and adenoidal tissues (Rowe et al. 1953). Infection in the tonsils can be persistent for many years without causing any symptoms of disease but Ad presence can be detected at the time of tonsillectomy in many patients undergoing this procedure. In addition, Ads may be shed in the feces for years after acute infection (Fox et al. 1977). Many of the Ad infections observed in immunocompromised hosts are probably secondary to reactivation of latent infection in the host or donor organ. The mechanisms involved in the establishment of persistent or latent infections are unknown. However, recent reports on the presence of Ad DNA in peripheral blood lymphocytes (Horvath et al. 1986) and the ability of Ads to replicate in B and T cell lines (Lavery et al. 1987) suggest a potential reservoir.

The Ad system has been extensively studied as a model of transformation by DNA viruses. Although these human viruses are not incriminated in human malignancies, Ads of group A (Ad12, 18, 31) are highly oncogenic in newborn hamsters (Huebner et al. 1962). These Ad serotypes and others can be made to transform cells in tissue culture. Both E1A and E1B have been shown to be involved in the transforming process (Graham et al. 1975). E1A has some regions of homology with myc and myb oncogenes (Ralston and Bishop, 1983). The E1B gene can be replaced by the ras oncogene which together with E1A can efficiently transform cells (Ruley, 1983). A large amount of material on the function of E1A has been concisely summarized (Moran and Mathews, 1987). A novel finding recently published indicates that an E1A polypeptide binds and activates the protein product of the host retinoblastoma (RB) resistance gene. The binding of the E1A polypeptide to RB functionally mimics the deletion of the RB gene which is characteristic of children with malignant retinoblastoma (Whyte et al. 1988). The RB gene in its normal functional state presumably acts as an anti-oncogene; thus, its deletion leads to the derepression of cellular oncogenes.

IMMUNE RESPONSES TO VIRAL INFECTION

Humoral and cellular effector mechanisms

Both cellular and humoral immune defense mechanisms participate in the host response to viral infections. Neutralizing antibodies can block viral infection by various mechanisms which include aggregating particles, induction of conformational changes, and interfering with virus attachment or release of infectious nucleic acid within the cell (Mandel, 1982). For Ads, antibodies to fiber as well as some antibodies to hexon block viral infection by interfering with virus uptake. In addition, antibody to penton base appears to block release of virus from endocytic vesicles without affecting virus uptake (Seth et al. 1984). Antibodies may also bind to viral antigens on the surface of infected cells and make them susceptible to complement-mediated or cell-mediated lysis. Formation of antibody-antigen complexes may also initiate inflammatory reactions.

Cell-mediated immune responses play a crucial role in dealing with virus infections, as with other intracellular pathogens. This is supported by observations in individuals with primary T cell immune defects and experiments in animals in which adoptive transfer of immune lymphocytes can alter the outcome of lethal infection (Lin and Askonas, 1981). Natural killer (NK) cells, which nonspecifically lyse viral-infected cells, may function in limiting the spread of viruses. Virus-specific cytotoxic T lymphocytes (CTL) can lyse infected cells which display viral surface antigens. By recognition of viral antigens on accessory cells in association with class II histocompatibility antigens, helper T cells can be activated to differentiate and elaborate lymphokines. Some of these lymphokines can amplify the T, B and NK cell responses described above. Other lymphokines can enhance the phagocytic activity of macrophages or have direct antiviral effects, such as gamma interferon (IFN).

MHC-restricted immune cytolysis in vitro and in vivo

The CTL response is felt to represent a primary mechanism in host recovery from acute viral infection (Zinkernagel and Althage, 1977). Virus-specific CTL have been demonstrated for a wide variety of viruses both *in vitro* and *in vivo*. Cytotoxic T cells recognize viral antigen in association with the MHC class I antigen (H-2 in mouse, HLA in humans) on the cell surface (Doherty and Zinkernagel, 1975). The mechanism is called H-2 or HLA restricted immune cytolysis. Class I histocompatibility antigens are composed of a 45kb heavy chain, consisting of 1 conserved (α1) and 2 variable domains (α2 and α3), that is noncovalently bound to an invariant 12kb light chain, ß2-microglobulin (Figure 2). MHC class I antigens exhibit homology to conserved domains present in all class I antigens and immunoglobulins (the "immunoglobulin superfamily") and to the Ad transmembrane glycoprotein E3-19K that will be described

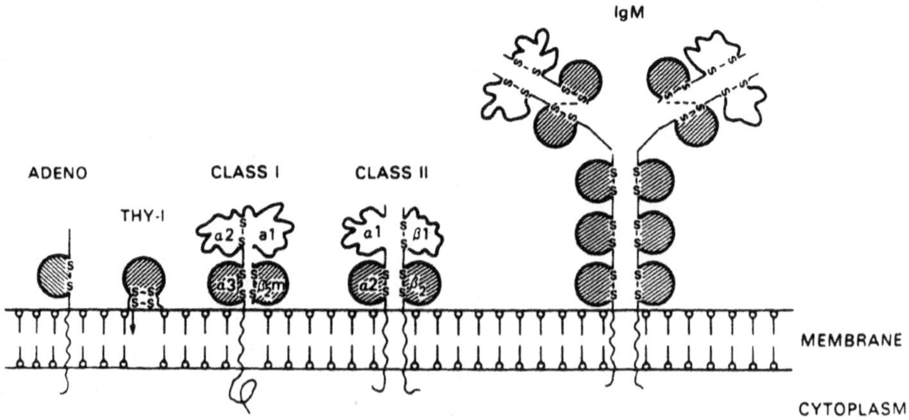

Figure 2. Homology between domains of MHC products, immunoglobulins, and the Ad E3-19K. Molecules of the immune system which exhibit homologous conserved domains (shaded areas) are shown schematically with their hydrophobic domains embedded in membrane. "Adeno" refers to the E3-19K glycoprotein; class I MHC is shown with the transmembrane portion of the heavy chain embedded (ß2 microglobulin has no transmembrane domain). For comparison, the class II MHC and IgM proteins are shown; Thy-1 is a glycoprotein found on the cell surface of rodent thymocytes. From Chatterjee and Maizel (1984).

below. There has been a great deal of experimentation on the nature of the viral target that binds to class I MHC, but no direct binding studies between MHC and the antigens residing in their CTL recognition site have been successful. Recently, however, x-ray crystallographic studies have revealed that only proteolytic products of antigens, representing 8-20 amino acids, are capable of fitting within the antigen-binding pocket that is formed by both the $\alpha1$ and $\alpha2$ domains of the heavy chain (Bjorkman et al. 1987a,b). This finding may explain the discovery of SV40 and influenza-specific CTL activity directed against internal viral proteins (Campbell et al. 1983; Townsend et al. 1984). For the Ad system, there is evidence that the CTL response to Ad-transformed rodent cells appears to recognize the nuclear proteins coded by either the E1A or E1B regions (Bellgrau et al. 1988; Fohring et al. 1983).

Effect of MHC on the immune response

It is likely that differences in MHC genotype can affect the ability of individuals to develop an immune response to a specific pathogen and affect disease outcome. In the mouse, the effect of different H-2K or D gene products in viral restriction has been demonstrated for several viruses. For example, Sendai virus, a natural pathogen of mice, is normally recognized in the C57BL/6 (B6) mouse by the cellular immune system in association with a K^b gene product. In one of the H-$2K^b$ mutant mouse strains in which the CTL response after Sendai infection was decreased, the virulence of the virus was enhanced and recovery from viral infection was adversely affected (Kast et al. 1986; Nairn et al. 1980). In humans, susceptibility to a wide variety of diseases has been observed to be associated with the MHC genotype (Brinton et al. 1984). The spondyloarthropathies show a very strong association with HLA-B27; a similar association has been shown for Reiter's syndrome which may be precipitated by enteric bacterial infections. Type I diabetes, which is associated with HLA-DR3 or DR4, appears to have an immunologic component which can even be ameliorated by immunosuppression with cyclosporin A (Bougneres et al. 1988). There is some evidence that the trigger is a viral infection such as coxsackie B or even rubella virus, based on the association of the congenital rubella syndrome with eventual diabetes in 25% of the surviving patients (Rubinstein et al, 1982; Yoon et al. 1979).

Virus-induced immunopathology

For many viral infections, the immune response may be an important cause of pathology in addition to direct cell lysis or inhibition of cell replication by viruses. There are examples in animals and humans in which prior exposure to an infectious agent in the presence of an intact immune system is detrimental at specific stages of the infectious process (Nash, 1984). Lymphocytic choriomeningitis virus (LCMV) of the newborn mouse is usually benign, although animals remain chronically viremic (Oldstone and Dixon, 1970). *In vitro* LCMV is relatively noncytopathic. However, transfer of immune serum or spleen cells from immunized animals to chronically infected mice resulted in worsening of disease and tissue injury in the latter. Moreover, LCMV can cause a fatal infection in adult mice, which appears to result from virus-specific CTL activity in brain tissue (Buchmeier et al. 1980). In humans, dengue virus can result in a severe hemorrhagic form of the infection upon reexposure to the virus (Halstead et al. 1970). The severity of dengue infection seems mediated by specific antiviral antibody which increases binding of virus to Fc receptors and enhances its replication in macrophages. In addition, immunization with either inactivated measles virus or respiratory syncytial virus vaccine (Martin et al. 1979; Nader et al. 1968) were both ineffective in preventing subsequent infections which were either more severe or presented with atypical clinical manifestations.

Modulation of Immune Response by Virus

Ads have developed a variety of mechanisms by which infected cells can evade the host immune system. Two unique mechanisms decrease MHC class I antigens on the cell surface, which may result in the inhibition of CTL responsiveness and alter the manifestations of acute disease. Downregulation of Ad-specific CTL activity may also play a role in establishing viral persistence and latency. In addition, Ads have specific mechanisms to resist the effects of interferons and the cytokine, TNF.

Other viruses have evolved different mechanisms to inhibit the immune response. For example, patients with acute measles or influenza infection exhibit evidence of mild immunosuppression, which is probably related to direct infection of immunocompetent cells with these viruses (Casali et al. 1984). More importantly, fatal immunodysfunction is caused by the direct infection and destruction of T4 cells by human immunodeficiency virus type 1 (HIV-1) (Ho et al. 1987). Some of the destructive effects of HIV-1 extend to uninfected cells (Germain, 1988; Siliciano et al. 1988). In addition, the HIV-1 envelope undergoes antigenic modulation which results in reduction of the HIV-1-specific immune response.

ADENOVIRUS PROTEINS THAT MODULATE THE IMMUNE RESPONSE

The early adenovirus transcription region 3 (E3)

The Ad E3 region has been found to play an important role in virus-host interaction during acute infection. We and others have provided evidence that the mechanisms to be described below are general for all Ad serotypes studied and involve a response mediated by recruitment of cells of the immune system. The E3 region was the focus of attention of virologists since the earliest attempts to grow sufficient virus to use for immunization of military recruits against Ad pneumonia (Hartley et al. 1956). When Ad was grown in monkey kidney cells to produce vaccine, it was realized subsequently that these lots of infected tissue culture fluid also contained simian virus 40 (SV40) (Rowe and Baum, 1965). When the Ad vaccine was passed in the presence of antibody to SV40, infectious SV40 was eliminated. However, with the discovery of the SV40 T antigen, it was realized that the Ad vaccine lots contained SV40 T antigen but no infectious SV40 virion (Lewis et al. 1966). The molecular event that occurred was the recombination of Ad7 and SV40 DNA, with the substitution of part of the Ad E3 region by a SV40 T antigen gene fragment. Subsequent constructs were able to demonstrate that the entire Ad E3 region could be deleted without any compromise of growth of Ad2 in tissue culture (Kelly and Lewis, 1973). However, because the E3 region had been conserved in many different Ads, the search for its potential *in vivo* function began.

An E3 glycoprotein can bind to the MHC and prevent its transport to the plasma membrane

The first function attributed to the Ad E3 gene family was discovered following the observation that a glycoprotein product of the 19kDa ORF (E3-19K) could be coimmunoprecipitated with the class I MHC protein in Ad2-infected human cells (Persson et al. 1979). The initial data suggested that E3-19K bound tightly to class I antigens on cell membranes and might serve as a target for recognition by cytotoxic T cells (Kvist et al. 1978). Subsequently, it was determined that a complex between the E3-19K and class I antigens formed and remained fixed in the endoplasmic reticulum (ER) (Anderson et al. 1985; Burgert and Kvist et al. 1985). The result was reduction of the amount of MHC class I antigen that was transported to the cell surface. Reduction of the number of class I molecules on the surface of Ad-infected cells is likely to inhibit recognition by CTL. Therefore, E3-19K may function to downregulate the CTL response and may affect patterns of acute disease (either enhancement or attenuation). The E3-19k may also facilitate the establishment of persistent infection or latency.

The E3 transcription region contains multiple open reading frames

Before further description of the E3-19k glycoprotein and its interaction with HLA, it should be noted that the E3 region between 76-86 map units (m.u.) in group C Ad2 and Ad5 contains many other ORFs that could code for proteins larger than 6kDa (Cladaras and Wold, 1985; Herisse et al. 1980; Herisse and Galibert, 1981) (Figure 3). An analysis of the complex transcription pattern of the Ad E3 region and an assessment of which of its ORFs are predicted to code for actual proteins has been recently published (Bhat and Wold, 1987). Some of the ORFs other than the

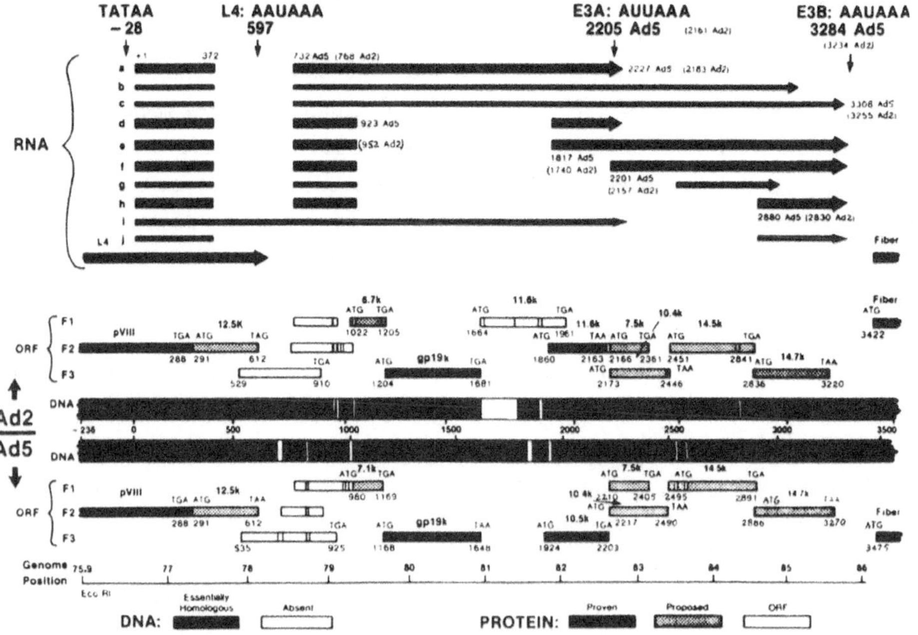

Figure 3. Schematic representation of the E3 transcription and translation units of Ad2 and Ad5. The genomic regions included in each of the mRNAs are shown by the bars at the top. The relative amounts of each mRNA are shown with the thickest bars representing the most prevalent mRNAs. mRNA control regions such as the TATAA box and 2 polyadenylation sites are shown. Overlap with the late region L4 on the left and the fiber on the right are included. The putative open reading frames (ORFs) are shown with various degrees of shading to indicate the probability that they code for actual proteins. Solid rectangles represent proven polypeptides; crosshatched are the proposed ones; whereas the unlikely coding ORFs are open rectangles. From Cladaras and Wold (1985).

one coding for the E3-19k glycoprotein will be discussed below. However, first, the many issues concerning the presence of the E3-19k protein amongst the various Ad serotypes, the physical basis for its interaction with the 45kDa heavy chain of the class I MHC, the structural features of the anchoring domain of the E3 19kDa protein and the role of the interaction between the E3-19K and class I antigens in pathogenesis will be summarized.

Characterization of the E3 MHC class I-binding glycoprotein

Common Structural Features of E3-19K Proteins from Subgroups B and C. Because the E3 region of the group C Ad2 was shown to be nonessential for replication in tissue culture, it was unclear initially whether the E3 region, including the E3-19K glycoprotein, would be preserved within other Ad serotypes. However, subsequent data confirmed that the analog of the Ad2 E3 MHC class I-binding glycoprotein was preserved in serotypes from group B, D, and E (Paabo et al. 1986). The group B Ad35 glycoprotein was extensively characterized at the protein and nucleic acid levels in our laboratory (Flomenberg et al. 1987b; 1988). It was shown to be a 166 amino acid polypeptide, containing typical signal sequence, transmembrane, and cytoplasmic tail domains and four N-linked glycosylation sites. The predominant carbohydrate moiety was a structure containing eight mannose residues; this indicates that transport of the protein from the ER to the Golgi, which would result in processing of the carbohydrate, did not occur. The Ad35 glycoprotein bound to HLA similar to its Ad2 counterpart containing 2 N-linked carbohydrates. However, the Ad35 polypeptide failed to react with polyclonal rabbit antibody which was generated to the purified Ad2 E3-19K (Figure 4). This observation is consistent with the limited sequence homology between the Ad2 and Ad35 glycoproteins (50% at the DNA level and 30% at the protein level). A comparison of the sequences of the class I antigen-binding glycoproteins from four serotypes- Ad2, Ad5, Ad35 and Ad3 - was performed in a search for a common binding site for HLA. We noted several structural similarities among the four serotypes such as five conserved cysteine positions and a 22 amino acid region in which 16 amino acids are identical and two others are similar (Figure 5). These structural features are currently being studied by mutagenesis for clues about the E3-19k binding site with HLA.

The interaction between E3-19K and class I histocompatibility antigens has been shown to occur in the ER and inhibits the transport of class I antigen to the cell surface. The transmembrane glycoprotein E3-

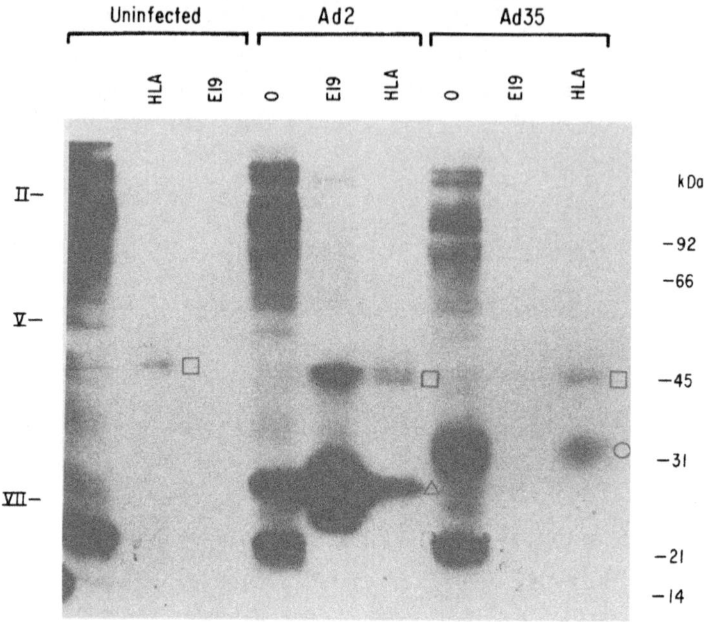

Figure 4. Co-immunoprecipitation of HLA and the E3 glycoproteins from Ads types 2 and 35. Immunoprecipitation of [^{3}H]mannose-labeled uninfected, Ad2, or Ad35 infected cell lysates with either a monoclonal antibody against HLA class I antigens or antibody against purified Ad2 E3-19K. Lanes marked "O" represent aliquots of the cell lysates before immunoprecipitation. A portion (10^6 cpm) of each [^{3}H]mannose-labeled lysate was solubilized in 2% Triton X-100 and incubated with antibody overnight, and immune complexes were precipitated with Staph A. Lanes marked HLA represent immunoprecipitates with the monoclonal antibody against HLA class I antigens, W6/27. Lanes marked E19 represent immunoprecipitates with the polyclonal antibody against the Ad2 E3-19K. The 45K class I heavy chain is indicated by squares. The Ad2 and Ad35 E3 region glycoproteins, which were coprecipitated with antibody to class I antigens, are indicated by the triangle and circle, respectively. Note that the Ad2 and Ad35 glycoproteins have mobilities of 25kDa and 29kDa, respectively, due to the presence of different amounts of oligosaccharide residues. From Flomenberg et al. (1987b).

19k is one of the few known examples of a viral-coded protein which remains in the ER. Its strong binding affinity to class I antigens may simply block further transport of class I antigens mechanically or perhaps cause conformational changes which inhibit specific transport mechanisms. Recent studies of Ad2 E3-19k mutants have indicated that a signal for retention in the ER is present in the cytoplasmic tail (Paabo et al. 1987). Since only the last three amino acids -Lys-Met-Pro- are conserved among the four glycoproteins sequenced, this domain is likely to be critical for ER targeting.

```
3  MGAILVVLALLSLLGLGSANLNPLDHDPCLDFDPENCTLTFAPDTSRLCGVLIKCGWDCRSVE
35 MGPILVLLVLLSLLEPGSAN-----YDPCLDFDPENCTLTFAPDTSRICGFLLSADGKCRSVE
2  M-RY-MILGLLALAAVCSA--------AKKVEFKEPACNVTFKSEANE-CTTLIKCTTEHEKLI
5  MIRY-MILGLLTLASAHGT--------TQKVDFKEPACNVTFAAEANE-CTTLIKCTTEHEKLL

3  ITHNNKTWNNTLSTTWEPGVPQWYTVSVRGPDGSIRISNNTFI|FSEMCDLAMFMSRQYDLWPP|
35 ITHNNKTWNNTLSTTWEPGVPEWYTVSVRGPDGSIRISNNTFI|FSEMCDLAMFMSKQYSLWPP|
2  IRHKDKIGKYAVYAIWQPGDTNDYNVTVFQGENR-KTFMYKFP|FYEMCDITMYMSKQYKLWPP|
5  IRHKNKIGKYAVYAIWQPGDTTEYNVTVFQGKSH-KTFMYTFP|FYEMCDITMYMSKQYKLWPP|

3  SKENIVAFSIAYCLVTCIITAIICVCIHLLIVIRPRQSNEEKE|KMP|
35 SKDNIVTFSIAYCLCACLLTALLCVCIHLLVTTRIKNAN-NKE|KMP|
2  QK--CLENTGTFCSTALLITALALVCTLLYLKYKSRRSFIDEK|KMP|
5  QN--CVENTGTFCCTAMLITVLALVCTLLYIKYKSRRSFIEEK|KMP|
```

Figure 5. Comparison of the sequence of E3-region glycoproteins from 4 Ad types: Conservation of cysteines and a 22 amino acid sequence. The sequence of the Ad3 and Ad35 (Group B) E3-19K proteins are compared to the same proteins from Ad2 and Ad5 (Group C). While there is considerable homology within subgroups, there is little homology between Ad2 and Ad35 except for a 22 amino acid span (denoted by the large rectangle) in which 16 amino acids are identical and 2 others are similar (I-L and F-Y within the boxed structure). The last 3 amino acids (KMP), which are within an 8 amino acid cytoplasmic domain, are also identical. The conserved cysteines are denoted by an asterisk. From Flomenberg et al. (1988).

MHC Class I Antigen Domains Which Bind to the Adenovirus E3-19K. The site on the MHC class I antigen to which the Ad glycoprotein binds has been determined to be the intraluminal α_1 and α_2 domains of the 45-kDa heavy chain (Figure 2) (Burgert and Kvist, 1987). The 12kDa β_2 microglobulin, which is also a component of the MHC class I antigen, is not part of the binding site to the viral glycoprotein nor is displaced during such binding (Signas et al. 1982). Exon shuffling experiments between mouse H-2 molecules such as K^k, which binds poorly to the Ad glycoprotein, and K^b, which binds well, have been used to localize the interaction between class I antigens and E3-19K. Differences in the affinity of binding of E3-19K to the HLA alleles A2 and B7 have also been reported (Severinsson et al. 1986). These data suggest that the ability of E3-19K to modulate class I antigens and its effect on disease outcome may vary based on the class I MHC genotype of the host.

Other E3 Genes Involved in Viral Pathogenesis. The E3 region appears to be dedicated to multiple interactions with the immune system, in addition to the E3-19k viral glycoprotein that modulates MHC class I antigen expression. A map of the various ORFs that potentially could code for proteins greater than 6kDa within the Ad2 E3 region has been constructed (Cladaras and Wold, 1985). In Figure 3, the complex splicing patterns of this region can be appreciated, in addition to the promoter sequences and the two polyadenylation sites that allow the expression of two families of polypeptides. The net result is nine ORFs that could code for proteins larger than 6kDa. Later in the infectious cycle, the x, y and z leaders are preferentially spliced to the ORF for fiber which is a late structural protein; this eliminates the E3 region and downregulates the expression of proteins from this early region. The Ad2 E3 19kDa, 11.6kDa, 14.7kDa, and recently the 10.4kDa ORFs have been demonstrated to code for actual proteins (Carlin et al. 1989; Wold et al. 1984). The 6.7kDa, 7.5kDa, and 14.5kDa ORFs are predicted to code for polypeptides by virtue of their codon usage. The 12.4kDa ORF is probably transcribed from the late leader promoter, unrelated to the E3 region, and another 11.6kDa ORF is predicted to be noncoding.

For group B Ads, three serotypes have been sequenced sufficiently in the E3 region to reach some consensus about the structure of this subgroup of the Ads. Again 9 ORFs have been recognized but there are three differences from the group C E3 region. In the group B Ad3 E3, a 16kDa ORF replaces the shorter 6.7kDa of Ad2 (Signas et al. 1986). Two unique ORFs of approximately 20kDa in the former replace the noncoding 11.6kDa ORF in the latter. In Ad35, the 20.3 and 20.6kDa ORFs are both predicted to be glycoproteins with four and five potential sites for glycosylation. Each has an amino terminal hydrophobic domain typical for a signal sequence, and the carboxy end contains a potential transmembrane domain and a hydrophilic cytoplasmic tail. It is not known definitively whether these group B 20kDa ORFs are expressed. However, Ad3 mRNAs from this region have been detected and both 20kDa ORFs are predicted to be coding. Sequence data on the Ad35 E3 from our laboratory demonstrated that there was no coding sequence for the polypeptide analogous to the coding 11.6kDa ORF of Ad2. An analogous 9kDa ORF is present in Ad3, and a related 7.7kDa ORF has been identified in the group B Ad7 (Hong et al. 1988). The significance of the unique absence of this region from the Ad35 E3 unit is unclear. Whether the loss of this ORF exerts an effect on the immune recognition of Ad35 infected cells or can explain the persistence of this virus in the urinary tracts of immunosuppressed hosts is unknown to us at this time.

Recently, the 14.7kDa protein at the 3' end of the E3 region has been shown to protect Ad infected cells from cytolysis by TNF (Gooding et al. 1988). TNF is a cytokine secreted by macrophages in response to inflammatory stimuli and has been associated with antitumor and antimetabolic activities (Beutler and Cerami, 1987; Old, 1985). Previously, TNF had been reported to inhibit the replication of some DNA and RNA viruses (Creasy et al. 1987; Wong and Goeddel, 1986). TNF has also been implicated in the inhibition of several other intracellular parasites such as chlamydia and rickettsia (Shemer-Avni et al. 1988). Most uninfected control cells are resistant to TNF lysis as are wild-type Ad-infected cells. However, cells infected with Ad mutants in which the E3 14.7kDa gene is deleted become susceptible to TNF. The susceptibility of Ad-infected cells to TNF lysis appears to be a property of the E1A gene because cells transfected with this isolated viral region become susceptible to TNF (Cook et al. 1988). The susceptibility of cells transformed by the Ad E1A to lysis by activated macrophages or natural killer (NK) cells may also be a property of TNF, which has been shown to be involved in the cytolysis by both macrophages and NK cells. Thus, both the E3 14.7kDa protein and E3-19kDa downregulate the immune response early in the Ad infectious cycle. These processes may serve to enhance Ad virulence by protecting infected cells against premature lysis before infectious viral progeny can be produced. Conversely, their function may be to attenuate acute disease by avoiding the development of immunopathologic phenomena.

Other adenovirus genes which affect immune responses

Adenovirus 12 E1A Suppresses MHC Transcripts in Transformed Cells. Associated with the Ad E1 region is a second mechanism that downregulates the cell surface expression of the class I MHC products. This mechanism is different than the ER trapping of class I antigens described above and is limited to the oncogenic viral serotypes. The product of the early gene Ad 12 E1A, which is a potent *trans*acting Ad gene and is important in cell transformation, can decrease the amount of class I MHC mRNA in Ad12 transformed cells (Schrier et al. 1983). This is due to either the specific inhibition of class I MHC transcription (Vasavada et al. 1986) or the post-transcriptional processing of the class I MHC mRNA (Vaessen et al. 1987). The lack of the class I MHC antigens on the cell surface is postulated to facilitate the survival of malignant Ad12 transformed cells in animals by decreasing MHC restricted CTL recognition and subsequent destruction of the abnormal cells. There is evidence that subsequent replacement of the MHC gene by transfection decreases the malignancy of Ad transformed cells *in vivo* (Tanaka et al. 1985). Thus, virus-host interactions of great significance to the persistence of viral transformed cells are revealed by these findings.

Adenovirus VAI RNA. The Ad VAI RNA is a small viral-coded transcript which may modulate the immune response by the direct inhibition of the action of interferons (IFNs) (Kitajewski et al. 1986). The VAI RNA is synthesized in large amounts late after infection from a gene located at about 30 map units on the Ad2 chromosome. Ads are generally unaffected by IFN, in contrast to many other viruses. One of the primary effects of IFN is induction of a ds RNA-dependent protein kinase that phosphorylates the protein initiation factor eIF2a and can inhibit viral protein synthesis. The VAI RNA specifically inhibits phosphorylation of eIF2a, allows viral protein synthesis to continue, and thus, permits the Ad to evade the inhibitory effects of IFN. In addition, there is recent evidence for modulation of IFN antiviral activity by the E1A region (Anderson and Fennie, 1987).

RODENT MODELS OF DISEASE INDUCED BY HUMAN ADENOVIRUS SEROTYPES

The functional roles of the downregulation of MHC class I antigen expression on the cell surface and the resistance to TNF and IFN in pathogenesis of Ad infection have not yet been determined. However, there are a number of observations in several rodent species that suggest that the E3 region proteins may alter the manifestations of Ad disease. This issue is of considerable significance, not only during Ad infection, but also because several groups of investigators are attempting to insert foreign genes into the Ad E3 region in order to use the virus as a vector for the expression of potentially immunizing polypeptides (see above). The use of Ads for such purposes is attractive, because types 4 and 7 have been used extensively as oral immunizing agents in young adults (Top et al. 1971). The experience with the safety, immunogenicity and route of administration of these two Ad serotypes would be an advantage in the choice of these viruses as immunization vectors. Although Ads with deletions which included the E3 region have been inadvertently administered to humans as described above, in the previous vaccine formulation, they were always given together with intact virus in a combination that required two hit kinetics for viable progeny. Thus, each infected cell had the Ad E3-19kDa glycoprotein. However, in the current vaccine model, the Ad deletion has been engineered to yield a replication competent virus lacking only the E3 region (Berkner and Sharp, 1982).

The approach to determine the role of the Ad E3 proteins in pathogenesis has employed a heterologous combination consisting of human Ad in a rodent host such as the cotton rat or hamster. The human Ads have been used because there are many E3 region mutants already available for human serotypes 2 and 5. Although the use of human Ads is attractive and they produce protein products from the early gene clusters in rodent cells, they generally replicate late genes poorly and produce little progeny virus in rodents. *In vivo* it is possible for Ad5 to be replicated in the cotton rat and hamster, but large amounts of input virus must be used; animals infected with wild type (wt) Ad usually show little or no histopathology or clinical disease (Morin et al. 1987; Pacini et al. 1984). Within these limitations, Ginsberg et al. (1987) have reported some intriguing results for Ad5 and its mutants when introduced into the lungs by intranasal inoculation of cotton rats. Wild type Ad5 produces low grade pulmonary pathology and no mortality in cotton rats but animals inoculated with E3 deletions of Ad5 show evidence of a more severe pneumonia with some accompanying mortality. It was suggested that the enhancement of pulmonary disease due to Ad5, in the absence of the E3 region, was the result of immunopathology. Whether these observations can be extrapolated to lytic Ad infections in humans is unclear.

As part of a strategy to use human Ads as a vector for delivering foreign polypeptides, E3 deletions of Ad5 have been inoculated into hamsters (Morin et al. 1987). In some experiments, these Ad5 deletions had inserts of hepatitis B surface antigen (HBsAg). The Ad4 and Ad7 strains that had been used in humans, as described above, do not infect hamsters. In the Ad5 model, antibody to HBsAg was produced by inoculation of greater than 10^7 PFU of virus. However, of more relevance to the role of E3 in pathogenesis, the Ad5 HBsAg recombinants representing E3 deletions did not enhance the pathology of the parental Ad5. This observation is at variance with data from the cotton rat model. In an attempt to answer whether alterations in the E3 region will affect the virulence or promote persistence of the Ads, a better model for Ad pathology must be developed. We are directing our efforts toward defining such a model.

ADENOVIRUS TYPE FL INFECTION OF MICE

Many mammals and avian species are infectable with Ads that are limited in their host range. For example, canine hepatitis is a well-known veterinary disease caused by canine Ad. Chickens are susceptible to the extensively studied chicken Ad designated chicken embryo lethal orphan (CELO) virus. Animal Ads of veterinary importance have been reviewed by Ishibashi and Yasue (1984). However, the animal with the best studied MHC genetics is the mouse and is the subject of our current studies. The AdFL or mouse adenovirus type 1 (MAV-1) is a natural infection of mice, that causes fatal infection in newborn and suckling mice (Hartley and Rowe, 1960). AdFL can cause asymptomatic persistent viruria in adult mice (Van der Veen and Mes, 1973). AdFL infections in different mouse strains are being analyzed, and the effect of MHC class I antigen variations on *in vivo* pathogenesis are currently being evaluated in our laboratory. In addition, the effects of various E3 mutants on *in vivo* pathogenesis will be examined. Because there have been few molecular studies of AdFL (Larsen et al. 1979), we are in the process of analyzing its E3 region with the aim of constructing the necessary mutants for the study of the E3-19kDa interaction with H2 and the function of other E3 ORFs.

SUMMARY

The Ad E3 region appears to play a role in *in vivo* pathogenesis by modulation of the host's immune response by at least two different mechanisms. The well-characterized E3-19kDa inhibits cell surface expression of MHC class I antigens, which presumably downregulates the CTL response. In addition, the E3 14.7kDa protein confers resistance of Ad-infected cells to TNF. It is likely that other E3 proteins also play a role in virus-host interaction. Other Ad mechanisms have been described for downregulation of MHC class I antigens and inhibition of the action of IFN. *In vivo*, the specific roles of these mechanisms in Ad pathogenesis have not been defined. However, in one animal model, deletion of the E3 region resulted in enhanced disease which may have been the result of immunopathology. Therefore, the present data suggest that modulation of the host immune response is beneficial to the survival of Ad.

It is clear that we are far from understanding the fine tuning of the immune response to viral infection or the evolution of viruses which has enabled them to subvert host defense functions. However, the E3 region of Ad seems to have clustered a number of these important functions within a single transcription unit. Elucidation of both the viral functions and the host responses to these activities should help us understand these complex interactions. Similar to the clustering of Ad functions essential for DNA replication in the E2A and E2B region and the grouping of transforming functions in the E1A and E1B region, the Ad E3 region has focused our attention on those host mechanisms that are important in both the manifestations of disease as well as recovery from viral infection.

REFERENCES

Anderson KP, Fennie EH (1987) Adenovirus early region 1A modulation of interferon antiviral activity. J Virol 61:787-795.

Anderson M, Paabo S, Nilsson T, Peterson PA (1985) Impaired intracellular transport of class I MHC antigens as a possible means for adenovirus to evade immune surveillance. Cell 43:215-222.

Bellgrau D, Walker TA, Cook JL (1988) Recognition of adenovirus E1A gene products on immortalized cell surfaces by cytotoxic T lymphocytes. J Virol 62:1513-1519.

Berkner KL, Sharp PA (1982) Preparation of adenovirus recombinants using plasmids of viral DNA. In: Gluzman Y (ed) Eukaryotic Viral Vectors. Cold Spring Harbor, New York. pp 193.

Beutler B, Cerami A (1987) Cachectin: more than a tumor necrosis factor. N Engl J Med 316:379-385.

Bhat BM and Wold WSM (1987) A small deletion distant from a splice or polyadenylation site dramatically alters pre-mRNA processing in region E3 of adenovirus. J Virol 61:3938-3945.

Bjorkman PJ, Saper MA, Sanraoui B, Bennett WS, Strominger JL, Wiley DC (1987a) Structure of the human class I histocompatibility antigen, HLA-A2. Nature 329:506-512.

Bjorkman PJ, Saper MA, Sanraoui B, Bennett WS, Strominger JL, Wiley DC (1987b) The foreign antigen binding site and T cell recognition regions of class I histocompatibility antigens. Nature 329:512-518.

Bougneres TF, Carel JC, Castano L, Coitard C, Garden SP, Landais P, Hors J, Mihatsch NS, Paillard M, Chauusain JL, Bach JF (1988) Factors associated with early remission of Type I diabetes in children treated with cyclosporine. N Eng J Med 318:663-670.

Brandt CD, Kim HW, Vargosdo AJ, Jeffries BC, Arrobio JO, Rindge B, Parrott RH, Chanock RM (1969) Infections in 18,000 infants and children in a controlled study of respiratory tract disease. I. Adenovirus pathogenicity in relation to serologic type and illness syndrome. Am J Epidemiol 90:484-500.

Brinton NA, Blank KJ, Nathanson N (1984) Host genes that influence susceptibility to viral diseases. In: Notkins AL, Oldstone MBA (eds) Concepts of Viral Pathogenesis. Springer Verlag, New York. pp 71-85.

Buchmeier MJ, Welsh RM, Dutko FS, Oldstone MBA (1980) The virology and immunobiology of lymphocytic choriomeningitis virus infection. Adv Immunol 30:275-332.

Burgert H, Kvist S (1985) An adenovirus type 2 glycoprotein blocks cell surface expression of human histocompatibility class I antigens. Cell 41:987-997.

Burgert H, Kvist S (1987) The E3/19K protein of adenovirus type 2 binds to the domains of histocompatibility antigens required for CTL recognition. EMBO 6:2019-2026.

Campbell AE, Foley FL, Tevethia SS (1983) Demonstration of multiple antigenic sites of the SV40 transplantation rejection antigen by using cytotoxic T lymphocytic clones. J Immunol 130:490-496.

Carlin CR, Tollefson AE, Brady HA, Hoffman BL, Wold WSM (1989) Epidermal growth factor receptor is down-regulated by a 10,4000 MW protein encoded by the E3 region of adenovirus. Cell 57:135-144.

Casali P, Rice GPA and Oldstone MBA (1984) Viruses disrupt functions of human lymphocytes. J Exp Med 159:1322-1337.

Chang C, Lepine P, Lelong M, Le-Tan-Vinn, Statge P, Virat J (1958) Severe and fatal pneumonia in infants and young children associated with adenovirus infection. Am J Hygiene 67:367-378.
Chardonnet Y, Dales S (1970) Early events in the interaction of adenovirus with Hela cells. I. Penetration of type 5 and intracellular release of the DNA genome. Virology 40:462-477.
Chatterjee D, Maizel JV Jr. (1984) Homology of adenoviral E3 glycoprotein with HLA-DR heavy chain. Proc Natl Acad Sci USA 81:6039-6043.
Cladaras C and Wold WSM (1985) DNA sequence of the early E3 transcription unit of adenovirus 5. Virology 140:28-43.
Cook J, May B, Chen MJ, Shalloway D, Walker T (1988) E1A gene expression in transformed cells induces cytolytic susceptibility by TNF-dependent and independent mechanisms. Abstract presented at the Tumor Virus Meeting on SV40, polyoma, and adenoviruses. Cold Spring Harbor Laboratory, Cold Spring Harbor, New York. p 246.
Creasy AA, Doyle LV, Reynonds MT, Jung T, Lin LS, Vitt CR (1987) Biological effects of recombinant tumor necrosis factor and its novel muteins on tumor and normal cell lines. Can Res 47:145-149.
Cutt JR, Shenk T, Hearing P (1987) Analysis of adenovirus early region 4 encoded polypeptides synthesized in productively infected cells. J Virol 61:543-552.
Dales S, Chardonnet Y (1973) Early events in the interaction of adenovirus with HeLa cells. IV. Association with microtubules and the nuclear core complex during vectorial movement in the inoculum. Virology 56:465-483.
deJong J, Wigand R, Kidd AH, Wadell G, Kapsenberg JG, Muzerie CJ, Wermenbol AG, Fritzlaff RG (1983) Candidate adenoviruses 40 and 41: Fastidious adenoviruses from human infant stool. J Med Virol II:215-231.
deJong PJ, Valderrama G, Spigland I, Horwitz MS (1983) Adenovirus isolates from the urines of patients with the acquired immunodeficiency syndrome. Lancet 1:1293-1295.
Doherty PC, Zinkernagel RM (1975) H-2 compatibility is required for T-cell mediated lysis of target cell infected with lymphocytic choriomengitis virus. J Exp Med 141:502-507.
Fitzgerald DJP, Padmanabhan R, Pastan I, Willingham MC (1983) Adenovirus-induced release of epidermal growth factor and Pseudomonas toxin into the cytosol of KB cells during receptor-mediated endocytosis. Cell 32:607-617.
Flomenberg PR, Chen M, Munk G, Horwitz MS (1987a) Molecular epidemiology of adenovirus type 35 infections in immunocompromised hosts. J Infect Dis 155:1127-1134.
Flomenberg PR, Chen M, Horwitz MS (1987b) Characterization of a major histocompatibility complex class I antigen-binding glycoprotein from adenovirus type 35, a type associated with immunocompromised hosts. J Virol 61:3665-3671.
Flomenberg PR, Chen M, Horwitz MS (1988) Sequence and genetic organization of adenovirus type 35 early region 3. J Virol 62:4431-4437.
Fohring B, Gallimore PH, Mellow GH, Raska K Jr (1983) Adenovirus type 12 specific cell surface antigen in transformed cells is a product of the E1b early region. Virology 13:463-472.
Fox JP, Hall CE, Cooney MK (1977) The Seattle Virus Watch. VII. Observations of adenovirus infections. Am J Epidemiol 105:362-386.
Germain RN (1988) Antigen processing and CD4+ T cell depletion in AIDS. Cell 54:441-444.
Ginsberg HS, Bello LJ, Levine AJ (1967) Control of biosynthesis of host macromolecules in cells infected with adenovirus. In: Colter JS, Paranchych W (eds.) The Molecular Biology of Viruses. Academic Press. New York. pp 547-572.
Ginsberg HS, Valdesuso J, Horswood R, Chanock RM and Prince G (1987) Adenovirus gene products affecting pathogenesis. In: Chanock R, Lerner RA, Brown F, Ginsberg HS (eds.) Vaccines 87. Cold Spring Harbor Laboratory. Cold Spring Harbor. New York. pp 322-326.
Gooding LR, Elmore LW, Tollefson AE, Brady HA, Wold WSM (1988) A 14,700 MW protein from the E3 region of adenovirus inhibit cytolysis by tumor necrosis factor. Cell 53:341-346.

Graham FL, Abrahams PS, Moulder C, Heijneker HL, Warnaar SO, deVries FAJ, Fiers W, van der Eb AJ (1975) Studies on *in vitro* transformation by DNA and DNA fragments of human adenoviruses and simian virus 40. Cold Spring Harbor Symp Quant Biol 39:673.

Halstead SB, O'Rourke EF (1977) Dengue virus and mononuclear phagocytes: I. Infection enhancement by non-neutralizing antibodies. J Exp Med 146:201.

Hartley JW, Heubner RJ, Rowe WP (1956) Serial propagation of adenovirus (APC) in monkey kidney tissue cultures. Proc Soc Exp Biol Med 92:667.

Hartley JW and Rowe WP (1960) A new mouse virus apparently related to the adenovirus group. Virology 11:645.

Hennache B, Boulanger P (1977) Biochemical study of KB cell receptor for adenovirus. Biochem J 166:237-247.

Herisse J, Courtois G, Galibert F (1980) Nucleotide sequence of the EcoRI D fragment of adenovirus 2 genome. Nucl Acids Res 8:2173-2192.

Herisse J, Galibert F (1981) Nucleotide sequence of EcoRI E fragment of adenovirus 2 genome. Nucl Acids Res 9:1229-1240.

Ho DD, Promeranz RJ, Kaplan JC (1987) Pathogenesis of infection with human immunodeficiency virus. N Engl J Med 317:278-286.

Hong JS, Mullis KG, Engler JA (1988) Characterization of the early region 3 and fiber genes of Ad7 (in press).

Horvath JL, Palkonyay L, Weber J (1986) Group C adenovirus DNA sequences in human lymphoid cells. J Virol 59:189-192.

Horwitz MS (1971) Intermediates in replication of type 2 adenovirus. Virology 8:675-683.

Horwitz MS (1985a) Adenoviral diseases. In: Fields BN (ed.) Virology. Raven Press. New York. pp 477-495.

Horwitz MS (1985b) Adenoviruses and their replication. In: Fields BN (ed.) Virology. Raven Press. pp 433-476.

Horwitz MS (1989) Adenoviruses and their replication. In: Fields BN (ed.) Virology (in press, volume 2).

Horwitz MS, Valderrama G, Korn R, Spigland I (1984) Adenovirus isolated from the urines of AIDS patients: characterization of group B recombinants. In: Gottlieb MS and Groopman JE (eds.) Acquired Immune Deficiency Syndrome. Alan R Liss Inc. New York. pp 187-207.

Huebner RJ, Rowe WP, Lane WT (1962) Oncogenic effects in hamsters of human adenovirus types 12 and 18. Proc Natl Acad Sci USA 48:2051.

Ishibashi M, Yasue H (1984) Adenoviruses of animals. In: Ginsberg HS (ed.) The Adenoviruses. Plenum Press. New York. pp 497-562.

Kast WM, Bronkhorst, deWaal LP, Melief CJM (1986) Cooperation between cytotoxic and helper T lymphocytes in protection against lethal Sendai virus infection. J Exp Med 164:723-738.

Kelly TJ, Lewis AM Jr. (1973) Use of non-defective adenovirus simian virus 40 hybrids for mapping the simian virus 40 genome. J Virol 12:643-652.

Kelsey DS (1978) Adenovirus meningoencephalitis. Pediatrics 61:291-293.

Kemp MC, Hierholzer JC, Cabradilla CP, Obijesti JF (1983) The changing etiology of epidemic keratoconjunctivitis: Antigenic and restriction enzyme analysis of adenovirus types 19 and 37 isolated over a 10 year period. J Infect Dis 148:29-33.

Kitajewski J, Schneider RJ, Safter B, Munemitsie SM, Samuel CE, Thimmappaya B and Shenk T (1985) Adenovirus VA1 RNA antagonizes the antiviral action of interferon by preventing activation of the interferon-induced elF-2α kinase. 45:195-200.

Koneru B, Jaffe R, Esquivel CO, Junz R, Tode S, Iwatsaki S, Starzl TE (1987) Adenoviral infections in pediatric liver transplant recipients. JAMA 258:489-492.

Kvist S, Ostberg L, Persson H, Philipson L (1978) Molecular association between transplantation antigens and a cell surface antigen in an adenovirus-transformed cell line. Proc Natl Acad Sci USA 75:5674-5678.

Landry ML, Long CKY, Neddermann K, Solomon L, Hsiung GD (1987) Disseminated adenovirus infection in an immunocompromised host. Pitfalls in diagnosis. Am J Med 83:555-559.

Larsen SH, Margolskee RF, Nathans D (1979) Alignment of the restriction map of mouse adenovirus FL with that of human adenovirus 2. Virology 97:406.

Lavery D, Lu SM, Lufka T, Chen-Kiang S (1987) Productive infection of cultured human lymphoid cells by adenovirus. J Virol 61:1466-1472.

Lewis AM Jr., Baum SG, Prigge KO, Rowe WP (1966) Occurrence of adenovirus SV40 hybrids among monkey kidney cell adapted strains of adenovirus. Proc Soc Exp Biol Med 122:214-218.

Lewis JB, Mathews M (1980) Control of adenovirus early gene expression: A class of immediate early products. Cell 21:303-313.

Lichy JH, Field J, Horwitz MS, Hurwitz J (1982) Separation of the adenovirus terminal protein precursor from its associated DNA polymerase: Role of both proteins in the initiation of adenovirus DNA replication. Proc Natl Acad Sci USA 79:5225-5229.

Lichy JH, Enomoto T, Field J, Friefeld BR, Guggenheimer RA, Ikeda J, Nagata K, Horwitz MS, Hurwitz J (1983) Isolation of proteins involved in the replication of adenovirus DNA *in vitro*. Cold Spring Harbor Symp Quanti Biol. 47:731-740.

Lin YL, Askonas BA (1981) Biological properties of an influenza A virus specific killer T cell clone. J Exper Med 154:225-234.

Mandel B (1982) Interaction of viruses with neutralizing antibodies. In Frankel-Conrat H, Wagner RR (eds.) Comprehensive Virology. Plenum Press. New York. pp 37-121.

Martin DB, Weiner LB, Nieburg PJ, Blair DC (1979) Atypical measles in adolescents and young adults. Ann Intern Med 90:877-881.

Moran E, Mathew MB (1987) Multiple functional domains in the adenovirus E1A gene. Cell 48:177-178.

Morin JE, Lubeck MD, Barton JE, Conley AJ, Davis AR, Hung PP (1987) Recombinant adenovirus induces antibody response to hepatitis B virus surface antigen in hamsters. Proc Natl Acad Sci USA 84:4626.

Mufson MA, Belshe RB, Horrigen TJ, Zollar LM (1973) Cause of acute hemorrhagic cystitis in children. Am J Dis Child 126:605-609.

Nader P, Horwitz MS, Rousseau P (1968) Atypical exanthem after exposure to measles in vaccinated children. J Ped 72:22.

Nairn R, Yamaga K, Nathenson SG (1980) Biochemistry of gene products from murine MHC mutants. Ann Rev Genet 14:241-277.

Nash AA (1984) Viruses as regulators of delayed hypersensitivity T-cell and supressor T-cell function. In: Notkins AL and Oldstone MBA (eds.) Concepts in Viral Pathogenesis. Springer Verlag. New York. pp 225.

Nevins JR (1982) Induction of the synthesis of a 70,000 dalton mammalian heat shock protein by the adenovirus E1A gene product. Cell 29:913-919.

Numazaki Y, Kumasaka T, Yano N, Yamanaka M, Miyazawa T, Tikai S, Ishida N (1973) Further study of acute hemorrhagic cystitis due to adenovirus type II. N Engl J Med 289:344-347.

Old L (1985) Tumor necrosis factor (TNF). Science 230:630-632.

Oldstone MBA, Dixon FJ (1970) Pathogenesis of chronic disease associated with persistent lymphocytic choriomeningitis viral infection. J Exp Med 131:1-14.

Paabo S, Bhat BM, Wold WSM, Peterson PA (1986) Adenoviruses of subgenera B,C,D and E modulate cell-surface expression of major histocompatibility complex class I antigens. Proc Natl Acad Sci USA 83:9665-9669.

Paabo S, Bhat BM, Wold WSM, Peterson PA (1987) A short sequence in the COOH-terminus makes an adenovirus membrane glycoprotein a resident of the endoplasmic reticulum. Cell 50:311.

Pacini DL, Duvobi EJ, Clyde WA, Jr. (1984) A new animal model for human respiratory tract disease due to adenovirus. J Infect Dis 150:92-97.

Pastan I, Seth P, Fitzgerald D, Willingham M (1987) Adenovirus entry into cells. Some new observations on an old problem. In: Notkins AL and Oldstone MBA (eds.) Concepts in Viral Pathogenesis. Springer Verlag. New York. pp 141-156.

Persson H, Kvist S, Ostberg L, Petersson PA, Philipson L (1979) Adenovirus early glycoprotein E3-19K and its association with transplantation antigens. Cold Spring Harbor Symp Quant Biol 44:509-517.

Persson H, Philipson L (1982) Regulation of adenovirus gene expression. Curr Top Microbiol Immunol 97:157-203.
Philipson L, Lonberg-Mola K, Petterson U (1968) Virus receptor interaction in an adenovirus system. J Virol 2:1064-1075.
Ralston R, Bishop JM (1983) The protein products of the myc and myb oncogenes and adenovirus Ela are structurally related. Nature (London) 306:803-806.
Rowe WP, Huebner RJ, Gillmore LK, Parrott RH, Ward TGH (1953) Isolation of a cytopathogenic agent from human adenoids undergoing spontaneous degeneration in tissue culture. Proc Soc Exp Biol Med 84:570.
Rowe WP, Baum SG (1965) Studies of adenovirus SV40 hybrid viruses. II. Defectiveness of the hybrid particles. J Exp Med 122:955-966.
Rubinstein P, Walker ME, Fedur B, Witt ME, Cooper LV, Ginsberg-Fellsner S (1982) HLA system in congenital rubella patients with and without diabetes. Diabetes 31:1088-1091.
Ruley H (1983) Adenovirus early region 1A enables viral and cellular transforming genes to transform primary cells in culture. Nature 304:602.
Schrier PI, Bernards R, Vaessen RTMJ, Houweling A, van der EB AJ (1983) Expression of class I major histocompatibility antigens switched off by highly oncogenic adenovirus 12 in transformed rat cells. Nature 305:771-775.
Seth P, Fitzgerald D, Ginsberg H, Williamham M, Pastan I (1984) Evidence that the penton base of adenovirus is involved in potentiation of toxicity of Pseudomonas exotoxin conjugated to epidermal growth factor. Molec Cell Biol 4:1428-1533.
Severinsson L, Martens I, Peterson PA (1986) Differential association between two human MHC class I antigens and an adenoviral glycoprotein. J Immunol 137:1003-1009.
Shemer-Avni Y, Wallach D, Sarov I (1988) Inhibition of *Chlamydia trachomatis* growth by recombinant tumor necrosis factor. Infect Immun 56:2503-2506.
Shields AF, Hackman RC, Fife KH, Corey L, Meyers JD (1985) Adenovirus infections in patients undergoing bone-marrow transplantation. N Engl J Med 315:529-533.
Signas C, Katze MG, Persson H, Philipson L (1982) An adenovirus glycoprotein binds heavy chains of class I transplantation antigens from man and mouse. Nature 299:175-177.
Signas C, Akusjarvi G, Pettersson U (1986) Region E3 of human adenoviruses; differences between the oncogenic adenovirus 3 and the non-oncogenic adenovirus 2. Gene 50:173-184.
Siliciano RF, Lawton T, Knall C, Karr RW, Berman P, Gregory T, Reinherz EL (1988) Analysis of host-virus interactions in AIDS with anti-gp120 T cell clones: Effects of HIV sequence variation and a mechanism for CD4+ cell depletion. Cell 54:561-575.
Similia SO, Ylikorkala O, Wasz-Huckert O (1971) Type 7 adenovirus pneumonia. J Pediatr 79:605-611.
Sissons JGP, Oldstone MBA (1985) Host responses to viral infections. In: Fields BN (ed.) Virology. Raven Press. New York. pp 265-279.
Stadler H, Hierholzer JC, Oxman MN (1977) New human adenovirus (candidate adenovirus type 35) causing fatal disseminated infection in a renal transplant recipient. J Clin Microbiol 6:257-265.
Svensson U, Persson H, Everitt E (1981) Virus-receptor interaction in the adenovirus system. I. Identification of virion attachment proteins of the HeLa cell plasma membrane. J Virol 38:70-81.
Tanaka K, Isselbacher KJ, Khoury G, Jay G (1985) Reversal of oncogenesis by the expression of a major histocompatibility complex class I gene. Science 228:26-30.
Top FH, Jr, Buescher EL, Bancroft WH, Russell PK (1971) Immunization with live types 7 and 4 adenovirus vaccines. II. Antibody response and protective effect against acute respiratory disease due to adenovirus type 7. J Infect Dis 124:155-160.
Townsend ARM, McMichael AJ, Carter NP, Huddleston JA, Brownlee GG (1984) Cytotoxic T cell recognition of the influenza nucleoprotein and hemagglutin expresses in transfected mouse L cells. Cell 39:13-25.

Vaessen RTMJ, Houweling A, van der Eb AJ (1987) Post-transcriptional control of class I MHC mRNA expression in adenovirus 12 transformed cells. Science 235:1486-1488.
Van der Veen J, Mess A (1973) Serological classification of two mouse adenoviruses. Arch Gesamte Virusforsch 42:235.
Vasavada R, Eager KB, Barbanti-Brodano G, Caputo A, Ricciardi RP (1986) Adenovirus type 12 early region 1A proteins repress class I HLA expression in transformed human cells. Proc Natl Acad Sci USA 83:5257-5261.
Wadell G (1984) Molecular epidemiology of human adenoviruses In: W. Doerfler (ed.) Current Topics in Microbiology and Immunology. Vol 110. Springer-Verlag. Berlin. pp 191-220.
Whyte P, Buchkovci KJ, Horowitz JM, Friend SH, Raybuck M, Weinberg R, Harlow E (1988) Association between an oncogene and an anti-oncogene: the adenovirus E1A proteins bind to the retinoblastoma gene product. Nature 34:124-128.
Wigand R, Butha A, Dreizin RS, Esche H, Ginsberg HS, Green M, Hierholzer JC, Kalter SS, McTerran JB, Petterson U, Russell WC, Wadell G (1982) Adenoviridae. Second report. Intervirology 18:169-176.
Wold WSM, Cladaras C, Magie SC, Yacoub N (1984) Mapping a new gene that encodes an 11,600 molecular weight protein in the E3 transcription unit of adenovirus 2. J Virol 52:307-313.
Wong GHW, Goeddel DV (1986) Tumour necrosis factors alpha and beta inhibit virus replication and synergize with interferon. Nature 323:819-822.
Yoon JW, Austin M, Onodera T, Notkins AL (1979) Virus-induced diabetes mellitus: isolation of a virus from the pancreas of a child with diabetes ketoacidosis. N Engl J Med 300:1173-1179.
Zahradnik JM, Spencer MJ, Porter DD (1980) Adenovirus infection in the immunocompromised patient. Am J Med 68:725-732.
Zinkernagel RM, Althage A (1977) Antiviral protection by virus-immune cytotoxic T lymphocytes infected target cells are lysed before infectious virus progeny is assembled. J Exp Med 145:644-651.

DISCUSSION

Gauntt C (University of Texas, San Antonio, TX):

Marshall, that was great. I enjoyed it very much. The question is, in the coxsackie models there is a big controversy now over whether coxsackie B-3, for example, persists in heart tissues. Compounding that is a recent discovery by Steve Tracy that there are human enteroviral-like sequences in normal cells. So, if one does hybridizations, you don't really know if you're picking up the "normal" cell sequences t .at are enteroviral-like, or if you're picking up the virus. Do you have similar problems or controversies in adeno? Does it persist? Are there sequences like adeno in cells?

Horwitz M:

Adenoviruses can persist. They persisted enough for Hartley and Rowe to first isolate adenoviruses at the time of tonsillectomies, from tonsils, presumably, not infected clinically. There's no question that in humans, other adenoviruses such as types 12 and 18 persist in the GI tract for long periods of time. In fact, the E19 mechanism may have something to do with persistence. I didn't get into that, but yes, that aspect of adeno is certainly shared with what you've mentioned for coxsackie B. The mouse adenoviruses can also persist either in urine for MAV-1 or in the stool for MAV-2. In terms of sequences shared between Ads and the host, although perhaps there are regions that could cause some confusion, it is really not a problem. A slide I didn't show demonstrates some distant homology between the E19, for instance, and some of the gammaglobulin genes. However, we don't have to worry about that because in this era of selecting the sequences you want to use as probes, there are adenovirus regions that absolutely do not react with normal tissue.

Gauntt C:

Do you know what product might be involved in altering the cells so that NK cells recognize an infected cells?

Horwitz M:

The only thing we know about NK, which is a mechanism I didn't discuss, is that the target seems to be a protein coded by E1A. The mechanism has nothing to do with the E19 glycoprotein or the E3 region.

Cesario T (University of California Irvine Medical Center, Orange, CA):

We know that in some cases, such as hepatitis, that the disease itself, the cytopathology is dependent upon not so much the virus, but rather the immune recognition of viral determinants. Is this the case with adenoviruses, that we can enhance the disease, if you will, by increasing expression of viral determinants on the cell surface? And if the viral proteins are held within the cells and, conversely, if they are not expressed well, is this a way of inhibiting the development of pathology in disease?

Horwitz M:

Adenovirus, as I mentioned, is not a membrane virus, so it puts very few of its structural proteins into the cell surface. However, let me qualify that a bit. The primary targets for cytotoxic T lymphocytes seem to be some of the E1A and E1B regulatory proteins which reach the cell surface, perhaps in processed forms. That doesn't answer your question though, in terms of enhanced pathology. In the cotton rat model, the Ad5 E19 deletion mutant causes the disease to worsen, presumably at a time when more MHC is on the cell surface. This would suggest that allowing the immune system to see the infected cell perhaps enhances the pathology and actually causes death in some cases. I think that once we get the mouse model to a corresponding stage, we will be able to answer the issue of immunopathology with far more precision.

Snurff D (California State Department of Health, Berkeley, CA):

Since you originally isolated your adeno 35 from AIDS patients, and you did see a fairly common isolation rate of the adeno 35-34 from AIDS patients, do you think that they have any significance in AIDS?

Horwitz M:

Well, we've obviously wondered about that. It's a key question. I think its significance is still theoretical. Dr. Mosca's talk yesterday very nicely illustrated that some of the adeno genes, like E1A from Ad2 and 5 do transactivate HIV genes. We have not done similar studies, in terms of transactivation, for Ad35. In the early days, we did do mixed infections to see what affect adeno would have on the growth of HIV in tissue culture. However, we wern't able to show any affect, either positive or negative. I look back on those studies and think, perhaps, we used too permissive a system for HIV. We used a tissue culture adapted HTLV-III isolate and the permissive H9 T4 lymphocyte cell line. Were we to do it today, we would have used a fresh HIV isolate without all of the adaptation to tissue culture cells, and we would have used a less permissive host cell, rather than H9. I don't know whether the answer that adeno and HIV did not affect each other would be the same.

McDaniel H (Dallas Fort Worth Medical Center, Grand Prarie, TX):

I wondered, while we were waiting for these proceedings and hopefully, with illustrations too, if you've written any reviews or know of any reviews of these mechanisms of viral attack against the immune system? Is there evidence that other viruses use similar mechanisms?

Horwitz M:

There is no other evidence that a viral gene product binds to MHC and prevents transport. There are reports of changing MHC levels after viral infections. The mechanisms haven't been worked out. There's been a report for herpes, for example. But, nothing that's similar to the adeno mechanism. However, I can't believe adeno is unique. I think we just don't know enough about how other viruses may affect and modulate MHC.

So, that still is an unknown, and hopefully, some people in the audience either know more about this from studying their particular virus, or will in the future. In terms of reviews, what I've put together today has not been well reviewed in toto. I can give you individual references, but not the picture that I've shown you today. But, you'll get the book.

Jones J (National Jewish Center for Immunology and Respiratory Medicine, Denver, CO):

Some of the other mechanisms responsible for expression of MHC have to do actually, with the IFN system, and upgrading and downgrading that particular problem. Is there any relationship in the adeno with the VA mRNAs and the E19 in that respect, say in the cotton rat model? It seems like IFN is pushing it up, yet the E19 is pushing it down.

Horwitz M:

That's a very good line of questioning. As I brought these separate topics together, I also thought of that question. I think it's something worth looking at. Because VA RNA can inhibit the action of IFN, it may block interferon-induced upregulation of MHC. Together with the E19-MHC interaction, the net effect would be downregulation of MHC in Ad-infected cells.

Jones J:

My second question has to do with persistence of adeno of any of these serotypes in the lung.

Horwitz M:

I don't know of its persistence in the lung. It certainly, in our immunosuppressed patients, persisted in the urinary tract. It's persistent in the tonsils. It's persistent in the GI tract, certainly, for years. Adeno can integrate, but in most of these cases, it seems that it's sort of a low level productive infection. There is a report form Canada that all of us have adeno 5 sequences in our lymphocytes that we acquire, in an epidemiologically understandable way, as we grow older. In other words, children don't have it and adults have sequences in their lymphocytes that persist. That study has not been confirmed by others, but that is an interesting observation. So, it's another site of persistence. I'm not aware, at least in the immunocompetent, of persistence in the lung.

Burnette G (Burnette Labs, Buena Park, CA):

My question, Dr. Horwitz, is a little bit concerning Dr. Cesario's question. Applications, what do we in the clinical lab tell a physician who has some question about the treatment? In the past, to prevent autoimmune disease as a sequence of a viral infection, some clinicians administered steroids to avoid that kind of destruction of any organ by these cytotoxic lymphocytes. Do you envision, in your studies, that when we do the typing of HLA in a patient, we could somehow suggest to a clinician what kind of patients will develop this kind of pathogenicity and what HLA types will be somehow protected?

Horwitz M:

I think the answer to those questions are many years off still. Just to reiterate though, our ability to treat adenoviruses are really woefully inadequate. In our other laboratory endeavors, we study the adeno DNA polymerase, and it would be really nice to get some good inhibitors of DNA polymerase, such as for the herpes viruses. But, we don't have any. People who develop keratoconjunctivitis suffer still with no help from antivirals. Some people use steroids at certain stages of the illness to ameliorate symptoms. When someone occasionally gets a severe adeno pneumonia, we again have nothing therapeutically to offer those patients in terms of antivirals.

Cesario T:

Is there any evidence that people with certain HLA types are more prone to severe adenovirus infections that others?

Horwitz M:

It's not been worked out for adeno. There are other models of disease where HLA typing perhaps does cluster with immunopathology. The majority of patients with Reiter's disease have HLA type B27. After a variety of infectious agents, those patients can go on to develop arthritis and a variety of mucosal inflammatory manifestations. That's one of the examples from humans. I am not aware of any attempts to correlate poor outcome from Ad infection with a specific HLA type. Presumably, there are variations and we would expect to show them first in mice, and perhaps get to that point in humans after the mouse studies.

SYNTHETIC PEPTIDES FROM A HYPOTHETICAL RECEPTOR BINDING SITE: A POSSIBLE APPROACH TO A RHINOVIRUS VACCINE?

GUDRUN WERNER and JOSEPH W. McCRAY

Sandoz-Forschungsinstitut GmbH, Brunnerstrasse 59 A-1235 Vienna, Austria

INTRODUCTION

Rhinoviruses are well established as the major causative agent of the common cold (Couch, 1984; Gwaltney, 1982). Since there are many methodological difficulties with diagnosis, estimates for the proportion of rhinoviruses among other virus families causing colds vary between 10 and more than 50% (Phillpotts and Tyrell, 1985). Although usually harmless in healthy children and adults, rhinovirus colds can induce severe complications in patients with chronic respiratory states like asthma or chronic bronchitis (Gregg, 1983). Rhinovirus has also been isolated from immunosuppressed infants with lower respiratory tract infections normally not caused by these viruses (Krilov et al. 1986). The economic burden of the common cold is enormous. According to a national health survey the common cold accounts for about 20% of all disabling conditions leading to roughly 30 million lost days of work per year in the United States alone (Couch, 1984). In contrast to the relative significance of the medical problem scientific efforts towards methods for control have been limited. There has been much optimism about antiviral drugs for prophylactic or therapeutic treatment of the disease. However, of the synthetic anti-rhinovirus compounds found to be active *in vitro* none proved effective in experimental or naturally occurring rhinovirus infections (Al-Nakib and Tyrell, 1987).

One major drawback to both therapeutic and vaccine anti-rhinovirus approaches is the great variability of the virus culminating in more than 120 distinct human serotypes. The possibility has been discussed by several groups that the number of serotypes predominating during an acute outbreak may be limited (Beem, 1969; Cooney et al. 1982; Fox, 1976; Mogabgao et al. 1975). If it is true that one has to protect against not more than a dozen serotypes, a conventional vaccine seems possible though very cost intensive to prepare. On the other hand, it has been observed that patterns of serotypes change over time, with higher numbered or non-typable strains accumulating (Fox, 1976). This suggests that an antigenic drift may exist comparable to influenza viruses. Therefore, an alternative approach to prevent virus infection seems to be desirable.

By using synthetic peptides representing indispensable and therefore unchangeable parts of the virus particle, it may be possible to overcome the virus's strategy for evading immune surveillance. We used synthetic peptides of rhinovirus type 14 (HRV 14), which are relatively conserved even among other picornaviruses, to prepare site specific antibodies. These antibodies turned out to neutralize 60% of a representative set of rhinovirus serotypes *in vitro* (McCray and Werner, 1987). The rationale behind our approach and the possibilities for a synthetic rhinovirus vaccine will be discussed in the present report.

THE THREE-DIMENSIONAL STRUCTURE OF RHINOVIRUS PARTICLES AND THE "CANYON" HYPOTHESIS

The elucidation of the three-dimensional structure of rhinovirus 14 (Rossmann et al. 1985) and of another picornavirus, poliovirus 1 (Hogle et al. 1985), by x-ray crystallography brought a breakthrough in our understanding of mechanisms for neutralization and escape therefrom. The original idea about the organization of the four capsid proteins VP1 to VP4 into a protomer, 60 of which are arranged into 12 pentamers, which in turn build up the icosahedral particle harboring the asymmetrical RNA genome can now be generally accepted (Medappa et al. 1971). Sequencing the genome of HRV 14 mutants which were selected for their resistance to several neutralizing monoclonal antibodies (Rueckert et al. 1986) has shown that mutations were clustered at just four sites with only very few amino acid substitutions compared to wild type HRV 14. In most cases only one substitution is sufficient to confer to the virus resistance to one particular antibody. These sites, therefore, are thought to be the major immunogenic sites or predominant neutralizing epitopes. From the three-dimensional structure the mutations were identified as parts of very flexible peptide loops forming large protrusions on the external surface of the virus particle. A sequence comparison with other picornaviruses revealed that especially these loops are the most variable regions. Thus, antigenic variability of rhinoviruses is easily explained, whereas other factors probably come into play to explain why polio viruses occur only as three serotypes.

The large surface protrusions create a deep narrow canyon around the pentamer axes. Sequences lining the canyon are relatively conserved among picornaviruses indicating that this site is under strong selective pressure. Due to its dimensions the canyon, especially its flor, should be inaccessible to antibodies. These findings led to the proposal that the canyon is the cell surface receptor binding site. The differences in amino acids contributing to the canyon walls should be sufficient to explain the fact that different picornavirus genera use different cellular receptors. According to competition experiments with antireceptor antibodies (Abraham and Colonno, 1984; Colonno et al. 1986) even rhinoviruses can be subdivided into two groups. HRV 14 belongs to the major group comprising 80% of serotypes and HRV 2 to the minor one. The canyon hypothesis was the basis of our considerations. It has been shown by many groups (Anderer, 1963; Arnon et al. 1971; Bittle et al. 1982; Emini et al. 1983a; Steward and Howard, 1987; Sutcliffe et al. 1983) that synthetic peptides representing partial sequences of a protein, exposed on its surface (Lerner, 1984), can induce antibodies which bind to the native protein. In addition, synthetic peptides can confer immunity to sites on a protein which are usually not dominant epitopes (Lerner, 1984; Walter, 1986). Considering the huge antibody repertoire, we thought it should be possible by using synthetic peptides to raise antibodies binding in the canyon.

ANTIBODIES TO PEPTIDE SEQUENCES PARTIALLY EXPOSED IN THE CANYON CROSS-NEUTRALIZE DIFFERENT RHINOVIRUS SEROTYPES

Mainly two peptide sequences MYVPPGAPNP (single letter code, starting at amino acid 151 of VP1) and AYTPPGARGP (starting at position 130 of VP3) of HRV 14 are conserved in rhino, polio and foot-and-mouth disease (FMD) viruses and are situated near the floor of the canyon (Luo et al. 1987; Rossmann et al. 1985). In order to be certain that the peptides would be immunogenic we extended both sequences beyond their N- and C-terminal residues according to the HRV 14 sequence and coupled them chemically to a carrier protein (key hole limpet hemocyanin, KLH). For that purpose a C-terminal cysteine residue was added. Both peptides, termed PVP1A and PVP3A, and their sequence variations among other picornaviruses are shown in Table 1. Peptide-KLH conjugates were injected into rabbits and the antibody titers induced were followed in an ELISA with the pure peptides as antigens. Because initial experiments with

TABLE 1. AMINO ACID SEQUENCES OF PEPTIDES PVP1A AND PVP3A OF HRV14 AND VARIATIONS THEREOF IN OTHER RHINO SEROTYPES AND OTHER PICORNAVIRUSES

	virus	references
PVP1A		
H-Val-Val-Gln-Ala-Met [Tyr] Val [Pro-Pro-Gly-Ala] Pro-Asn [Pro] Lys-Glu-Cys-OH	Rhino 14	Callahan et al.,1985
H- -Met- -Tyr- -Ile- -Thr- -OH	Rhino 1B	Hughes et al.,1988
H-Thr-Met- -Tyr- -Val- -Asn-Ser- -OH	Rhino 2	Skern et al.,1985
H- -Leu- -Phe- -Val- -Glu-Lys- -OH	Rhino 89	Duechler et al.,1987
H- -Tyr- -Ile- -Val- -Glu-Lys- -OH	Polio 1	Kitamura et al.,1981
H- -Tyr- -Ile- -Ile- -Thr- -Ser- -OH	Polio 3	Stanway et al.,1984a
PVP3A		
H-Lys-Leu-Ile-Leu-Ala [Tyr] Thr [Pro-Pro-Gly-Ala] Arg-Gly [Pro] Glu-Asp-Cys-OH	Rhino 14	Stanway et al.,1984b
H- -Leu- -Ile-Asp-Lys- -Ala-Thr- -OH	Rhino 1B	
H- -Thr- -Ile-Ala-Glu- -Thr-Thr- -OH	Rhino 2	
H- -Leu-Ile- -Val-Gly-Lys- -Lys-Ser- -OH	Rhino 89	
H- -Leu-Val-Ser- -Asp-Pro- -Lys-Lys- -OH	Polio 1	Racaniello, 1981
H- -Ile-Leu-Val- -Ala- -Glu-Pro- -Thr-Ser- -OH	Polio 3	
H-Arg-Tyr-Met-Val- -Ile- -Val-Glu-Thr- -Pro-Gln- -OH	FMDV	Carrol et al.,1984
H- -Phe-Leu-Ile- -Gly-Lys- -Thr-Ser- -OH	EMCV	Palmenberg,1984
Control peptides		
H-Ala-Pro-Ala-Arg-Ser-Pro-Ser-Pro-Ser-Thr-Gln-Pro-Trp-Cys-OH	CSF-1	Wong et al.,1985
H-Asp-Ala-Lys-Val-Lys-Leu-Ile-Lys-Gln-Glu-Leu-Asp-Lys-Tyr-Lys-Asp-Cys-OH	RSF-1	Collins et al.,1984

animal sera showed only marginal neutralization at very low titers we purified the antipeptide antibodies by affinity chromatography on the corresponding peptide affinity gels (McCray and Werner, submitted). These antibodies now clearly neutralized HRV 14 in a microneutralization test (MNT, Table 2). Antibodies against both peptides were equally active. In control experiments antibodies against an unrelated peptide (CSF-1) or a commercial rabbit IgG preparation did not neutralize HRV 14. Since we expected that our antibodies should neutralize other serotypes as well, we tested a representative panel of serotypes in the *in vitro* neutralization test. As shown in Table 3 roughly 60% of serotypes tested were neutralized, although to a varying extent. Unfortunately, no sequence information on the canyon site besides that from HRV 14, HRV 2 and HRV 89 is available at present. Therefore, it is not known whether sequences equivalent to PVP1A and PVP3A of HRV 14 are different for serotypes not neutralized by the antipeptide antibodies. On the other hand, it could well be that the three-dimensional structure of those serotypes is somewhat different at this site as has been found for mengo virus (Rossmann et al. 1987). Here a set of five deep pits instead of a canyon surrounds each fivefold axis. Perhaps such sites are, in fact, not accessible to antibody molecules. Alternatively, neutralization may depend on the affinity of the cell surface receptor binding site to the receptor counteracting the binding of the antipeptide antibodies. It is well known that antipeptide antibodies recognize the native protein with lower affinity, possibly due to the high conformational flexibility of peptides (Hirayama et al. 1985). Since the antibodies are polyclonal, it

TABLE 2. MICRONEUTRALIZATION TEST (MNT) OF HRV 14 WITH ANTIPEPTIDE ANTIBODIES

Antibody Preparation	Concentration of Antibodies[a] (µg/ml)	Activity in MNT (Neutralizing Titer)
Anti-PVP1A	18	+ (1:64)
Anti-PVP3A	12	+ (1:64)
Anti-PVP1A-PVP3A (flow-through fraction)	120	-
Anti-PVP1A-PVP3A (eluate fraction)	7	+ (1:32)
Rabbit IgG	300	-

[a]Values indicate minimum concentrations of antibodies needed to neutralize HRV 14 in MNT or the maximum concentration tested if no neutralization was selected. Reprinted with permission from McCray and Werner (1987).

TABLE 3. NEUTRALIZATION OF HUMAN RHINOVIRUSES BY ANTIPEPTIDE ANTIBODIES

Neutralizing Activity	Serotypes Tested
Weak (<1:8)	9[a],10,16,40[a],45[a],68[b],71[a]
Normal (1:8-1:32)	17[a],24,26[a],28,32[a],36,51,58[a],72[b]
Strong(>1:32)	3[b],5[a],6,14,23,27,35[a],37[a],48[a],50[b],64
None	*1a*[c],*2*,4,7,8,13,15,18,19,20,21,22,25,33, 38,45,*47*[b],*62*[b],75,85 (polio-1,polio-2,echo-9,coxsackie-B3)

[a]These viruses are negative with the anti-HRV 14-hyperimmune serum.
[b]These serotypes were weakly positive with a rabbit hyperimmune serum against HRV 14 whole virus particles. Titers ranged between 1:20 (HRV 3) and 1:640 (HRV 55) compared to 1:80,000 for HRV 14.
[c]Viruses in italics belong to a minor group of rhinoviruses using another receptor for cell penetration. Reprinted from McCray and Werner (1987) with permission.

would be difficult to interpret kinetic studies necessary to determine affinity constants and we have not yet performed such experiments.

Interestingly, many serotypes not neutralized by the antipeptide antibodies belong to the minor receptor group of rhinoviruses, for example HRV 2. We also tested antibodies to the authentic HRV 2 sequence corresponding to PVP1A (Table 1) in a neutralization test. These antibodies neutralized HRV 14 out not HRV 2. Polio-1, echo-9 and coxsackie B3, selected members of other picornavirus genera, were also not neutralized by any of the antipeptide antibodies. They, too, recognize other cellular receptors than HRV 14. The somewhat unexpected results with HRV 2 are discussed in a later section.

Based on estimated concentrations of virus and antibodies in the neutralization test, we can calculate that the antibodies only neutralize in large molecular excess. Therefore, it appears likely that only a small proportion of the antibodies is neutralizing. We do not know at present whether one antibody molecule per virus particle is sufficient to induce a conformation change which then would hinder early steps in the virus replication cycle such as attachment to the cell, penetration or uncoating. A conformational change is one of the often discussed mechanisms for virus neutralization (Dimmock, 1987). If the antipeptide antibodies interfere with attachment to the cellular receptor, it may be necessary to block each binding site on all 12 pentamers. In a preliminary experiment (Figure 1) with radiolabeled virus particles anti-PVP3A antibodies inhibited adsorption to the cells in contrast to unrelated antipeptide antibodies. Thus, we assume that the antipeptide antibodies in fact bind to or near the cellular receptor binding site of rhinovirus 14 and related serotypes. However, this assumption should be tempered by the observations that there are antipeptide antibodies of polio-1 which neutralize the virus although the immunizing sequences according to the three-dimensional structure are paradoxically found in the interior (Chow et al. 1985; Emini et al. 1983a). When these antibodies are indeed binding only to the cognate sequences in virus particles, these results require more complex explanations, for example, a conformational change during isoelectric transition of the virus (Emini et al. 1983b). We, therefore, did several experiments to test the site-specificity of our antipeptide antibodies.

THE SPECIFICITY OF THE ANTIPEPTIDE ANTIBODIES

In an immunoblotting assay with purified rhinovirus particles anti-PVP1A and anti-PVP3A antibodies reacted strongly with the corresponding capsid proteins VP1 and VP3, respectively, as had been expected (Figure 2). A very weak crossreactivity of anti-PVP1A antibodies with VP3 and inversely anti-PVP3A antibodies with VP1, but not with VP2 was also observed. A similar crossreactivity was found in an ELISA-test with the

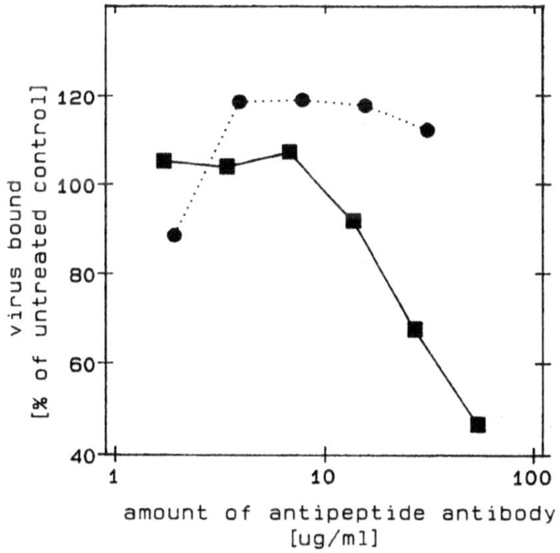

Figure 1. Inhibition of HRV 14 binding to HeLa cells by anti-PVP3A antibodies. (■ , anti-PVP3A; ● , anti-CSF-1).

peptides as antigens (Figure 3). The antipeptide antibodies did not bind to unrelated control peptides such as CSF-1, a 13 amino acid sequence of human granulocyte-macrophage colony stimulating factor or RSF-1, a partial sequence of the fusion protein of human respiratory syncytial virus. The crossreactivity of the antipeptide antibodies observed in immunoblotting and ELISA-tests can easily be explained by the core sequence PPGA common to PVP1A and PVP3A peptides (Table 1).

In an ELISA-test with purified virus particles fixed to the plates antibodies against PVP1A and PVP3A, but not anti-CSF-1 or anti-RSF-1 antibodies were bound. Furthermore, the binding of the antipeptide antibodies to the virus could be competed in a concentration-dependent manner by the corresponding free peptides as shown in Figure 4. This indicates that antibodies may indeed bind to the site of the virus represented by the peptides. Certainly, the virus may be denatured during binding or fixation to the plates, thereby making this site easily accessible to the antibodies. Therefore, we cannot know for sure, whether on native virus particles the same site is recognized by the antipeptide antibodies. However, since 60% of the serotypes were neutralized, we assume that the antipeptide antibodies can bind to a conserved site of native virus particles.

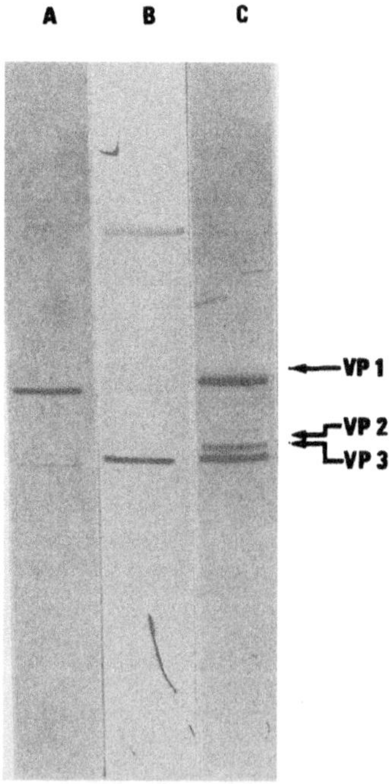

Figure 2. Immunoblotting of HRV 14 capsid proteins with the antipeptide antibodies: anti-PVP1A antibodies (a) at 40 μg/ml, anti-PVP3A antibodies (b) at 60 μg/ml and an anti-HRV 14 (whole virus) hyperimmune serum at 1:60. Reprinted from McCray and Werner (1987) with permission.

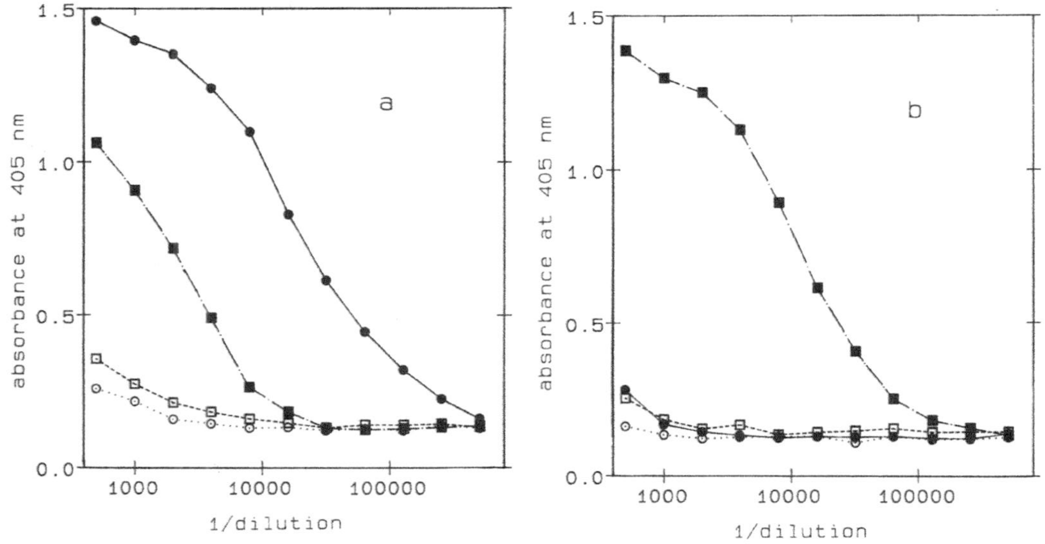

Figure 3. Reactivity of the antipeptide antibodies in ELISA with different peptides: anti-PVP1A-(a) and anti PVP3A-(b) antibodies with PVP1A (●), PVP3A (■), CSF-1 (○) and RSF-1 (□).

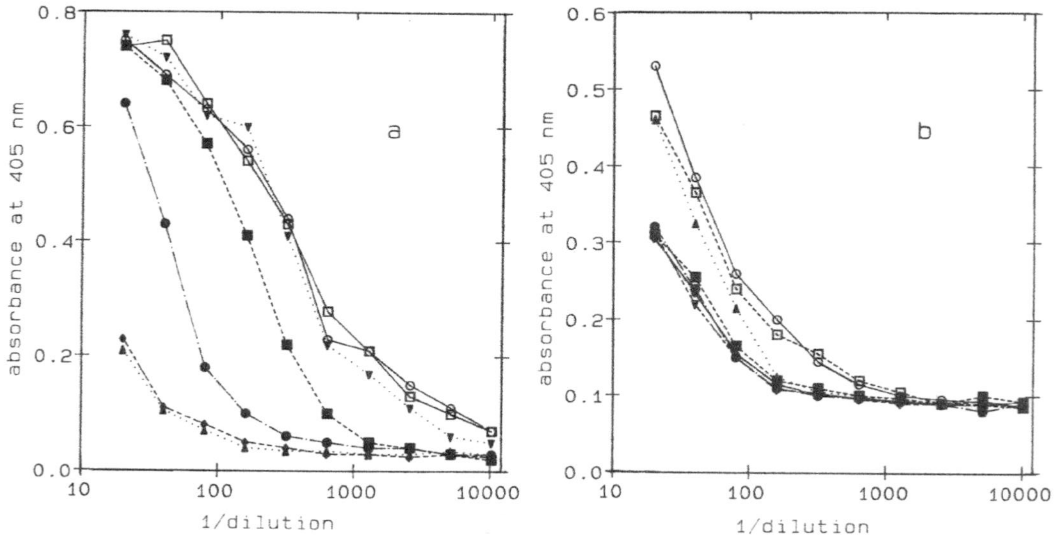

Figure 4. Binding of the antipeptide antibodies to virus particles in ELISA and its competition by the free peptides: anti-PVP1A antibodies (a) without peptide (○), with 10 µg/ml CSF-1 (□) and with PVP1A at 1 ng/ml (▼), at 10 ng/ml (■), at 100 ng/ml (●), at 1 µg/ml (♦) and at 10 µg/ml (▲); anti-PVP3A antibodies (b) without peptide (○), with 50 µg/ml CSF-1 (□) and with PVP3A at 10 ng/ml (▲), at 100 ng/ml(■), at 1 µg/ml (●), at 10 µg/ml (▼) and at 50 µg/ml (♦). Adapted from McCray and Werner (1987) with permission.

FURTHER CHARACTERIZATION OF THE NEW NEUTRALIZING EPITOPE

The peptides PVP1A and PVP3A selected for immunization of rabbits are both 17 amino acids long. Since we would expect from the three-dimensional structure of the virus particle that the N-terminus, being buried in the interior, would not be accessible, the neutralizing antipeptide antibodies should bind preferentially to the core sequence PPGA or to the more C-terminal amino acids of the peptides. The latter possibility seemed rather unlikely, because these amino acids are more variable between rhino serotypes and other picornaviruses. Essentially, those antibodies crossreacting with both peptides were thought to recognize the core sequence PPGA. Therefore, we enriched this fraction by purifying the antipeptide antibodies through two consecutive affinity columns. Anti-PVP1A antibodies were first purified on a PVP1A affinity column, followed by a PVP3A affinity chromatography step. An inverse protocol was performed on anti-PVP3A antibodies. In both cases, antibodies bound to and eluted from the second column neutralized HRV 14 much better than antibodies from the first column (Table 2). Interestingly, antibodies depleted of this crossreacting fraction but probably still containing the activities directed against the C-terminal parts of the peptides did not or only very weakly neutralize HRV 14. This is, indeed, surprising because those sequences ascending along the walls to the rim of the canyon should be recognized by antibodies much more easily. Perhaps the proportion of such antibodies in our preparation is too small, since the C-terminus of the peptides was used for chemical coupling to the carrier protein, or antibodies binding to it have an altered specificity because they also recognize spacer molecules or amino acids of the carrier.

Taking these results together, we suggest that the neutralization of HRV 14 and related serotypes we observed is caused by antibodies primarily directed against the sequence PPGA, which is partially exposed at the canyon floor. The cross neutralization of different serotypes and the preliminary observation of inhibition of adsorption of the virus to cells by the antipeptide antibodies support Rossmann's et al. (1985) view of the canyon being involved in receptor binding. Of course, we have no direct proof that the antipeptide antibodies bind in the native canyon which would contradict the expectation that the canyon is too narrow for the entry of antibodies. Since the antibody repertoire is nearly unlimited and up to now the three-dimensional structure of just a few Fab-fragments has been determined, antibodies with more convex antigen binding domains able to fit into rather concave narrow structures might exist

(Chang et al. 1985). Alternatively, our antipeptide antibodies may bind to an altered conformation of this site which could be induced during attachment of the virus to the cell. Knowing that the rate of attachment of HRV 14 and of other members of this group to the cellular receptor is very low compared to HRV 2 (Colonno et al. 1986; Lonberg-Holm and Korant, 1972) favors the latter possibility and may also explain why HRV 2 is not neutralized by any of the antipeptide antibodies, not even by the HRV 2 PVP1A sequences. Therefore, to obtain antipeptide antibodies which can neutralize HRV 2 may require peptides with more rigid conformations simulating the shape of the canyon. Such peptides might induce antibodies able to bind to the canyon sequences with sufficient affinity to compete with the cellular receptor.

Nevertheless, we think that our results open the way to a rhinovirus vaccine protecting against many and hopefully one day all serotypes. The more practical requirements of such an approach are discussed in the following section.

IS A SYNTHETIC RHINOVIRUS VACCINE POSSIBLE?

As mentioned earlier, a conventional vaccine, for example live attenuated virus, inactivated virus, or subunit structures of the virus's outer shell is not a realistic possibility for rhinoviruses due to the existence of so many serotypes. A more promising approach would be to immunize against conserved parts of the virus particle such as sequences lining the canyon. Synthetic or molecularly cloned peptides are a first step in that direction, the goal of which would be to outwit viral mechanisms for escaping immune surveillance or, in other words, to redirect the immune system to attack the invader's most vulnerable sites.

ADVANTAGES AND DISADVANTAGES OF SYNTHETIC PEPTIDE VACCINES

An ideal vaccine has to fulfill the following criteria: immunity induced should be effective and durable, no or only minimal side effects should be evoked and the vaccine should be cheap, stable and easy to administer. The last points are particularly important for a rhinovirus vaccine and could be most likely and most easily achieved by synthetic peptides. As far as the immunogenicity of synthetic peptides is concerned, the task is difficult (Zanetti et al. 1987). Many synthetic immunogens have been tested for many different pathogens such as viruses, bacteria or parasites. Table 4 gives a summary of synthetic peptides evaluated as vaccine candidates for humans so far. Very encouraging

TABLE 4. POTENTIAL PEPTIDE VACCINES FOR HUMAN USE

	REFERENCES
Anti-bacterial vaccines	
Streptococcal M protein	Dale and Beachey (1982)
Diphteria toxin	Audibert et al. (1982)
Cholera toxin	Jacob et al. (1984), Guyon-Gruaz et al. (1986)
E. coli enterotoxin	Klipstein et al. (1986)
Gonococcal pili	Rothbard and Schoolnik (1985)
Anti-parasitic vaccines	
P. falciparum circumsporozoite (CS) protein	Good et al. (1987), Herrington et al. (1987)
P. falciparum merozoite protein	Cheung et al. (1986), Pattarroyo et al. (1988)
Anti-viral vaccines	
Influenza virus hemagglutinin	Wilson et al. (1984), Shapira and Arnon (1986)
Hepatitis B virus HBsAg	Dreesman et al. (1982), Gerin et al. (1983), Przewlocki et al. (1986) Itoh et al. (1986)
Poliovirus 1 capsid protein	Emini et al. (1983a)
Rhinovirus capsid protein	McCray and Werner (1987), Francis et al. (1987a)
HIV envelope protein	Ho et al. (1988) Chanh et al. (1986)
HIV gag proteins	Sarin et al. (1986)
Herpes simplex glycoprotein D	Watari et al. (1987)
Contraceptive vaccines	
Human chorionic gonadotropin (hCG)	Stevens (1986)

results were obtained in volunteers with, surprisingly enough, an oral synthetic immunogen of *E. coli* enterotoxin (Klipstein et al. 1986). A major disadvantage of peptides as immunogens normally deduced from the primary structure of the protein is that the corresponding antibodies recognize only sequential epitopes on the native structure of the protein. Antibodies induced by the native protein and neutralizing its activity, however, in most cases bind to residues which are not contiguous in the sequence, but are brought together by the folding of the polypeptide chain. For instance, antigenic site 2 of poliovirus is made up of residues from the structural proteins VP1 and VP2 (Hogle et al. 1985). In order to elicit an adequate level of protective immunity it may be necessary to mimic the correct tertiary conformation of even several neutralizing epitopes. On the other hand, experiments with synthetic peptides of VP1 of FMDV have shown that a sequence comprising amino acids 141-160 induces a better protective immune status than the molecularly cloned whole VP1 (Bittle et al. 1982).

For the induction of an adequate humoral immune response, a cooperation between B and T cells is required, meaning that the antigen should contain T cell epitopes in addition to the B cell epitopes. The covalent coupling of the peptide, having a relatively small size similar to haptens, to a carrier protein usually guarantees the attachment of sufficient T cell sites for the stimulation of T cell help and explains earlier observations that chemical haptens in order to become immunogenic have to be conjugated to a protein (Landsteiner, 1946; Mitchison, 1971). Unlike B cells, T cells cannot recognize the native tertiary structures of proteins (Ashwell and Schwarz, 1986), instead antigens must be presented to the T cell receptor in the form of linear peptides bound to MHC class II molecules (Allen, 1987; Guillet et al. 1987). This is accomplished by a proteolytic processing of the antigen in particular antigen presenting cells such as monocytes, dendritic cells or the B cells themselves (Pierce and Margoliash, 1988). T cell stimulatory peptides tend to be amphipathic α-helical structures with the more hydrophobic amino acids binding to the MHC molecules at one side of the helix and the hydrophilic residues interacting with the T cell receptor complex at the opposite side (Delisi and Berzofsky, 1986). Furthermore, such peptides must be sufficiently different from self structures to be recognized by the T cell receptor (Guillet et al. 1987).

Recently the nonresponsiveness to the previously mentioned (Bittle et al. 1982) synthetic VP1 peptide of FMDV in a genetically restricted mouse strain has been overcome by adding foreign T cell epitopes from ovalbumin or sperm whale myoglobin (Francis et al. 1987a). It has to be noted that only the addition of certain T cell sites induced high neutralizing antibody titers in those mice indicating that the T cell help stimulated by the different T cell epitopes is B cell clone specific.

Therefore, being dependent upon genetic factors such as the individual makeup of the highly polymorphic MHC structures, the available repertoire of T and B cells, the operation of the idiotype network, etc., the immunogenicity of peptides and particularly the spectrum of the induced antibodies is difficult to predict for a large population to be vaccinated (Steward and Howard, 1987). Nevertheless, even if it may be difficult to raise high affinity antibodies neutralizing the invading pathogen at high titers by using synthetic peptide vaccines, a priming of the immune system, as it has been demonstrated with peptide sequences of polio-1 and FMDV (Emini et al. 1983a; Francis et al. 1985) might be sufficient to suppress virus replication enough to ameliorate or block completely the overt symptoms of infection. The possibility of priming the immune system for an antibody response against structures of the canyon proposed to be the cell surface receptor binding site of rhinoviruses seems particularly attractive.

Antibodies induced by the linear peptides PVP1A or PVP3A probably bind to the canyon with low affinity. The reason may be steric hindrance or simply that the number of contact residues in the native canyon offered by the peptides is too low. Peptides mimicking the conformation of the canyon would be difficult to design. However, if the cause of the neutralization of the different rhinovirus serotypes observed *in vitro* is antipeptide antibodies blocking the canyon, such antibodies may be increased upon contact with the native structure of the virus.

Furthermore, somatic mutation in B cell clones induced by the peptides may create antibodies upon exposure to the native virus which fit better into the natural conformation of the hypothetical receptor binding domain (Cohen, 1971; Milstein, 1986). Therefore, one might be able to instruct the immune system to recognize structures of the virus to which it normally does not react.

IMPORTANCE OF A LOCAL IMMUNE RESPONSE FOR PROTECTION AGAINST RHINOVIRUSES

As has been shown earlier (Buscho et al. 1972; Knopf et al. 1970) a crucial step for a potential rhinovirus vaccine to be effective is the induction of a secretory IgA response in the upper respiratory tract, the primary target for rhinovirus replication. This has been achieved, for example, by an intranasal instillation of volunteers with an inactivated HRV 13 vaccine, which conferred good protection against infection and illness on a subsequent challenge with homologous live virus (Knopf et al. 1970). The same vaccine given by a parenteral route induced a serum antibody response not correlating with resistance towards disease at all. Therefore, we tested a scheme of intranasal immunization of rabbits with the peptide-KLH conjugates, which had been shown to induce high antipep-

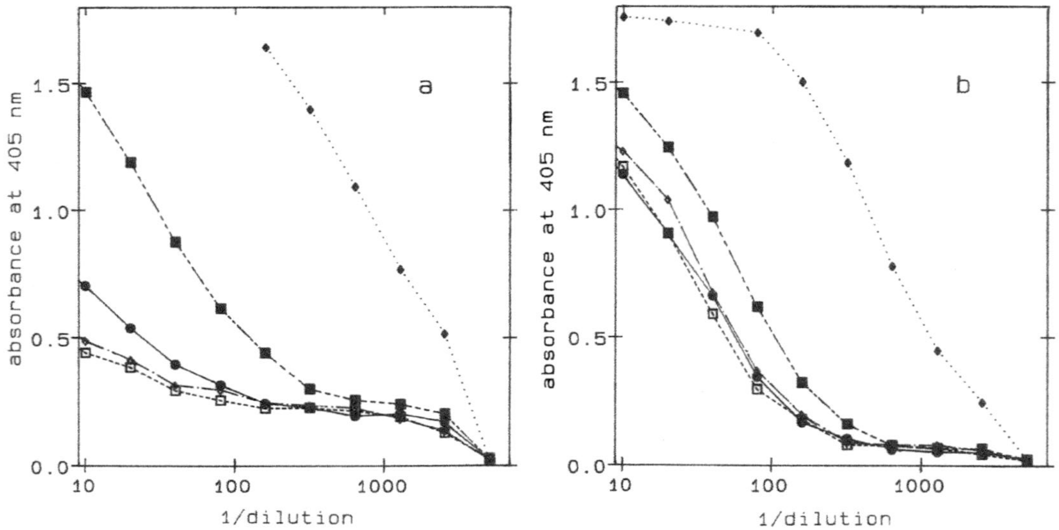

Figure 5. Intranasal immunization of rabbits with peptide-KLH-conjugates (PVP1A-KLH (a) and PVP3A (b)). Rabbits were immunized on days 0, 7, 19. Control antimals received phosphate buffered saline only. Antibody activity in serum samples against the corresponding peptides from days 0 (immunized, ●), 19 (■ ; control □), 26 (◆ , ◇) measured in ELISA.

tide titers when given to rabbits in an intradermal and subcutaneous hyperimmunization scheme with complete Freund's adjuvant (CFA), an incomplete Freund's adjuvant (IFA). Even without using an adjuvant, intranasal infusion of the peptide-KLH conjugates in a buffered saline suspension induced a remarkable antipeptide titer in the serum of the animals (Figure 5). Unfortunately, attempts to measure a local peptide specific IgA response in nasal washings have failed so far, presumably due to technical difficulties. On the other hand, the earlier antirhinovirus vaccine studies had shown that the local IgA titers were paralleled by a comparable raise in serum titers against the immunogen (Knopf et al. 1970). Therefore, we would expect that the serum antibody titers found in our intranasally immunized rabbits reflect also a secretory IgA response.

CARRIER-COUPLED VERSUS FREE PEPTIDES AS IMMUNOGENS

In our opinion, KLH is not a suitable carrier for use in humans and particularly not for a potential rhinovirus vaccine. Likewise, there are a number of problems inherent in the use of a carrier protein, for example, a potential sensitization towards the carrier protein eventually followed by allergic side effects (Audibert, 1986). Another problem may be the possibility of raising a specific epitopic suppression when using a carrier such as tetanus toxoid which most humans are immunized against. This phenomenon is well described in animal experiments (Herzenberg et al. 1980; Schutze et al. 1985), but it is not certain whether it also exists in humans.

In some cases it has been shown that free peptides or oligomers thereof, especially when potential T cell antigenic peptides were linked to them, were able to induce efficient immunity (Dreesman et al., 1982; Francis et al., 1987a; Good et al., 1987; Leclerc et al., 1987). Interestingly, two separate antigenic determinants (residues 141 to 158 and 200 to 213) of VP1 of FMDV coupled together to a 40 amino acid peptide by a central ProProSer spacer elicited high levels of neutralizing antibodies and protected cattle against intradermolingual challenge with infectious virus (DiMarchi et al., 1986).

These results led us to try dimeric forms of both peptides (PVP1A and PVP3A) for an intranasal application in rabbits. We prepared disulfide linked dimers by oxidation of the C-terminal cysteines with hydrogen peroxide and formulated them in buffered salt solutions or suspensions with or without detergent. By using a suspension of dimeric PVP3A with 0.1% nonionic detergent as adjuvant we were able to induce a measurable

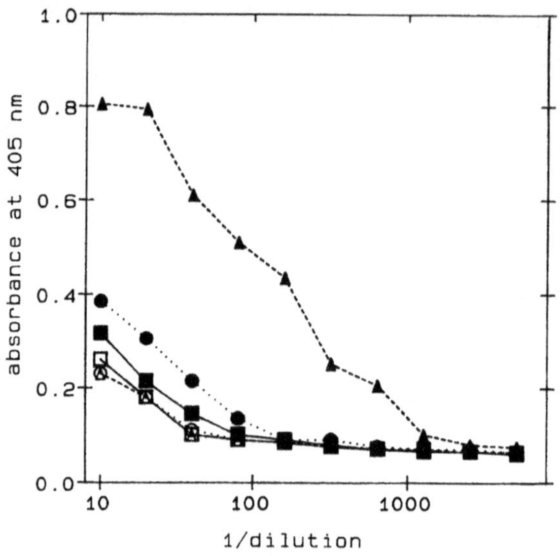

Figure 6. Intranasal immunization of rabbits with PVP3A dimers. Animals were immunized on days 0, 10, 18, and 24. Control rabbits received only buffer containing detergent. Antipeptide antibody activity in serum samples from days 10 (immunized, ■ ; control, □), 18 (●; ○), and 24 (▲ ; △), measured in ELISA.

specific antibody response in serum (Figure 6), although not in all animals immunized and the titers were substantially lower than those obtained with peptides coupled to KLH (Figure 5). In a preliminary experiment, we gave two subsequent intranasal applications of purified HRV 14 in buffered saline to two of those rabbits which did not manifest a measurable antipeptide antibody titer in serum after 3 intranasal dosages of PVP3A dimers. After this virus challenge one animal developed a measurable anti-PVP3A titer concomitant to an anti-virus titer in serum. Interestingly, control rabbits which received virus exclusively did not respond with an anti-PVP3A titer although antibodies binding to other sites of the virus were clearly measurable. These observations indicate that it may in fact be possible to prime for the canyon site using our peptides.

On the other hand, results of the intranasal application of peptide dimers let us hope that we can increase the immunogenicity of the peptides by using better adjuvants (Arnon, 1986). For example, murabutide or liposomes have proven to be effective adjuvants for peptide antigens (Chedid et al., 1982; Hopp, 1984; Watari et al., 1987).

Unfortunately, besides infection of chimpanzees (Jennings and Dick, 1987) and of mice with a mouse adapted strain of HRV 2 (Yin and Lomax, 1983), there is no animal model available for human rhinovirus infections. Therefore, the feasibility of our peptide vaccine could probably most easily be tested in humans. Even if we cannot expect our peptide dimers to be immunogenic in every individual of the large human population because the peptide probably does not contain T cell epitopes for each individual genetic makeup of MHC II molecules, a recent observation (Milich et al., 1987) and our own first experiments with virus challenge made us optimistic that we perhaps need not attach T cell sites for so many different genetic backgrounds (Rothbard, 1987). Since T cells naturally recognize peptide antigens, it is not surprising that one can stimulate T cell help by using isolated peptides. However, unexpectedly the authors (Milich et al., 1987) were able to prime with a single T cell determinant from the nucleocapsid of hepatitis virus for an antibody response, after a subsequent boost with the intact virus, against a second protein comprising the viral envelope. This led to the conclusion that immunogens such as viruses, being aggregates of many different structural proteins, are treated by the immune system as single molecules. B cells internalize the whole virus particles and present T-helper determinants from each of the viral proteins on their surfaces in association with MHC II antigens. Therefore, we would expect that after priming by the peptides for the conformation of the canyon, upon contact with the virus, displaying many different T cell epitopes on its capsid proteins, a sufficient stimulation of T-helper cells could be induced.

CONCLUSIONS

The recent elucidation of the three-dimensional structure of rhinovirus leading to the identification of a potential cell surface receptor binding site and the progress with synthetic peptides as potential vaccines stimulated us to evaluate an approach to a synthetic rhinovirus vaccine. We prepared site specific antibodies in rabbits against two peptides PVP1A and PVP3A of HRV 14 which neutralized roughly 60% of 48 different serotypes *in vitro*, although the corresponding sites on native virus particles were suggested to be inaccessible to antibody molecules for steric reasons. Several indications led us to assume that the antipeptide antibodies indeed neutralize by blocking the cell surface receptor binding site, but a direct proof can only be performed by X-ray crystallography of monoclonal antibodies (Fab fragments) bound to virus particles. First attempts towards an intranasally applicable vaccine form by using peptide dimers to avoid problems inherent to carrier proteins led to the induction of low neutralizing titers of antipeptide antibodies in the serum of immunized rabbits. Therefore, we assume that it might be possible to achieve a more immunogenic form by using more potent adjuvants which could more strongly induce the local secretory immune response thought to be important for protection against rhinovirus infections.

REFERENCES

Abraham G, Colonno RJ (1984) Many rhinovirus serotypes share the same cellular receptor. J Virol 51:340-345.

Allen PM (1987)Antigen processing at the molecular level.Immunology Today 8:270-273.

Al-Nakib W, Tyrell DAJ (1987) A 'new' generation of more potent synthetic anti-rhinovirus compounds: comparison of their MICs and their synergistic interactions. Antiviral Res 8:179-188.

Anderer FA (1963) Preparation and properties of an artificial antigen immunologically related to tobacco mosaic virus. Biochim Biophys Acta 71:246-248.

Arnon R (1986) Peptides as immunogens: prospects for synthetic vaccines. Current Topics Microbiol Immunol 130:1-12.

Arnon R, Maron E, Sela M, Anfinsen CB (1971) Antibodies reactive with native lysozyme elicited by a completely synthetic antigen. Proc Natl Acad Sci USA 68:1450-1455.

Ashwell DJ, Schwartz RH (1986) T-cell recognition of antigen and the Ia molecules as a ternary complex. Nature 320:176-179.

Audibert F (1986) Carriers and adjuvants for the development of synthetic vaccines. In 1st Forum in Virology.Paris: Ann Inst. Pasteur/Virol. 137E:pp.514-517.

Audibert F, Jolivet M, Chedid L, Arnon R, Sela M (1982) Successful immunization with a totally synthetic diphtheria vaccine.Proc Natl Acad Sci USA 79:5042-5046.

Beem M (1969) Rhinovirus infections in nursery school children (abstract). J Pediatr 74:818-820.

Bittle JL, Houghten RA, Alexander H, Shinnik TM, Sutcliffe JG, Lerner RA, Rowlands DJ, Brown F (1982) Protection against foot-and-mouth-disease by immunization with a chemically synthesized peptide predicted from the viral nucleotide sequence. Nature 298:30-33.

Buscho RF, Perkins JC, Knopf HLS, Kapikian AZ, Chanock RM (1972) Further characterization of the local respiratory tract antibody response induced by intranasal instillation of inactivated rhinovirus 13 vaccine. J Immunol 108:169-177.

Callahan PL, Mizutani S, Colonno RJ (1985) Molecular cloning and complete sequence determination of the RNA genome of human rhinovirus type 14. Proc Natl Acad Sci USA 82:732-736.

Carrol AR, Rowlands DJ, Clarke BE (1984) The complete nucleotide sequence of the RNA coding for the primary translation product of foot and mouth disease virus. Nucl Acids Res 12:2461-2472.

Chang CH, Short MT, Westholen FA, Stevens FJ, Wang BC, Furey Jr W, Solanai A, Schiffer M (1985) Novel arrangement of immunoglobulin variable domains: X-ray crystallographic analysis of the R-chain dimer Bence-Jones protein loc. Biochemistry 24:4890-4897.

Chanh TC, Dreesman GR, Kanda P, Linette GP, Sparrow JT, Ho DD, Kennedy RC (1986) Induction of anti-HIV neutralizing antibodies by synthetic peptides. EMBO J 5:3065-3071.

Chedid LA, Parant MA, Audibert FM, Riveau GJ, Parant FJ, Lederer E, Choay JP, LeFrancier PL (1982) Biological activity of a new synthetic muramyl peptide devoid of pyrogenicity. Infect Immun 35:417-424.

Cheung A, Leban J, Shaw AR, Merkli B, Stocker J, Chizzolini C, Sander C, Perrin LH (1986) Immunization with synthetic peptides of a *Plasmodium falciparum* surface antigen induces antimerozoite antibodies. Proc Natl Acad Sci USA 83:8328-8332.

Chow M, Yabrov R. Bittle J, Hogle J, Baltimore D (1985). Synthetic peptides from four separate regions of the poliovirus type 1 capsid protein VP1 induce neutralizing antibodies. Proc Natl Acad Sci USA 82:910-914.

Cohen M (1971) The take-home lesson. Ann N.Y. Acad Sci 190:529-584.

Collins PL, Huang YT, Wertz GW (1984) Nucleotide sequence of the gene

encoding the fusion (F) glycoprotein of human respiratory syncytial virus. Proc Natl Acad Sci USA 81:7683-7687.
Colonno RJ, Callahan PL, Long WJ (1986) Isolation of a monoclonal antibody that blocks attachment of the major group of human rhinoviruses.J Virol 57:7-12.
Cooney MK, Fox JP, Kenney GE (1982) Antigenic groupings of 90 rhinovirus serotypes. Infect Immun 37:642-647.
Couch RB (1984) The common cold: control? J Infect Dis 150:167-173.
Dale JB, Beachey EH (1982) Protective antigenic determinant of streptococcal M protein shared with sarcolemma membrane protein of human heart. J Exp Med 156:1165-1176.
Delisi C, Berzofsky J (1985) T-cell antigenic sites tend to be amphipathic structures. Proc Natl Acad Sci USA 82:7048-7052.
DiMarchi R, Brooke G, Gale C, Crackwell V, Doel T, Mowat N (1986) Protection of cattle against foot-and-mouth disease by a synthetic peptide.Science 232:639-641.
Dimmock NJ (1987) Multiple mechanisms of neutralization of animal viruses. Trends Biochem Sci 12:70-75.
Dreesman GR, Sanchez Y, Ionescu-Matin I, Sparrow JT, Six HR, Peterson DL, Hollinger FB, Melnick JL (1982) Antibody to hepatitis B surface antigen after a single inoculation of uncoupled synthetic HBsAg peptides. Nature 295:158-160.
Duechler M, Skern T, Sommergruber W, Neubauer C, Grundler P, Fogy I, Blaas D, Kuechler E (1987) Evolutionary relationships within the human rhinovirus genes: Comparison of serotypes 89,2 and 14. Proc Natl Acad Sci USA 84:2605-2609.
Emini EA, Jameson BA, Wimmer E (1983a) Priming for and induction of anti-polio-virus neutralizing antibodies by synthetic peptides. Nature 304:699-703.
Emini EA, Show YK, Lewis AJ, Crainic R, Wimmer E (1983b) Functional basis of poliovirus neutralization determined with monospecific neutralizing antibodies. J Virol 46:466-474.
Fox JP (1976) Reviews and commentary-is a rhinovirus vaccine possible? Am J Epidemiol 103:345-354.
Francis MJ, Fry CM, Rowlands DJ, Brown F, Bittle JL, Houghten RA, Lerner RA (1985) Immunological priming with synthetic peptides of foot-and-mouth disease virus. J Gen Virol 66:2347-2354.
Francis MJ, Hastings GZ, Syred AD, McGinn B, Brown F, Rowlands DJ (1987a) Non-responsiveness to a foot-and-mouth disease virus peptide overcome by addition of foreign helper T-cell determinants. Nature 300:168-170.
Francis MJ, Hastings GZ, Sangar DV, Clark RP, Syred A, CIarke BE, Rowlands DJ, Brown F (1987b) A synthetic peptide which elicits neutralizing antibody against human rhinovirus type 2. J Gen Virol 68:2687-2691.
Gerin JL, Alexander H, Shih JWK, Purcell RH, Dapolito G, Engle R, Green N, Sutcliffe JG, Shinnick TM, Lerner RA (1983) Chemically synthesized peptides of hepatitis B surface antigen duplicate the d/y specificities and induce subtype-specific antibodies in chimpanzees. Proc Natl Acad Sci USA 80:2365-2369.
Good MF, Maloy WL, Lunde MN, Margalit H, Cornette JL, Smith GL, Moss B, Miller LH, Berzofsky JA (1987) Construction of a synthetic immunogen: use of new T-helper epitope on malaria circumsporozoite protein. Science 235:1059-1062.
Gregg I (1983) Provocation of airflow limitation by viral infection: implication for treatment. Eur J Resp Dis (suppl) 128:369-379.
Guillet JG, Lai MZ, Briner TJ, Buus S, Sette A, Grey HM, Smith JA, Gefter ML (1987) Immunological self, nonself discrimination. Science 235:865-870.
Guyon-Gruaz A, Delmas A, Pedoussaut S, Halimi H, Milhaud G, Raulais D, Rivaille P (1986) Oral immunization with a synthetic peptide of cholera toxin B subunit. Eur J Biochem 159:525-528.
Gwaltney JM, Jr (1982) Rhinoviruses. In Evans AS (ed.), Viral Infections in Humans: Epidemiology and Control. New York: Plenum Publishing Corporation, pp. 491-514.
Herrington DA, Clyde DF, Losonsky G, Cortesia M, Murphy JR, Davis J, Baqar S, Felix AM, Heimer EP, Gillessen D, Nardin E, Nussenzweig RS,

Nussenzweig V, Hollingdale MR, Levine MM (1987) Safety and immunogenicity in man of a synthetic peptide malaria vaccine against *Plasmodium falciparum* sporozoites. Nature 328:257-259.
Herzenberg LA, Tokuhisha T, Herzenberg LA (1980) Carrier priming leads to hapten-specific suppression. Nature 285:664-667.
Hirayama A, Takagaki Y, Karush (1985) Interaction of monoclonal antibodies with lysozyme. J Immunol 134:3241-3247.
Ho DD, Kaplan JC, Rackaukas IE, Gurney M (1988) Second conserved domain of gp 120 is important for HIV infectivity and antibody neutralization. Science 239:1021-1023.
Hogle JM, Chow M, Filman DJ (1985) Three-dimensional structure of poliovirus at 2.9 A resolution. Science 229:1358-1365.
Hopp TP (1984) Immunogenicity of a synthetic HBsAg peptide: enhancement by conjugation to a fatty acid carrier. Mol Immunol 21:13-16.
Hughes PJ. North C, Jellis CH, Minor PD, Stanway G (1988) The nucleotide sequence of human rhinovirus 1B: molecular relationships within the rhinovirus genus. J Gen Virol 69:49-58.
Itoh Y, Takai E, Ohnuma H, Kitajima K, Tsuda F, Machida A, Mishiro S, Nakamura T, Miyakawa Y, Mayumi M (1986) A synthetic peptide vaccine involving the product of the pre- S(2) region of hepatitis B virus DNA: protective efficacy in chimpanzees. Proc Natl Acad Sci USA 83:9174-9178.
Jacob CO, Pines M, Arnon R (1984) Neutralization of heat-labile toxin of *E. coli* by antibodies to synthetic peptides derived from the B subunit of cholera toxin. EMBO J 3:2889-2893.
Jennings LC, Dick EC (1987) Transmission and control of rhinovirus colds. Eur J Epidemiol 3:327-335.
Kitamura N, Semler BL, Rothberg PG, Larsen GR, Adler CJ, Dorner AJ, Emini EA, Hanecak R, Lee JJ, van der Werf S, Anderson CW, Wimmer E (1981) Primary structure, gene organization and polypeptide expression of poliovirus RNA. Nature 291:547-553.
Klipstein FA, Engert RF, Houghton RA (1986) Immunization of volunteers with a synthetic peptide vaccine for enterotoxigenic *Escherichia coli*. Lancet 1:471-472.
Knopf HLS, Perkins JC, Bertran DM, Kapikian AZ, Chanock RM (1970) Analysis of the neutralizing activity in nasal wash and serum following intranasal vaccination with inactivated type 13 rhinovirus. J Immunol 104:566-573.
Krilov L, Pierik L, Keller E, Mahan K, Watson D, Hirsch M, Hamparian V, McIntosh K (1986) The association of rhinoviruses with lower respiratory tract disease in hospitalized patients. J Med Virol 19:345-352.
Landsteiner K (1946) The specificity of immunological reactions. Harvard University Press.
Leclerc C, Przewlocki G, Schutz MP, Chedid L (1987) A synthetic vaccine constructed by copolymerization of B and T cell determinants. Eur J Immunol 17:269-273.
Lerner RA (1984) Antibodies of predetermined specificity in biology and medicine. Adv. Immunol 36:1-44.
Lonberg-Holm K, Korant BD (1972) Early interaction of rhinoviruses with host cells. J Virol 9:29-40.
Luo M, Vriend G, Kamer G, Minor I, Arnold E, Rossmann MG, Boege U, Scsaba G, Duke GM, Palmenberg AC (1987) The atomic structure of Mengo Virus at 3.0 A resolution. Science 235:182-191.
McCray JW, Werner G (1987) Different rhinovirus serotypes neutralized by antipeptide antibodies. Nature 329:736-738.
Medappa KC, McLean C, Rueckert RR (1971) The structure of human rhinovirus 1A. Virology 44:259-270.
Milich DR, McLachlan A, Thornton GB, Hughes JL (1987) Antibody production to the nucleocapsid and envelope of the hepatitis B virus primed by a single synthetic T cell site. Nature 329:547-549.
Milstein C (1986) From antibody structure to immunological diversification of immune response. Science 231:1261-1268.
Mitchison NA (1971) The carrier effect in the secondary response to hapten-protein conjugates. I. Measurement of the effect with transferred cells and objections to the local environment hypothesis. Eur J Immunol 1:10-19.
Mogabgab WJ, Holmes BJ, Pollack B (1975) Antigenic relationships of com-

mon rhinovirus types from disabling upper respiratory illnesses. Dev Biol Stand 28:400-411.
Palmberg AC, Kirby EM, Janda MR, Drake NL, Duke GM, Potrak KF, Collett MS (1984) The nucleotide and deduced amino acid sequences of the encephalomyocarditis viral polyprotein coding region. Nucl Acids Res 12:2969-2985.
Patarroyo ME, Amador R, Clavijo P, Moreno A, Guzman F, Romero P, Tascon R, Franco A, Murillo LA, Ponton G, Trujillo G (1988) A synthetic vaccine protects humans against challenge with asexual blood stages of *Plasmodium falciparum* malaria. Nature 332:158-161.
Phillpotts RJ, Tyrell DAJ (1985) Rhinovirus colds. Br Med Bul 41:386-390.
Pierce SK, Margoliash E (1988) Antigen Processing: an interim report. Trends Biochem Sci 13:27-29.
Przewlocki G, Audibert F, Jolivet M, Chedied L, Kent SBH, Neurath AR (1986) Production of antibodies recognizing a hepatitis B virus (HBV) surface antigen by administration or murabutide associated to a synthetic pre-S HBV peptide conjugated to a toxoid career. Biochem Biophys Res Commu 140:557-564.
Racaniello VR, Baltimore D (1981) Molecular cloning of poliovirus cDNA and determination of the complete nucleotide sequence of the viral genome. Proc Natl Acad Sci USA 78:4887-4891.
Rossmann MG, Arnold E. Erickson JW, Frankenberger EA, Griffith JP, Hecht HJ, Johnson JE, Kamer G, Luo M, Mosser AG, Rueckert RR, Sherry B, Vreind G (1985) Structure of a human common cold virus and functional relationship to other picornaviruses. Nature 317:145-153.
Rossmann MG, Arnold E, Griffith JP, Kamer G, Luo M, Smith TH, Vriend G, Rueckert RR, Sherry B, McKinley MA, Diana G, Otto M (1987) Common cold viruses. Trends Biochem Sci 12:313-318.
Rothbard J (1987) Synthetic peptides as vaccines. Nature 330:106-107.
Rothbard JB, Schoolnik GK (1985) The primary sequence and antigenic structure of gonococcal pilin: Approaches towards a gonococcal vaccine. Adv Exp Med Biol. 185:247-273.
Rueckert RR, Sherry B, Mosser A, Colonno R, Rossmann M (1986) Location of Four Neutralization Antigens on the Three-Dimensional Surface of a Common-Cold Picornavirus, Human Rhinovirus 14. In Crowell RL, Lonberg-Holm K (ed.), Virus Attachment and Entry into Cells. Washington D.C.: American Society for Microbiology, pp. 21-27.
Sarinn PS, Sun DK, Thornton AH, Naylor pH, Goldstein AL (1986) Neutralization of HTLV-III/LAV replication by antiserum to thymosin $\alpha 1$. Science 232:1135-1137.
Schutze MP, Leclerc C, Jolivet M, Audibert F, Chedid L (1985) Carrier-induced epitopic suppression, a major issue for future synthetic vaccines. J Immunol 135:2319-2322.
Shapira M, Arnon R (1986) In Kandell AP, Patriarca PA (eds.),. Options for the control of influenza. Lissabon, pp 391-406.
Skern T, Sommergruber W, Blass D, Grundler P, Fraundorferr F, Pieler C, Fogy I, Kuechler E (1985) Human rhinovirus 2: complete nucleotide sequence and proteolytic processing signals in the capsid region. Nucl Acids Res 13:2111-2126.
Stanway G, Hughes PJ, Mountford RC, Reeve P, Minor PD, Schild GC, Almond JW (1984a) Comparison of the complete nucleotide sequences of the genomes of the neurovirulent poliovirus P3/Leon/37 and its attenuated Sabin vaccine derivative P3/Leon 12a1b. Proc Natl Acad Sci USA 81:1539-1543.
Stanway G, Hughes PJ, Mountford RC, Minor PD, Almond JW (1984b) The complete nucleotide sequence of a common cold virus human rhinovirus 14. Nucl Acids Res 12:7859-7875.
Stevens VC, (1986) Current status of antifertility vaccines using gonadotropin immunogens. Immunol Today 7:369-374
Steward MW, Howard CR (1987) Synthetic peptides: A next generation of vaccines? Immunol Today 8:51-58.
Sutcliffe IG, Shinnick TM, Green N, Lerner RA (1983) Antibodies that react with predetermined sites on proteins. Science 219:660-666.
Walter G (1986) Production and use of antibodies against synthetic peptides. J Immunol Methods 88: 149-161.
Watari E, Dietzschold B, Szokan G, Heber-Katz E (1987) A synthetic pep-

tide induces long-term protection from lethal infection with herpes-simplex virus 2. J Exp Med 165:459-470.
Wilson IA, Niman HL, Houghton RA, Cherenson AR, Connolly ML, Lerner RA (1984) The structure of an antigenic determinant in a protein. Cell 37:767-778.
Wong GG, Witek JS, Temple PA, Wilkens KM, Leary AC, Luxenberg DP, Jones SS, Brown EL, Kay RM, Orr EC, Shoemaker C, Golde DW, Kaufman RJ, Hewick RM, Wang EA, Clark SC (1985) Human GM-CSF: molecular cloning of the complementary DNA and purification of the natural and recombinant proteins. Science 228:810-814.
Yin FH, Lomax NB (1983) Host range mutants of human rhinoviruses in which non-structural proteins are altered. J Virol 48:410-418.
Zanetti M, Sercarz E, Salk J (1987) The immunology of new generation vaccines. Immunol Today 8:18-25.

DISCUSSION

***Cesario T (University of California Irvine Medical Center, Orange, CA)*:**

What do we know about the body's reaction to rhinoviruses? We know that alpha IFN is useful and can be protective, but what do we know of the body's cellular immune response, for example, to rhinoviruses?

***Werner G*:**

I think the cellular immune response in rhinovirus infections is not important. It is only the local IgA titer which can prevent infection.

***Cesario T*:**

That leads me to my next question, and that is also my assumption. Since the late hypersensitivity takes two to three weeks or perhaps more to develop, its unlikely to play an important role in current developments. On the other hand, if we give a vaccine to rhinoviruses, is it possible that we might induce delayed hypersensitivity, but then on exposure to the live virus could, in fact, have a deleterous result. The reason I say that is because of our experience with parainfluenza vaccines lead us in that direction.

***Werner G*:**

Yes, this is a possibility. Maybe we sensitize for those sites and induce a cellular immune response.

***Cesario T*:**

I think that is a question that will come up increasingly with all types of vaccines. It might be useful in these kinds of studies to see, at some point, if one does in fact induce a significant delayed hypersensitivity reaction. Because it certainly looks, at least on the surface, from your work and from some of the earlier work from Colonno that this has enormous clinical potential to exploit this.

***Werner G*:**

I think normally, the epitopes for CD8 positive cytotoxic T-cells are different from those for T helper cells. So, one should collect more data about this.

***Cesario T*:**

Yes, the CD4 cells, I think, rely more on Class II histocompatibility, and the CD8 on Class I.

***Werner G*:**

The peptides seem to be different. They are created, probably, in different compartments of the cell. The association with Class I molecules, which is important for induction of cytotoxic T cells, is different from that with Class II molecules. At least, that's the model at present.

***Barshinger C (CFS Research Foundation, Inc., Denver, CO)*:**

You mentioned that the binding cavity is not accessible to the antibodies, yet somehow you're getting results. Have you considered the possibility of pulling back or opening up the binding cavity and using the virus itself.

***Werner G*:**

We believe that upon binding to the cellular receptor, a conformational change of the virus particle may be induced and then this side may be accessible to antibodies. That's one explanation. We have no direct proof that our antipeptide antibodies bind in the canyon. But that would be one possibility, how to explain the neutralization results.

***Jones J (National Jewish Center for Immunology and Respiratory Medicine, Denver, CO)*:**

Two other theoretical problems with peptide vaccines are one, if you make an antibody that will eventually induce an anti-idiotype, would that anti-idiotype then bind to the original receptor and alter cell function in some way that may or may not be expected?

***Werner G*:**

Yes. An anti-idiotype then would mimic the receptor binding site, and you are thinking that you would induce altered immune responses?

***Jones J*:**

That's one possibility. The other is that your antibody then will also bind to the original ligand directed against the receptor. Could that alter in another way? Are these things being looked at, not only by you, but by others who are looking at peptide vaccines?

***Werner G*:**

We have isolated mutants of rhino 14 which are resistant to these neutralizing antipeptide antibodies. But, we have not yet analyzed on a molecular level these mutants.

ANTIVIRAL CYTOTOXIC T-LYMPHOCYTES AND VACCINES

STEPHEN MARTIN AND BARRY T. ROUSE

Department of Microbiology, College of Veterinary Medicine, University of Tennessee, Knoxville, TN 37996-0845 (USA)

INTRODUCTION

A characteristic of antigen recognition by T lymphocytes is restriction by the gene products of the major histocompatibility complex (MHC) (Benacerraf, 1981; Zinkernagel and Doherty, 1979). Until 1983, class I MHC molecules (H-2K, D, and L in the mouse and HLA-A, B, and C in man) were considered as the sole restriction elements for virus-specific cytotoxic T lymphocytes (CTL) (Biddison, 1982; McMichael, 1980), but recently class II MHC molecules (Ia in the mouse and HLA-D and DR in man) were shown to restrict the recognition of antigen by some antiviral CTL (Jacobson and Biddison, 1986). The significance of class II MHC restricted CTL, however, remains unknown and, with the exception of measles virus (Jacobson et al. 1984, 1985), the majority of antiviral CTL appears to be restricted by class I MHC gene products. Therefore, our comments in this review will concentrate on class I MHC restricted CTL and, in addition, will emphasize responses against herpes simplex virus type 1 (HSV-1).

A consequence of MHC restriction is that T cells usually respond only to antigen upon the surface of an autologous cell. After virus specific CTL bind to viral antigen and MHC on the surface of a cell there follows a series of poorly defined events which result in cell lysis. Several studies have shown that CTL mediated lysis of infected cells *in vitro* can occur during the eclipse phase of virus replication (Zinkernagel and Anthage, 1977) leading to cell lysis prior to the release of infectious virus progeny. Whether CTL function *in vivo* via a similar lytic mechanism or via the release of immunomodulatory lymphokines such as gamma interferon (IFN-γ) remains undefined with most virus systems. It may well be that a combination of these events are necessary for the successful action of antiviral CTL.

ACTIVITY OF CTL *IN VIVO*

There is now compelling evidence that CTL play an important role in the control of virus infections. Thus, with many virus infections in the murine system, adoptive transfers of defined lymphocyte populations with specific CTL activity could control a subsequent infection. For example, Lin and Askonas (1981) demonstrated that influenza A virus-specific CTL clones mediated a reduction of virus titer in infected lungs, and in the lymphocytic choriomeningitis (LCMV) system CTL clones were able to clear virus from the infected spleen (Byrne and Oldstone, 1984). Moreover, virus specific CTL may also serve to protect mice against lethal HSV (Sethi et al. 1983), or lethal challenge of mice with simian virus 40 (SV40) transformed syngeneic cells (Gooding, 1982). In none of these instances was the lytic function of CTL shown to be responsible for the protective function. Indeed, the idea has arisen that only CTL clones also capable of IFN-γ production act in a protective fashion *in vivo* (Askonas and Pala, 1985). However, in one instance a role for the lytic function of CTL was clearly implicated (Lukacher et al. 1984). Thus Lukacher et al. (1984) demonstrated that an influenza A subtype-specific CTL clone could only reduce the titer of homologous virus subtype in the lungs of mice when infected concomitantly with both homologous and heterologous subtypes. In contrast, the adoptive transfer of a CTL clone which recognized cross reactive determinants reduced the titer of both virus subtypes in the doubly infected lungs. The interpretation of these data was that CTL mediated immunity required specific cell contact and may preclude the idea that CTL function *in vivo* by the release of nonspecifically acting lymphokines.

In the HSV system, studies by ourselves and others indicate that antiviral immunity is "incomplete" in the absence of a functional CTL response. This was first suggested by the early work of Howes et al. (1979) in which they showed that immune lymphocytes restricted to both class I and II MHC antigens were needed for resistance against lethal HSV challenge, a finding later confirmed by both Nash et al. (1981) and ourselves (Larsen et al. 1984). This concept was taken one step further by studies which showed that mice depleted of CD8 ($Lyt2^+$) bearing cells (in mice CTL have the $Lyt1^-2^+,3^+$ or CD8 phenotype) could survive a challenge of 10^4 plaque forming unit (PFU) of virus inoculated into the ear pinna, but were totally susceptible to a challenge of 10^5 PFU (Wildy and Gell, 1985). The interpretation of these findings was that whereas smaller challenge doses of HSV were controlled by CD4+ lymphocytes, complete protection against larger challenge doses required a functional CTL population. Recently, we provided further support for the above from studies on the levels of protection induced by recombinant vaccinia virus vectors in mice (Martin and Rouse, 1987b).

We showed that infection of mice expressing the H-2^k MHC haplotype with vaccinia virus recombinants which expressed genes coding for HSV-1 glycoprotein B (gB) (Martin et al. 1988a) or glycoprotein D (gD) (Martin et al. 1987a), but not a recombinant expressing the influenza A hemagglutinin (VHa) molecule, stimulated many aspects of anti-HSV T cell immunity with the notable exception of virus specific CTL. Moreover, whereas mice immunized with either the gB (Figure 1) (Martin and Rouse, 1987b) or the gD recombinant were immune to challenge with a low dose of virus, they failed to resist challenge with a high dose. Experiments currently in progress suggest that this situation can be reversed by the supplementation of gD or gB immune mice with HSV immune lymphocyte populations enriched for anti-HSV CTL activity (Martin et al. 1988a).

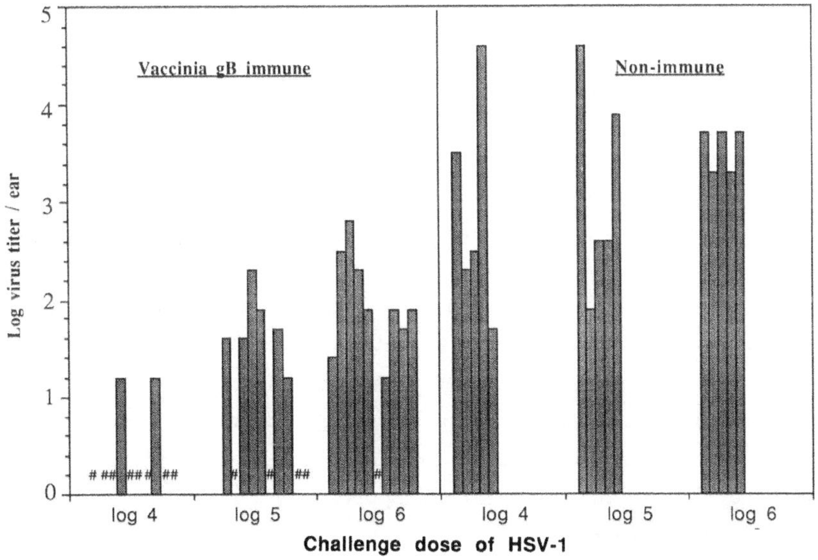

Figure 1. Quantitative analysis of ear clearance ability. Fourteen days after subcutaneous inoculation (left pinna) with the viruses shown groups of C3H/TEN were challenged in the right pinna with the dose of HSV-1 shown. Four days after challenge, the right pinnae were removed and the infectious virus titers (low_{10} $TCID_{50}$) determined on Vero cell monolayers. Results are shown for individual mice with animals represented as (#) having a virus titer of less than 0.47 logs of virus/ear.

The studies of Nash et al. (1981) and ourselves have used the model of HSV clearance from the skin of mice infected in the pinna since, like the clinical disease in man, this infection is usually non-lethal and localized to a small region of the epithelium (Hill et al. 1975). In this model, as with infection of man, axonal translocation along the sensory nerves innervating the infected neurodermatome may be as important as cell to cell contact for the cutaneous spread of a primary HSV infection (Hill et al. 1975; Wildy, 1985). Thus, the successful control of a primary infection would involve both the clearance of replicating virus at the site of challenge and the prevention of virus infecting the sensory nerves which innervate the site of epithelial virus replication, thereby inhibiting spread of virus to new areas of susceptible host tissue.

Based upon this model, one likely explanation for the inability of mice lacking CTL activity to clear high challenge doses of virus is that CTL independent immune mechanisms are involved only in the clearance of the epithelium infection, whereas CTL are involved in the control of the infection of the sensory nervous system. Thus, presumably CTL independent immune mechanisms can clear low doses of HSV from the challenge site with sufficient rapidity to prevent virus spread into the nervous system. However, such a mechanism cannot control a large challenge dose of HSV with sufficient rapidity to prevent a subsequent infection of the innervating sensory nerves. In the absence of a functional HSV-specific CTL response the infection of the nervous tissues continues unabated and virus spreads to new areas of the epithelium within the infected pinna.

This interpretation of the role of antiviral CTL in controlling infections gains some credibility in light of the recent studies of Nash and colleagues (1987). These authors demonstrated that CD8 depleted mice show a severe infection of the nervous system upon primary infection with HSV when compared to normal "CTL competent" mice. This is not, however, the only possible interpretation of the data, and the notion reviewed by Doherty (1985) that nerve tissues do not normally express class I MHC antigens would argue against CTL controlling virus infections of the CNS.

An alternative interpretation supported by our own data is that CTL also serve their function by controlling the epithelial infection of viruses such as HSV, but for the reasons outlined below their actions only become necessary in the face of large challenge doses of virus. In other words, class II MHC restricted responses control HSV challenge by inhibiting virus replication and the cell to cell spread of infection by mechanisms which, due to the lack of Ia expression murine epithelial cells, are largely non-focused and nonspecific. These might include lymphokine release (e.g., IFNs, tumor necrosis factors, chemotactins, etc.) together with the recruitment and activation of indiscriminate killer cells such as macrophages, polymorphonuclear leukocytes, and natural killer cells. However, since these mechanisms tend to show an unfocused, almost nonspecific activity (i.e., their recruitment is antigen specific, but not their action), they are unable to channel their full antiviral potential upon infected cells. A consequence of this would be that their actions are only truly effective when the viral challenge is small and involves only a small number of infected cells.

It thus follows that when the viral challenge is large, a more effective, more focused antiviral response is necessary. Our own studies which assessed the ability of antiviral lymphocyte populations to control a localized pinna infection with a relatively non-neurovirulent strain of HSV-1 lends credence to this interpretation (Martin and Rouse manuscript in preparation). Thus, mice immunized with either vacc gB or vacc gD are unable to rapidly clear a high challenge dose (6 logs of virus) of HSV-1 KOS despite its inability to infect the local sensory nerves and utilize axonal translocation as a primary mechanism of spread. This failure to clear a localized epithelial infection, despite obvious levels of antiviral antibody and virus recognizing CD4+ T cells, could be attributed to the inability of these constructs to induce HSV-1 specific CTL with the CD8+ phenotype.

As discussed previously, due to their restriction by class I MHC, antiviral CTL represent an exquisitely focused immune response and lyse the target cell they recognize. It would seem realistic to assume that if CTL recognition was to occur during the eclipse phase of viral replication, i.e., subsequent to infection but prior to the production of new infectious progeny, then one would predict that the resulting lysis would be effective in inhibiting virus spread (Zinkernagel and Anthage, 1977) and as a consequence would prevent the virus reaching the local sensory nerve endings. In order for CTL to recognize infected cells during this eclipse phase they would be expected to recognize either the exogenously acquired viral glycoproteins derived from the infecting virus, or the gene products which are the first to be synthesized after infection (alpha or immediate early genes) and are involved in the subsequent production of new infectious viral progeny (Roizman and Batterson, 1985).

In our studies with HSV we have found that target cells pretreated with actinomycin D to prevent viral gene expression and then infected with high multiplicities (>10) of HSV-1 bear serologically detectable levels of HSV glycoproteins (data not shown). Such target cells, however, act as very poor targets for murine HSV-specific memory CTL populations (Table 1). It is demonstrated in Table 2 that autologous L cells allowed to express only the immediate early viral gene products following infection with HSV-1 are good targets for HSV-specific memory CTL populations (Martin et al. 1988b). Quantitation of these HSV-1 specific CTL revealed that between 25% and 35% of CTL present in the draining lymph node of acutely infected mice recognize the viral immediate early gene products (Martin et al. 1988b). Some of these immediate early or alpha gene products of HSV-1 are thought to be transactivating factors (Everett, 1984) and their expression is a prerequisite to the subsequent expression of later gene classes which code for the structural components of HSV (Roizman and Batterson, 1985). Thus, the alpha gene specific CTL would be expected to recognize infected cells in the eclipse phase of replication and hence destroy them before infectious progeny are produced. The possibility of these CTL acting in a similar fashion *in vivo* needs to be investigated but would seem likely in view of the recent studies of Jonjic et al., 1988.

TABLE 1. LIMITED RECOGNITION OF HSV-1 STRUCTURAL ANTIGENS BY HSV-1 SPECIFIC MEMORY CTL POPULATIONS.

	Effector to Target Ratio			
Target Cell	100:1	50:1	25:1	12.5:1
L Cell ($H-2^k$)	10.7 ± 1.1[b]	8.7 ± 0.8	7.3 ± 1.2	4.0 ± 1.4
L Cell + ActD[a] + HSV-1	18.2 ± 1.5	15.1 ± 1.8	10.7 ± 1.1	7.4 ± 1.4
L Cell + HSV-1	75.5 ± 3.5	77.4 ± 8.3	66.8 ± 3.0	35.5 ± 6.6
EL 4 + HSV-1 ($H-2^b$)	10.5 ± 1.6	11.23± 1.4	3.9 ± 1.6	1.0 ± 2.0

Immune splenocytes derived from C3H/TEN ($H-2^k$) mice were stimulated *in vitro* with UV-inactivated HSV-1 for 5 days. The lymphocytes were then washed and mixed with ^{51}Cr-labeled target cells at various effector to target ratios and incubated for four hours in the presence of actinomycin D (5 μg/ml). Target cells had been previously infected with HSV-1 (MOI = 10) and simultaneously labeled with ^{51}Cr for a period of 4 hours.

[a] Target cells pretreated with actinomycin D (5 μg/ml) prior to infection with HSV-1 do not allow subsequent viral gene expression and are ideal targets for assessing the contribution of input viral structural antigens to the formation of CTL target antigen complexes.

[b] Percent specific lysis plus one standard deviation.

TABLE 2. RECOGNITION OF HSV-1 IMMEDIATE EARLY GENE PRODUCTS BY HSV-1 SPECIFIC MEMORY CTL POPULATIONS.

TARGET CELL (Pretreatment (Regime)	Effector to Target Ratio 40:1	20:1	10:1
L Cell ($H-2^k$)			
-a	5.8 ± 47[a]	3.9 ± 2.9	2.6 ± 3.4
-b			
L Cell + HSV-1			
ActD (5 μg/ml)	4.9 ± 4.9	3.1 ± 2.0	1.0 ± 0.5
L Cell + HSV-1	36.4 ± 4.2	28.9 ± 1.2	18.7 ± 1.3
L Cell			
Cycloheximide (75 μg/ml)			
ActD (75 μg/ml)	7.0 ± 6.2	4.6 ± 2.6	2.4 ± 1.3
L Cell + HSV-1			
Cycloheximide (75 μg/ml)			
ActD (5 μg/ml)	26.2 ± 5.2	25.8 ± 1.1	14.9 ± 1.1

CTL were generated as outlined in the legend for Table 1 and tested at the effector to target ratios shown in a four-hour chromium release assay. The rationale for the various inhibitors combinations is as follows: the first step (a) includes the addition of cycloheximide (translation inhibitor) or actinomycin D (transcription inhibitor) 1 hour prior to the infection of the cells with HSV-1 (M.O.I. 5). The cells are then cultured for 8-12 hours in the continued presence of the inhibitor. The second step involves the removal of cycloheximide by continued washing in the presence of actinomycin D. Cells are then cultured for a further 2 hours in the presence of actinomycin D before being used in the assay. This sequential treatment of HSV-1 infected cells allows the expression of viral immediate early genes (α genes) in the absence of the early (β) and late (γ) genes (Roizman and Batterson, 1985). The Actinomycin D pretreated cells do not express any viral genes act as controls for the contribution of input viral antigens to the formation of CTL target complexes.

[a] Results are expressed as mean percent specific lysis plus one standard deviation.

These authors use the infection of mice with murine cytomegalovirus (MCMV) as a model system with which to analyze the protective function of antiviral T-lymphocytes and to determine which viral antigens they recognize. With this model system they have shown that immune T lymphocytes bearing the CD8 phenotype limit viral spread, prevent tissue destruction, and protect against lethal disease (Reddehase et al. 1985, 1987a, 1987b, 1988). This population of antiviral CD8+ T lymphocytes are the cells which show CTL activity, the majority of which are reported to be specific for the major immediate early (IE) phosphoprotein of MCMV (Reddehase and Koszinowski 1984). This groups most convincing demonstration of the importance of CTL which recognize viral IE antigens was with studies of mice vaccinated with a vaccinia recombinant expressing the pp 89 gene of MCMV (Jonjic et al. 1988). This recombinant vector induced MCMV specific CD8+ T lymphocytes which were able to limit challenge MCMV replication in host tissues and consequently prevent mortality.

At the present time we lack sufficient data to similarly evaluate the precise mechanism of action of HSV-specific CTL and are actively investigating the ear clearance model to gain more insight into this problem. One aspect of research into antiviral mechanisms highlighted by these studies is the necessity to evaluate lymphocyte effector functions in a model which mimics as closely as possible the disease seen in the natural host. Thus HSV in man is a non-lethal epithelial disease which can involve infection of sensory nerves and their ganglia, and it is thought that axonal translocation is one of the major routes of virus spread to new uninfected epithelia. The mouse ear model mimics closely this pathogenesis of primary disease (Hill et al. 1975; Wildy, 1985). In contrast the lethal intraperitoneal infections used by previous workers to evaluate HSV-specific CTL (Sethi et al. 1983) involves an atypical pathogenesis, including nephritis, viremia (Renis et al. 1976) and the induction of suppressor cells specific for DTH (Lathey et al. 1987; Nash and Ashford, 1982). Consequently, an interpretation that CTL subserve protection only if they also release IFN- γ may only hold true in the intraperitoneal infection model, but it may not be the mechanism by which virus-specific CTL control infections of epithelia or sensory nerves.

Despite the problems in interpretation of the role of CTL in some models, there remains ample evidence to demonstrate the obvious importance of CTL for the efficient control of virus infection. We, therefore, use the subsequent sections to review our current knowledge concerning CTL induction, with particular emphasis on the aspects important to vaccine construction.

CTL INDUCTION

The basic steps involved in the generation of lytically active CTL from CTL-precursors (CTL-p) are depicted in Figure 2. In this simplified model, the activation process is divided into three steps, the last two of which can occur concurrently. In the first instance, the lytically inactive CTL-p recognize their target antigen on the surface of an

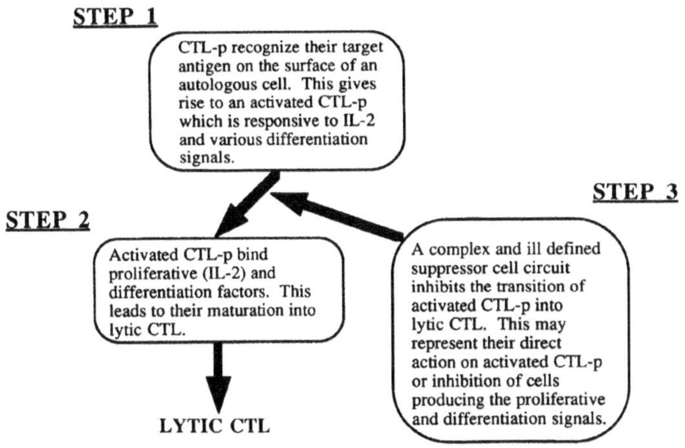

Figure 2. Schematic presentation of CTL induction

autologous cell. This binding of antigen/MHC induces the expression of lymphokine receptors on the surface of the CTL-p and the transition into an activated state (CTL-p*). In the second step the various growth and differentiation factors bind to their specific receptors expressed on the surface of the CTL-p* and trigger differentiation into a functionally lytic CTL. In the third step, ill-defined immunoregulatory suppressor cells and their products inhibit the transition of CTL-p* to CTL.

A closer analysis of these three steps allows us to gain an insight into the requirements of a successful CTL inducing vaccine.

ACTIVATION OF CTL-P (CTL TARGET ANTIGENS)

The essence of CTL-p activation is the recognition of specific target antigen together with class I MHC via the T cell receptor (TCR), hence a logical start to the design of CTL inducing vaccines is to identify the nature of the antigenic determinants recognized by CTL-p and their lytic progeny the CTL.

Many of the initial studies on the nature of viral determinants recognized by CTL dealt with an evaluation of the structural glycoproteins which are usually expressed in large amounts on the infected cell surface. This reasonable assumption that extracellular viral antigens would be recognized by the CTL TCR has been verified in numerous viral systems. In the case of anti-reovirus CTL the same antigenic domain of the viral hemagglutinin (HA) is recognized by both neutralizing antibody and CTL populations (Finberg et al. 1982). In contrast, the regions of the various HA molecules recognized by anti-influenza A and Sendai virus specific CTL's appear to be distinct from those recognized by antiviral antibodies (Burakoff et al. 1984; Mitchell et al. 1985) as are the CTL determinants on the G protein of vesicular stomatitis virus (Lefrancios and Lyles, 1983; Zinkernagel and Rosenthal, 1981).

Several workers have suggested that HSV glycoproteins act as CTL target antigens (Carter et al. 1982; Lawman et al. 1980) since the addition of glycosylation inhibitors such as tunicamycin to HSV-infected cells prevented their ability to act as targets for the majority of anti-HSV CTL. Later studies with mutant HSV viruses indicated that glycoprotein C (gC) of HSV-1 accounts for the specificity of up to 75% of anti-HSV CTL (Glorioso et al. 1985). Nevertheless, there are notable exceptions in the repertoire of extracellular or extravirion virus specified glycoproteins recognized by antiviral CTL. In our laboratory we have been unable to demonstrate the recognition of gD and are able to show very poor recognition of gB by anti-HSV CTL, despite the abundant expression of both glycoproteins on the surface of infected cells (Martin et al. 1987a, 1987b). This appears to be a real phenomenon rather than a laboratory artifact since the efforts of workers with transfected L cells expressing gB (Rosenthal et al. 1987), gD (Blacklaws et al. 1987; Martin et al. 1987a, 1988a; Rosenthal et al. 1987) or vaccinia recombinants expressing gD (Martin et al. 1987a, Yewdell, personal communication) have failed to detect their recognition by anti-HSV CTL. Experiments in our laboratory have excluded trivial explanations such as insufficient expression of glycoprotein by the various vectors, insufficient expression of MHC antigens by the target cells, inefficient stimulation of the necessary accessory cells and proliferation factors involved in CTL differentiation, and altered kinetics of induction of secondary gD specific anti-HSV CTL (Martin et al. 1987a).

This apparent failure to demonstrate gB or gD reactive CTL could have a number of explanations. These glycoproteins may not be able to associate with the class I MHC molecules expressed by the mouse strains so far investigated. Alternatively, the nonresponding strains may generate a potent gB or gD specific suppressor cell cascade which inhibits the generation of HSV CTL specific for these glycoproteins. Concordant with the latter idea, we could detect potent gD specific suppressor cells and factors capable of inhibiting HSV-specific CTL induction *in vitro* (Martin et al. 1987a), but it remains uncertain if such suppressor systems are generated sufficiently early during the immune response to turn off CTL induction. In support of the failure of H-2k restricted CTL to specifically recognize gD, or gB, limiting dilution studies which investigate CTL specificity at the clonal level have still been unable to detect gB or gD specific CTL (Martin et al. 1988, submitted) despite the fact that this technique usually dilutes suppressor and effector cells apart allowing the detection of both (Lefkovits and Waldmann, 1979). Another line of experimentation which failed to demonstrate a role for gD-specific suppressor cells *in vivo* were the co-immunization studies of mice with HSV-1 and vaccinia gD or gB in which normal frequencies of HSV-specific-CTL still developed in the draining lymph nodes (Martin and Rouse, unpublished observations).

In view of the present evidence, although still preliminary, we favor the interpretation that gD and gB fail to act as targets for HSV-specific-CTL as opposed to their preferential induction of suppressor cells which inhibit the induction of gD or gB specific CTL. In many respects it seems remarkable that such large, gD 63,000 daltons and gB 115,000 daltons, (Roizman and Batterson, 1985) extracellular molecules are not recognized by CTL. Although some discrepancy exists within the literature, it would appear that the neuraminidase (NA) of influenza A viruses, a large tetrameric molecule also found abundantly on the surface of virions and infected cells (Yewdell and Gerhard, 1981) also fails to act as a target antigen for antiviral CTL (Ada and Jones, 1986). These exceptions are surprising in view of the recent demonstration that even intracellular viral proteins which lack an amino terminus leader sequence which permits membrane translocation and surface expression can act as CTL target antigens. Examples of these are the alpha-genes of HSV (Table 2) and CMV (Reddehase and Koszinowski 1984), the nucleoproteins of influenza A and respiratory syncytial virus (Pala et al. 1986; Yewdell et al. 1985), the gag antigen of some retroviruses (Plata et al. 1987), the matrix protein, non-structural proteins, and RNA polymerases of influenza A viruses (Bennink et al. 1982, 1987) and the large T antigen of SV40 transformed cells (Tevethia, 1986). Furthermore, in the case of some viruses such as influenza A, a higher proportion of the antiviral CTL population recognizes internal antigens such as the nucleoprotein (30%) than the major extracellular antigen, namely HA (10%) (Andrews et al. 1987).

At face value these findings would seem confusing, but recent observations are beginning to provide clues to their understanding. For instance, Morrison et al. (1986), in studies on the nature of antigen recognition by anti-influenza CTL clones specific for HA, demonstrated that class I restricted CTL only recognized HA endogenously produced by the target cell. In contrast, class II restricted HA specific CTL recognized antigen exogenously acquired by the target cell. Furthermore, whereas the recognition by class II restricted CTL was impaired by lysomotropic agents which inhibit the lysosomal pathway of antigen processing (Anderson and Abraham, 1980; Kurlander and Barther 1982), the class I CTL recognition was unaffected. This failure of exogenously acquired target antigens to be recognized by the majority of class I restricted CTL is also found in the HSV system using a polyclonal CTL population (Table 1). The inability of autologous cells expressing passively acquired HSV antigens to act a CTL targets is even more surprising in the light of reports that 50-75% of HSV-specific CTL recognize glycoprotein C (gC) (Glorioso et al. 1985). However, this molecule is serologically detectable on the surface of such target cells (data not shown). This further substantiates the idea that CTL recognize viral antigens endogenously produced by the target cells and not merely passively acquired.

Although endogenous synthesis for CTL recognition may be the general rule, there are situations where exogenously acquired proteins can act as CTL target antigens. Examples include inactivated Sendai virus and certain virus antigen incorporated into liposomes (Burakoff et al. 1984; Lawman et al. 1981). These exceptions may be associated with the ability of the preparations used to fuse with the target cell. In accordance with this, Sendai virus grown in cells which inhibit the maturation of their functional fusion protein impairs the ability to passively sensitize cells for CTL recognition (Hale et al. 1980).

At first glance, the next studies discussed would appear to contradict the requirement for CTL antigen to be either fusion-active or endogenously produced by the target cell. Thus, the groups of Townsend et al. (1986a) and McMichael et al. (1986) were able to passively sensitize cells to act as CTL targets by pretreatment with synthetic peptides corresponding to the carboxy terminus one-third of influenza A nucleoprotein. In order to unify these apparently contradictory observations, we (Rouse et al. 1988) and others (Germain, 1986) have postulated that there exists a ubiquitous cellular function which performs the task of altering or processing a minor proportion of the host cell's endogenously produced proteins such that they will then associate with class I MHC and be recognized by the appropriate CTL.

ACTIVATION OF CTL-P (ANTIGEN PROCESSING)

"CTL processing" is a hypothetical cellular function which reconciles most of the anomalous reports concerning the nature of the viral determinants recognized by CTL. In essence, we imagine that during a protein's synthesis it may be recognized by the "CTL processing" system. If this happens, then the protein will be altered, degraded or its tertiary conformation unfolded in such a manner that it will be able to associate with class I MHC antigens while concomitantly exposing CTL target determinants. The physical location of this "processing" system within the cell could be within specialized areas of the rough endoplasmic reticulum (RER), or the Golgi apparatus. Consistent with this possibility are reports of compartmentation within these organelles (Kelly, 1987; Pryme, 1986). Alternatively or concomitantly, CTL processing could be associated with the cellular membrane. This would help to explain why fusion active preparations of virus, which are able to insert themselves into the host membrane, also act as CTL target antigens. The recent reports of membrane enzymes which acylate some proteins, together with the finding that a minor proportion of the T antigen in SV40 transformed cells exists in a membrane bound acylated form (Klockman and Deppert, 1983), may suggest that acylation of antigens is directly or indirectly associated with the formation of a CTL target antigen complex.

Even in this simple form, the "CTL processing" system explains many of the puzzling phenomena concerning the recognition of target cells by CTL. Thus, if this process requires continual production of various host proteins, then the ability of many viruses to inhibit host cell transcriptional and translational processes would explain why some viruses form poor CTL targets. For example, viruses such as HSV-2, and Moloney's leukemia virus (MoLV) inhibit host cell syntheses of molecules such as MHC heavy chains and are also only weakly recognized by antiviral CTL (Flyer et al. 1985; Jennings et al. 1985). Moreover, the inhibition of host cell functions by viral proteins produced late in the replication cycle of vaccinia may possibly explain why vaccinia virus late proteins do not form CTL target antigens (Mallon et al. 1985); in other words, they can no longer be appropriately "Processed" due to the viral inhibition of essential host cell functions. This would also explain why some vaccinia recombinants carrying their cloned genes under the control of a late promoter fail to act as CTL targets when vaccinia recombinants controlling the same cloned gene with an early promoter are recognized by the appropriate antiviral CTL (Coupar et al. 1986). Since "CTL processing" is an intracellular mechanism presumably associated with the normal mechanism of translation, it follows that most proteins would need to be endogenously produced to act as CTL target antigens and this would also predict that the nucleoprotein peptides of Townsend et al. (1986a) and McMichael et al. (1986) mimic "CTL processed" nucleoprotein of an influenza A virus infected cell. Such an interpretation seems even more likely in view of this groups latest findings which demonstrate that whereas the appropriate nucleoprotein peptide can sensitize a cell to become susceptible to lysis by influenza A specific CTL, the whole purified nucleoprotein molecule cannot do so (Bastin et al. 1987).

In its present form, the hypothesis is unable to explain why some endogenously produced molecules fail to act as CTL target antigens in most MHC I haplotypes tested even when found at the cell surface, for example, HSV-1 gB or gD, or influenza A NA. One clue to understanding the failure could come from recent knowledge about signal peptides and their role in trafficking proteins to appropriate locations within cells. A few brief examples of such internal regulatory sequences are given below. For instance, studies with proteins which translocate into the RER membrane prior to secretion or cell surface expression have led to the identification of several vital intermolecular signal region which dictate the fate of proteins (Hortsch and Meyer, 1984; Melancon and Garoff, 1986; Walter et al. 1984). One such vital signal recognition sequence which is present in the amino-terminus of secretory proteins, and is involved in the translocation of molecules across the RER membrane, is characterized by a long stretch of hydrophobic amino acid residues (usually containing Asn, Try, and Pro) (Deber et al. 1986) which are removed during the translocation process. In the case of transmembrane proteins, Blobel (1980) postulated the existence of a stop transfer signal present within the transmembrane region of the molecule. Accordingly, the orientation of complex multi-spanning membrane proteins can be logically explained by the sequential expression of signal recognition peptides (SRP) which initiate (and reinitiate) and then stop protein translocation (Blobel, 1980; Rapoport, 1985).

Another such peptide recognition sequence within proteins has been described by Ruoslahti and Pierschbacher (1986). These authors demonstrated that the tripeptide Arg-Gly-Asp (RGD) is crucial for the interaction of the 2,500 amino acid molecule fibronectin with its cell surface receptor. Subsequent analysis of other molecules which bind to cell surface receptors have implicated RGD in their recognition sequence (Ruoslahti and Pierschbacher, 1986). A similarly well characterized peptide sequence is Asn-xxx-Ser/Thr which, depending upon its location with the tertiary structure of the protein, acts as an N-linked glycosylation site for the molecule (Pless and Lennarz, 1977).

Finally, there is the intriguing demonstration that glycoproteins contain intrinsic structural features (SRF) which direct their transport from the trans-golgi apparatus to different areas within the cell (Kelly, 1987). This has also been shown to be true for viral glycoproteins, their SRF's determine whether they are transported to either the apical or basolateral surface of infected polarized cells (Boulau and Pendergast, 1980; Compans and Klenk, 1979).

In summary, it has become evident from several approaches investigating different molecules that short stretches of amino acids within the SRF are extremely important in determining the fate of that protein within the cell. More importantly to this discussion, these SRF's are being identified which direct the transport of proteins from the Golgi-apparatus to different areas within the cell (Kelly, 1987). Thus, it is a possibility that most proteins contain a SRF which directs their transport to the "CTL processing" compartment within the cell. It is here that a proportion of the molecules, will be appropriately "processed" and presented on the cell surface in the context of class I MHC. It is necessary to assume that only a proportion of the molecules are CTL-processed, because if this critical event involved the degeneration of the molecule to peptides of similar size to those utilized by Townsend et al. (1986a), then a totally efficient processing system may destroy all native molecules, thereby precluding their recognition in or upon virus infected cells by serological techniques.

A logical consequence of the CTL "processing" SRF hypothesis is that molecules lacking the appropriate SRF which allows them to be transported to the "CTL processing" compartment and appropriately modified could not act as a CTL target. This offers one explanation as to why some prominent extracellular viral glycoproteins may fail to be recognized by CTL, whereas intracellular molecules which do not contain a leader sequence allowing their expression upon the surface of the cell (Yewdell et al. 1985), or truncated glycoproteins which do not have membrane expressed molecules (Townsend et al. 1986b) can form CTL target complexes.

Another important consideration pertaining to the antigen recognition by CTL is the profound influence of host MHC antigens upon epitope selection for presentation to suitable T cells. The recent x-ray crystallographic resolution of the structure of the human class I MHC molecule has suggested that there exists a single specialized cleft within the molecules tertiary structure which is involved in the presentation of epitope to its corresponding TCR (Bjorkman et al. 1987a, 1987b). This model whereby class I MHC molecules position a single protein fragment (or peptide) for presentation to CTL is supported experimentally by the study of Maryanski et al. (1988), in which they show that various variant synthetic peptides (differing by a single amino acid) corresponding to the region of a HLA molecule recognized by allogeneic CTL could compete for recognition apparently at the level of the target cell. The existence of only a single binding site upon MHC molecules (Bjorkman et al. 1987a, 1987b; Brown et al. 1988) presumably necessitating a newly generated peptide to have a higher binding affinity for the MHC "cleft" than the currently residing peptide, together with the limitations of processing a molecule to provide suitable epitopes (Sercarz, 1986) would readily contribute to the apparent paucity of CTL epitopes upon many of the virus antigens analyzed.

EXPANSION OF CTL-P* (STEP 2)

This section is dealt with in an eclectic nature and for more detail the reader is referred to the comprehensive review of Gately et al. (1986) that has appeared recently. Antigen-activated CTL-p (CTL-p*) require a source of lymphokines before they can mature into functionally lytic CTL. These lymphokines can be added exogenously and are usually derived from the supernatant fluids of activated T helper cells. The complete array of lymphokines necessary for the maturation of CTL-p* are unknown, but it is generally accepted that IL-2 is mandatory for the proliferation of CTL-p*. However, the factors involved in CTL-p* differentiation factors are still the subject of some debate. Potential candidates for this function include IL-3, IFN-γ, and the less well defined T cell differentiation factors (TCDF's) (Gately et al. 1986). From the vaccinologist's point of view, the cloudy issue of which lymphokines are necessary for CTL-p* differentiation is not a great hinderance to the development of a CTL-inducing vaccine. The necessary lymphokines are non-antigen specific and their production can be induced from lymphocytes of many different antigenic specificities just by including an appropriate stimulator. Substances which are being considered to be an appropriate stimulator include immunomodulators such as muramyl dipeptide or tripeptide whose mechanisms of action are uncertain or purified protein derivative (PPD) of *Mycobacterium tuberculosis* which is a powerful T cell stimulant and activates the memory cells within BCG immunized (e.g., the majority of Europeans) individuals (Lachmann et al. 1986).

An unexpected observation concerning the provision of "help" for the development of a CTL response in many viral systems was the demonstration of their independence of CD4+ T lymphocytes, the classical helper cell. To date, this phenomenon has been demonstrated for the herpesviruses, HSV-1 (Newell et al. 1988, submitted and Nash et al. 1987), MCMV (Jonjic et al. 1988), the arenavirus, LCMV, and the poxvirus Ectromelia (Buller et al. 1987). It would, therefore, appear to be a widespread occurrence and is likely to be demonstrated in many more viral systems. The existence of CD4 independent antiviral CTL responses raises several interesting new considerations concerning the choice of CTL-inducing vaccines. For instance, not all viruses have a CD4 independent CTL response with SV40 specific CTL demonstrating classical helper cell dependence (S. Jennings personal communication). Consequently, this raises the possibility that for a given virus, such as HSV, some molecules may induce CTL in a CD4+ independent mechanism whilst others would be dependent upon classical helper cell participation. Since the work of Nash and colleagues (1987) has suggested that CD4+ cells may actually inhibit the HSV CTL response, a wise vaccine candidate may be one which induces CTL independently of CD4+ T lymphocytes. In view of the findings of Schmitt et al. (1988) which reveal clonal segregation between CD8+ T lymphocytes which are cytolytic (CTL) or which secrete IL-2 (helpers?), it is unlikely that CD4+ independent CTL are autonomous and the need for an additional helper cell stimulating moiety, albeit appropriate for a CD8+ T lymphocyte still needs to be considered for an efficient vaccine.

In summary, if suitable vaccine formulations can be identified which stimulate CTL-p, it will be necessary to attach them to appropriate carrier molecules, either a nonspecific molecule such as PPD which will stimulate the lymphokines necessary for CTL-p* differentiation and expansion, or a specific epitope recognized by CD8+ CTL helper cells.

SUPPRESSION OF CTL-P* (STEP 3)

Current literature would suggest that there are very few exceptions to the biological equivalent of Newton's third law which states "for every action there must be a reaction". In this case, the action can be considered to be the generation of CTL, with the reaction being the subsequent induction of suppressor T cells (Ts). The control of HSV-specific CTL by Ts involves a complex interaction of lymphocytes, antigen presenting cells, and soluble suppressor factors, the absolute nature of which we are only just beginning to understand (Rouse and Horohov, 1986). Essentially there appears to be three levels of T cells which interact via antigen presenting cells (APC) and soluble factors. The three T cells have been termed by others Ts1, Ts2, and Ts3 or more descriptively, suppressor-inducer, suppressor-transducer, and suppressor-acceptor cells, respectively (Asherson et al. 1986). Basically, the suppressor-inducer cell interacts with antigen presented on an IJ+ APC and produces an IJ+ idiotypic factor. This is presented to anti-idiotypic suppressor-transducer cell that, in turn, produces an IJ+ anti-idiotypic factor. We have identified this factor together with a non-antigen specific factor acting in the HSV-CTL induction system (Horohov et al. 1986). The former factor proved to be a 68,500 dalton two chain structure that expressed IJ^k determinants and anti-idiotypic activity. This specific suppressor factor appeared to act by causing cyclophosphamide sensitive IJ^+ $lyt1^+$ antigen specific acceptor cells to produce non-antigen specific suppressor factors. In other systems, these factors are assumed to operate by interfering with the activity of cytokines such as IL-1 (Lombardi et al. 1985), and IL-2 (Colizzi et al. 1984; Kramer and Kosinowski, 1982).

Although suppressor cells probably evolved, in part, to curtail exuberant positive immune responses, some viruses such as HSV appear to have exploited their existence and can preferentially stimulate them so as to dampen the magnitude of the antiviral CTL response (Rouse and Horohov, 1986). Therefore, it is crucial to understand the mechanistic details of suppressor cell circuits since this may provide clues as to ways of inhibiting or amplifying the circuit as the circumstances demand.

SUMMARY: IMPLICATIONS FOR VACCINES WHICH INDUCE A CTL RESPONSE

Based upon our present knowledge concerning the induction of antiviral CTL, it is possible to identify improvements in vaccine technology which will help to generate more effective vaccines.

In the previous section we have given brief attention to the role of helper cells in the generation of CTL. Although this area is poorly understood, we are in the fortunate position of being able to globally stimulate the plethora of nonspecific factors needed for step 2 of CTL activation (Figure 2). Thus, rather than identify suitable viral antigens that will co-stimulate CTL-p and lymphokine producing cells, we are able to include more generally acting molecules or adjuvants to more efficiently achieve the same purpose. One such potential molecule is PPD, a powerful T cell stimulant able to enhance cell mediated immune responses in animals previously sensitized with BCG (Lachmann et al. 1986), and this occurring apparently without the undesirable side effects such as toxicity, allergy, sickness, etc. of previously tried adjuvants.

Unfortunately, our knowledge concerning the nature of Ts inducing antigenic determinants is not very advanced. It is known that some carrier molecules preferentially favor suppressor cell responses when coupled to haptens (Schutze et al. 1985), and should not be included in future vaccines. This problem of avoiding suppressogenic determinants is further complicated by the demonstration that some B cell determinants are also recognized by Ts (Sakato and Fujio, 1986). Consequently, the avoidance of suppression may be accompanied by a failure to induce antibody, a disadvantage with diseases such as influenza (Mitchell et al. 1985) and HSV (Wildy, 1985). At the present time, our lack of knowledge concerning suppressogenic determinants means that vaccines must be constructed empirically with the hope of avoiding exuberant suppressor cell generation.

Our current knowledge concerning CTL recognizable determinants is a little more advanced, and the current line of research in many laboratories promises to thoroughly define such epitopes in the near future. It is obvious that many current inactivated or sub-unit vaccines will fail to induce a CTL response, due to a lack of fusion activity and consequently the inability to induce target cells to "CTL process" the relevant antigens. Perhaps one obvious improvement for such preparations would be to include them in fusion-active liposomes thereby mediating CTL target generation. The use of liposomes as vaccines, however, is fraught with problems concerning their trafficking to the appropriate areas of the body and their propensity to be ingested by the reticuloendothelial system. Therefore, current knowledge would dictate that successful CTL-inducing vaccines be either relevant synthetic peptides or replicating recombinant virus vectors.

The demonstration that CTL can recognize small synthetic peptides which presumably mimic a "CTL processed" epitope has important ramifications to vaccine generation. Thus, the ability to mass produce a small peptide recognized by antiviral CTL and couple it to a suitable carrier molecule lacking suppressogenic epitopes would be very advantageous in a vaccine development program. Unfortunately, there still remains major empirical problems to overcome when using peptide vaccines. For example, peptide vaccines usually only induce a quality antiviral immune response when given with medically unacceptable adjuvants (DiMarchi et al. 1986). Consequently, new methods of immunological enhancement are desperately required.

Perhaps the major limitation of subunit vaccines is the marked influence MHC genes have on peptide recognition by the immune system. Thus, presumably in accordance with the situation for class II MHC restricted responses, an antigen will only be recognized by CTL if it contains the suitable epitope, along with the agretope which allows association with class II MHC molecules, and does not contain a suppressogenic determinant (Sercarz, 1986). Thus, a suitable subunit vaccine must contain a multitude of CTL epitopes plus the appropriate agretopes if it is to stimulate a CTL response in the majority of individuals of an outbred (MHC heterozygous) population. It is well documented that the magnitude of the antiviral response is controlled by the MHC haplotype of the host organism. This was elegantly demonstrated by Ahmed et al. (1984a) when they reported that the fine antigenic specificity of murine CTL populations which were cross-reactive for differing strains of LCMV was determined by the H-2 haplotype of the host mouse. More recently, studies with vaccinia virus recombinants coding for the nucleoprotein, matrix protein, non-structural protein, and RNA polymerases of influenza A viruses indicate that low or high responsiveness of CTL to these individual molecules mapped to the class I MHC genes of man (Gotch et al. 1987) or mouse (Benninck et al. 1987).

Accordingly, it is likely that some viral molecules will fail to act as CTL target antigens because of their failure to associate with class I MHC molecules. The fact that MHC region heterozygotes will express several different HLA antigens does not appear to minimize the problem, since an influenza A nucleoprotein peptide described by McMichael et al. (1986) was only recognized by the virus-specific CTL of one MHC haplotype. In short, our lack of knowledge concerning the precise nature of CTL epitopes and their relevant agretopes make the task of identifying suitable subunit vaccines extremely difficult in the near future.

Despite the problems of MHC restriction, recombinant virus vectors carrying cloned genes whose products stimulate an antiviral CTL response show considerable promise as future vaccines. It is possible to generate recombinant viruses which are innocuous to the host and which target to the area in the body where an antiviral CTL response would be most beneficial. Moreover, since all antigens synthesized by vectors will be endogenously produced in cells, if they contain an appropriate SRF they should be CTL processed. Such vectors can carry a myriad of CTL-inducing viral antigens thereby minimizing non-responsiveness due to MHC restriction upon the CTL generation in response to some individual viral components. Finally, if the "CTL processing" signal sequences can be identified then chimeric viral genes could be constructed so as to include more of such sequences thereby allowing more efficient expression of relevant CTL epitopes. Our present state of knowledge would suggest that recombinant virus vectors may represent the better vaccines of the future for some viral diseases due to their superior ability to generate antiviral CTL responses in heterozygous populations.

It would not be unreasonable to suspect that persistent virus infections such as those caused by herpesviruses, and maybe even the human immunodeficiency virus type 1 (HIV-1), have intrinsic methods of eluding eradication by CTL. An example of this horrifying scenario has been demonstrated by Ahmed and colleagues (1984b) in which a variant of LCMV arises during a persistent infection. Although this LCMV mutant is itself capable of being recognized by antiviral CTL, it fails to induce (or may even suppress) a CTL response within the infected host. Surprisingly, the mutation responsible for this "suppression" causes a single amino acid change within the viral polymerase (Oldstone personal communication). The mechanisms by which this and other "mutant" viruses can avoid the antiviral CTL response are largely unknown but possibilities are the preferential induction of CTL inhibiting suppressor cells, the removal of "CTL processing" SRF's from viral molecules, and the inhibition of host cell functions vital to the generation of CTL-inducing molecules. Such avoidance methods would presumably help the viruses to reside persistently within the host. However, armed with a detailed knowledge of CTL induction, suitable vaccine constructs which induce herpesvirus or HIV-1 CTL in the face of CTL avoidance mechanisms and depleted helper cells (necessary for step 2, Figure 2) is not conceivable since numerous recent reports have demonstrated classical helper cell independent CTL induction (Buller et al. 1987; Nash et al. 1987; Sprent and Schafer, 1987).

A word of caution, however, is that vaccines selected solely for their ability to stimulate CTL responses in the absence of other antiviral responses may well be as ineffective as many current vaccines which induce antiviral responses in the absence of CTL. For example, Andrews et al. (1987) have shown that a vaccinia recombinant virus which was a potent inducer of CTL specific for influenza A nucleoprotein CTL was only able to induce low levels of protection against lethal challenge. In contrast, a vaccinia HA recombinant which was a poor inducer of CTL but also concomitantly induced other aspects of antiviral immunity as well as CTL elicited sound protection against lethal challenge.

The notion that antiviral CTL play a pivotal role in the control of many virus infections has over the years become well established. Regrettably, despite this knowledge very few viral vaccine preparations induce antiviral CTL, and consequently engender the host with a poor level of immunity and occasionally no protective immunity. The recent advances in our understanding of the TCR, T cell epitopes, particularly those recognized by CTL, and the MHC molecules, coupled with our progress in delineating the important aspects of antiviral immunity permit us to start and rationally design successful vaccine formulations.

In this review we have concentrated upon those recent advances which pertain to vaccines capable of inducing an antiviral CTL response. Accordingly, we have discussed the nature of viral antigens recognized by CTL, how such antigens need to be presented in order to activate CTLp, and mechanisms of nonspecifically inducing the necessary T cell help whilst concomitantly avoiding T cell suppression. In spite of the recent progress in our understanding of CTL biology, we still fall short of a sound insight into CTL induction and regulation. Hopefully, continued research by investigators around the world will prove fruitful and these deficiencies will soon become extinct.

ACKNOWLEDGEMENTS

We would like to thank Mrs. Marty Danilchuk and Ms. Paula Keaton for their patience and excellent typing of this manuscript.

REFERENCES

Ada GL, Jones PD (1986) The immune response to influenza infection.Current Topics Microbiol Immunol 128:1-54.

Ahmed R, Byrne JA, Oldstone MBA (1984a) Virus specificity of cytotoxic T lymphocytes generated during acute lymphocytic choriomeningitis virus infection: Role of the H-2 region in determining cross-reactivity for different lymphocytic choriomeningitis virus strains. J Virol 51:34-41.

Ahmed R, Salmi A, Butler LD, Chiller JM, Oldstone MBA (1984b) Selection of genetic variants of lymphocytic choriomeningitis virus in spleens of persistently infected mice. Role in suppression of cytotoxic T lymphocyte response and viral persistence. J Exp Med 60:521-540.

Anderson CL, Abraham GN (1980) Characterization of the Fc receptor for IgG on a human macrophage cell line, U937. J Immunol 125:2735-2741.

Andrews ME, Coupar BEH, Boyle DB, Ada GL (1987) The roles of influenza virus hemagglutinin and nucleoprotein in protection: Analysis using vaccinia virus recombinants. Scand J Immunol 25:21-28.

Asherson GL, Colizzi V, Zembala M (1987) An overview of T-suppressor cell circuits. Ann Rev Immunol 4:37-68.

Askonas BA, Pala P (1985) Human T cell clones.Influenza specific cytotoxic T cell clones and immune interferon. In: Human T Cell Clones. Humana Press, New Jersey, pp 381-388.

Bastin J, Rothbard R, Davey J, Jones I, Townsend A (1987) Use of synthetic peptide of influenza nucleoprotein to define epitopes recognized by class I restricted cytotoxic T lymphocytes. J Exp Med 165:1508-1523.

Benacerraf B (1981) Role of MHC gene products in immune regulation.Science 212:1229-1234.

Bennink JR, Yewdell JW, Gerhard W (1982) A viral polymerase involved in recognition of influenza virus-infected cells by a cytotoxic T cell clone. Nature 296:75-76.

Bennink JR, Yewdell JW, Gerhard W (1987) Anti-influenza cytotoxic T lymphocytes recognize the three viral polymerases and a non-structural protein: Responsiveness to individual viral antigens is major histocompatibility complex controlled. J Virol 61:1098-1102.

Biddison WE (1982) The role of the human major histocompatibility complex in cytotoxic T cell responses to virus infected cells. J Clin Immunol 2:1-9.

Bjorkman PJ, Saper MA, Samraoui B, Bennett WS, Strominger JL, Wiley DC (1987a) Structure of the human class I histocompatibility antigen HLA-A2.Nature 329:506-512.

Bjorkman PJ, Saper MA, Samraoui B, Bennett WS, Strominger JL, Wiley DC (1987a) The foreign antigen binding site and T cell recognition regions of class I histocompatibility antigens. Nature 329:512-518.

Blacklaws BA, Nash AA, Darby G (1987) Specificity of the immune response of mice to herpes simplex virus glycoproteins B and D constitutively expressed on L cell lines. J Gen Virol 68:1103-1114.

Blobel G (1980) Intracellular protein topogenesis.Proc Natl Acad Sci USA 77:1496-1500.

Boulau ER, Pendergast M (1980) Polarized distribution of viral envelope proteins in the plasma membrane of infected epithelial cells. Cell 20:45-54.

Brown JH, Jardetsky T, Saper MA, Samraoui B, Bjorkman PJ, Wiley DC (1988) A hypothetical model of the foreign antigen binding site of class II histocompatibility molecules. Nature 332:845-850.

Buller RML, Holmes KL, Hugin A, Frederickson TN, Morse HC III (1987) Induction of cytotoxic T cell responses *in vivo* in the absence of CD4 helper cells. Nature 328:77-80.

Burakoff SJ, Weinberger O, Krensy AM, Reiss CS (1984) A molecular analysis of the cytolytic lymphocyte T response. Adv Immunol 36:45-85.

Byrne JA, Oldstone MBA (1984) Biology of cloned cytotoxic T lymphocytes specific for lymphocytic choriomeningitis virus: Clearance of virus *in vivo*. J Virol 51:682-686.

Carter VC, Rice PL, Tevethia SS (1982) Intratypic and intertypic specificity of lymphocytes involved in the recognition of herpes simplex virus glycoproteins. Infect Immun 37:116-121.

Colizzi V, Ferluga J, Garreau F, Malkovsky M, Asherson GL (1984) Suppressor cells induced by BCG release nonspecific factors *in vitro* which inhibit DNA synthesis and IL-2 production. Immunology 51:65-71.

Compans RW, Klenk H (1979) Primary, secondary, tertiary, and quaternary structures. In: Fraenviel-Conrat H, Wagner RR (eds), Comprehensive Virology. Plenum Press, Vol. 13, pp 293-407.

Coupar BEH, Andrew ME, Both GW, Boyle DB (1986) Temporal regulation of influenza hemagglutinin expression in vaccinia virus recombinants and effects on the immune response. Europ J Immunol 16:1479-1487.

Deber CM, Brandl CJ, Deber RB, Hsu LC, Young XK (1986) Amino acid composition of the membrane and aqueous domains of integral membrane proteins. Arch Biochem Biophys 251:68-76.

DiMarchi R, Brooke G, Gale C, Crackwell V, Doel T, Mowat N (1986) Protection of cattle against foot-and-mouth disease by a synthetic peptide. Science 232:639-641.

Doherty PC (1985) Cell-mediated immunity and the CNS as a key role for interferon. TINS Feo. 41-42.

Everett RD (1984) Transactivation of transcription of herpes virus products: Requirements for two HSV-1 immediate-early polypeptides for maximum activity. EMBO J 3:3135-3141.

Finberg R, Spriggs DR, Fields BN (1982) Host immune response to reovirus: CTL recognize the major neutralization domain of the viral hemagglutinin. J Immunol 129:2235-2238.
Flyer DC, Burakoff SJ, Faller DV (1985) Retrovirus-induced changes in major histocompatibility complex antigen expression influence susceptibility to lysis by cytotoxic T lymphocytes. J Immunol 135:2287-2292.
Gately MK, Jenson JC, Benjamin WR (1986) Cytokines in T and B cell mediated immunity. Concepts Immunopathol 3:74-108 (Karger, Basel).
Germain RN (1986) The ins and outs of antigen processing and presentation. Nature 322:687-689.
Glorioso J, Kees U, Kumel G, Kirchner H, Krammer P (1985) Identification of herpes simplex type 1 (HSV-1) glycoprotein C as the immunodominant antigen for HSV-1-specific memory cytotoxic T lymphocytes. J Immunol 135:575-582.
Gooding LR (1982) Characterization of a progressive tumor from C3H fibroblasts transformed *in vitro* with SV40 virus. Immunoresistance *in vivo* correlates with the phenotypic loss of H-$2K^k$. J Immunol 129:1306-1312.
Gotch F, McMichael A, Smith G, Moss B (1987) Identification of viral molecules recognized by influenza-specific human cytotoxic T lymphocytes. J Exp Med 165:408-416.
Hale AH, Lyles DS, Fan DP (1980) Elicitation of anti-Sendai virus cytotoxic T lymphocytes by viral and H-2 antigens incorporated into the same lipid bilayer by membrane fusion and by reconstitution into liposomes. J Immunol 124:724-731.
Hill TJ, Field JH, Blyth WA (1975) Acute and recurrent infection with herpes simplex virus in the mouse. A model for studying latency and recurrent disease. J Gen Virol 28:341-347.
Horohow DW, Wyckoff J, Moore RN, Rouse BT (1986) Regulation of herpes simplex virus specific cell-mediated immunity by a specific suppressor factor. J Virol 58:331-338.
Hortsch M, Meyer DI (1984) Pushing the signal hypothesis: What are the limits? Biol Cell 52:1-8.
Howes EL, Taylor W, Mitchison NA, Simpson E (1979) MHC matching shows that at least two T cell subsets determine resistance to HSV. Nature 277:66-68.
Howes EL, Taylor W, Mitchison NA, Simpson E (1979) MHC matching shows that at least two T cell subsets determine resistance to HSV. Nature 277:66-68.
Jacobson S. Nepom GT, Richert JR, Biddison WE, McFarland HF (1985) Identification of a specific HLA-DR2 Ia molecule as a restriction element for measles virus-specific HCA class II-restricted cytotoxic T cell clones. J Exp Med 161:263-268.
Jacobson S, Biddison WE (1986) Virus-specific HLA class II-restricted cytotoxic cells. In: Concepts in Viral Pathogenesis II. Springer Verlag, NY, pp 187-192.
Jennings SR, Rice PL, Koszewski ED, Anderson RW, Thompson DL, Tevethia SS (1985) Effect of herpes simplex virus types 1 and 2 on surface expression of class I major histocompatibility antigens of infected cells. J Virol 56:757-766.
Jonjic S, Del Val M, Keil GM, Reddehase MJ, Koszinowski UH (1988) A non-structural viral protein expressed by recombinant vaccinia virus protects against lethal cytomegalovirus infection. J Virol 62:1653-1658.
Kelly RB (1987) From organelle to organelle. Nature 326:14-15.
Kockman H, Deppert W (1983) Acylated simian virus 40-specific proteins in the plasma membrane of HeLa cells infected with adenovirus 2-simian virus 40 hybrid virus Ad2+ ND2. Virology 126:717-720.
Kramer M, Koszinowski U (1982) T cell-specific suppressor factor(s) with regulatory influenza on interleukin 2 production and function. J Immunol 128:784-790.
Kurlander RJ, Batker J (1982) The binding of human immunoglobulin G1 monomer and small, covalently crosslinked polymers of immunoglobulin G1 to human peripheral blood monocytes and polymorphonuclear leukocytes. J Clinic Inv 69:1-8.

Lachmann PJ, Strangeways L, Vyakarnam A, Evans G (1986) Raising antibodies by coupling peptides to PPD and immunizing BCG-sensitized animals. In: Synthetic Peptides as Antigens. Wiley, Chichester, Ciba Foundation Symposium 119, pp 25-27.

Larsen HS, Feng MF, Horohov DW, Moore RN, Rouse BT (1984) Role of T lymphocyte subsets in recovery from herpes simplex virus infection. J Virol 50:56-59.

Lathey JL, Martin S, Rouse BT (1987) Suppression of delayed type hypersensitivity to herpes simplex virus type one following immunization with anti-idiotypic antibody - an example of split tolerance. J Gen Virol 68:1093-1102.

Lawman MJP, Courtney RJ, Eberle R, Shaffer PA, O'Hara MK, Rouse BT (1980) Cell-mediated immunity to herpes simplex virus: Specificity of cytotoxic T cells. Infect Immun 2:451-461.

Lawman MJP, Maylor PT, Huang L, Courtney RJ, Rouse BT (1981) Cell-mediated immunity to herpes simplex virus induction of cytotoxic T lymphocyte responses by viral antigens incorporated into liposomes. J Immunol 126:304-308.

Lefkovits I, Waldman H (1979) Statistical tests, limitations, reproducibility. In: Limiting Dilution Analysis of Cells in the Immune System. Cambridge University Press, Cambridge.

Lefrancois L, Lyses DS (1983) Antigenic determinants of vesicular stomatitis virus: Analysis with antigenic variants. J Immunol 130:394-398.

Lin YL, Askonas BA (1981) Biological properties of an influenza A virus-specific killer T cell clone. Inhibition of virus replication *in vivo* and induction of delayed-type hypersensitivity reaction. J Exp Med 154:225-234.

Lombardi G, Vismara D, Piccolella E, Colizzi V, Asherson GL (1985) A nonspecific inhibitor produced by *Candida albicans*-activated T cell impairs cell proliferation by inhibiting interleukin-1 production. Clin Exp Immunol 60:303-310.

Lukacher AE, Braciale VL, Braciale TJ (1984) *In vivo* effector function of influenza virus specific cytotoxic T lymphocyte clones is highly specific. J Exp Med 160:814-826.

Mallon VR, Domoer EA, Holowczak JA (1985) Vaccinia virus proteins on the plasma membranes of infected cells. II. Expression of viral antigens and killing of infected cells by vaccinia virus-specific cytotoxic T cells. Virology 145:1-23.

Martin S, Moss B, Berman PW, Laskey LA, Rouse BT (1987a) The mechanisms of antiviral immunity induced by a vaccinia virus recombinant expressing herpes simplex virus type 1 glycoprotein D: Cytotoxic T cells. J Virol 61:726734.

Martin S, Rouse BT (1987) The mechanisms of antiviral immunity induced by a vaccinia virus recombinant expressing herpes simplex virus type 1 glycoprotein D: Viral clearance. J Immunol 138:3431-3437.

Martin S, Cantin E, Rouse BT (1988a) Cytotoxic T lymphocytes: Their relevance in herpesvirus infections. Cytotoxic T cells: Biology and relevance to disease. N Y Acad Sci (in press).

Martin S, Courtney RJ, Fowler G, Rouse BT (1988b) Herpes simplex virus type 1-specific cytotoxic T lymphocytes recognize virus non-structural proteins. J Virol 62:2265-2273.

Maryanski JL, Pala P, Cerottini JC, Corradin G (1988) Synthetic peptides as antigens and competitors in recognition by H-2 restricted cytolytic T cells specific for HLA. J Exp Med 167:1391-1405.

Melanon P, Garoff H (1986) Reinitiation of translocation in the Semliki Forest virus structural polyprotein: Identification of the signal for the E1 gly coprotein. EMBO 5:1551-1560.

Mitchell DM, McMichael AJ, Lamb JR (1985) The immunology of influenza. Brit Med Bull 41:80-85.

Morrison LA, Lukacher AE, Braciale VL, Pan DP, Braciale TG (1986) Differences in antigen presentation to MHC class I and class II restricted influenza virus-specific cytolytic T Iymphocyte clones. J Exp Med 163:903-921.

McMichael AJ (1980) HLA restriction of human cytotoxic T cells. Springer Semin Immunopathol 3:3-22.

McMichael AJ, Gotch FM, Rotchbard J (1986) HLA B37 determines an influenza A virus nucleoprotein epitope recognized by cytotoxic T lymphocytes. J Exp Med 164:1397-1406.
Nash AA, Phelan J, Wildy P (1981) Cell-mediated immunity in herpes simplex virus infected mice: H-2 mapping of the delayed type hypersensitivity response and the antiviral T cell response. J Immunol 126:1260.
Nash AA, Ashford PN (1982) Split T cell tolerance in herpes simplex virus-infected mice and its implication for antiviral immunity.Immunol 45:761-767.
Nash AA, Jayasuriya A, Phelan J, Cobbold SP, Waldmann H, Prospero T (1987) Different roles for L3T4+ and Lyt2+ T cell subsets in the control of an acute herpes simplex virus infection of the skin and nervous system. J Gen Virol 68:825-833.
Pala P, Townsend AR, Askonas BA (1986) Viral recognition by influenza A virus cross-reactive cytotoxic T cells: The proportion of Tc that recognize nucleoprotein varies between individual mice. Europ J Immunol 6:193-198.
Plata F, Langlade-Demoyen P. Abastado JP, Berbar T, Kourilsky P (1987) Retrovirus antigens recognized by cytocytic T lymphocytes activate tumor rejection *in vivo*. Cell 48:231-240.
Pless DD, Lennarz WJ (1977) Enzymatic conversion of proteins to glycoproteins.Proc Natl Acad Sci USA 74:134-138.
Pryme IF (1986) Compartmentation of the rough endoplasmic reticulum. Molec Cell Biochem 71:3-19.
Rapoport TA (1985) Extensions of the signal hypothesis-sequential insertion model versus amphipathic tunnel hypothesis. FEBS Lett 187:1-10.
Reddehase MJ, Koszinowski UH (1984) Significance of herpesvirus immediate-early gene expression in cellular immunity to cytomegalovirus infection. Nature 312:369-371.
Reddehase MJ, Weiland F, Munch K, Jonjic S, Luske A, Koszinowski UH (1985) Interstitial murine cytomegalovirus pneumonia after irradiation: Characterization of cells that limit viral replication during established infection of the lungs. J Virol 55:264-273.
Reddehase MJ, Mutter W, Koszinowski UH (1987a) *In vivo* application of recombination interleukin-2 in the immunotherapy of established cytomegalovirus infection. J Exp Med 165:650-656.
Reddehase MJ, Mutter W, Munch K, Buhring J-J, Koszinowski UH (1987b) CD8+ positive T lymphocytes specific for murine cytomegalovirus immediate-early antigens mediate protective immunity. J Virol 61:3102-3108.
Reddehase MJ, Jonjic S. Weiland F, Mutter W, Koszinowski UH (1988) Adoptive immunotherapy of murine cytomegalovirus adrenalitis in the immunocompromised host: CD4-helper-independent antiviral function of CD8-positive memory T lymphocytes derived from latently infected donors. J Virol 62:1061-1065.
Renis HE, Eidson EE, Mathews J, Gray JE (1976) Pathogenesis of herpes simplex virus types 1 and 2 in mice after various routes of inoculation. Infect Immun 24:571-578.
Roizman B, Batterson W (1985) Herpesviruses and their replication. In: Fields BN, et al. (eds) Virology. Chapter 25, Raven Press, NY, pp 497-526.
Rosenthal KL, Smiley JR, South S, Johnson DC (1987) Cells expressing herpes simplex virus glycoprotein C but not gB, gD or gE are recognized by murine virus specific cytotoxic T lymphocytes. J Virol 61:2438-2447.
Rouse BT, Horohov DW (1986) Immunosuppression in virus infections. Rev Infect. Dis 8:850-873.
Rouse BT, Norley S, Martin S (1988) Antiviral cytotoxic T lymphocyte induction and vaccination. Rev Infect Dis 10:16-23.
Ruoslahti E, Pierschbacher MD (1986) Arg-Gly-Asp:A versatile cell recognition signal. Cell 44:517-518.
Sakato N, Fujio H (1986) Suppression of the delayed-type hypersensitivity response to hen egg white lysozyme (hel) by hel peptides in a genetically high responder mouse strain: Evidence of requirement of the loop structure for induction of suppressor T cells. Cell Immunol 100:66-78.

Schmitt J, Wagner H, Heeg K (1988) Reactivity of Ly2+ T cells against 2,4,6-trinitrophenyl (TNP) modified syngeneic stimulator cells: Specificity, frequency of interleukin-2 producing Ly2+ helper T cells and clonal segregation from Ly2+ cytotoxic T lymphocytes. Eur J Immunol 18:325-332.
Schutze MP, Leclerc C, Jolivet M, Audibert F, Chedid L (1985) Carrier-induced epitopic suppression: A major issue for future synthetic vaccines. J. Immunol 135:2319-2322.
Sercarz EE (1986) Hierarchies of epitope preference and the problem of Ir gene control. Concepts Immunopathol 3:61-67.
Sethi KK, Omata Y, Scheneis KE (1983) Protection of mice from fatal herpes simplex virus type 1 infection by adoptive transfer of cloned virus-specific and H-2-restricted cytotoxic T lymphocytes. J Gen Virol 64:443-447.
Sprent J, Schafer M (1986) Capacity of purified Lyt2- T cells to mount primary proliferative and cytotoxic responses to Ia- tumor cells. Nature 322:541-544.
Tevethia SS (1986) Heterogeneity of CTL reactive antigenic sites on SV40 tumor antigen. In: Notkins AL, Oldstone MBA (eds) Concepts in Viral Pathogenesis II. Springer Verlag, pp 182-187.
Townsend ARM, Rothbard J, Gotch FM, Bamadur G, Wraith D, McMichael AJ (1986a) The epitopes of influenza nucleoprotein recognized by cytotoxic T lymphocytes can be defined with short synthetic peptides. Cell 44:959-968.
Townsend ARM, Bastin J, Gould K, Brownlee GG (1986b) Cytotoxic T lymphocytes recognize influenza haemagglutinin that lacks a signal sequence. Nature (London) 234:575.
Walter P, Gilmore R, Blobel G (1984) Protein translocation across the endoplasmic reticulum. Cell 38:5-8.
Wildy P (1985) Herpesviruses: A background. Brit Med Bull 41:339-344.
Wildy P, Gell PGH (1985) The host response to herpes simplex virus. Brit Med Bull 41:86-91.
Yewdell JW, Gerhard W (1981) Antigenic characterization of viruses by monoclonal antibodies. Ann Rev Microbiol 35:185-206.
Yewdell JW, Bennink JR, Smith GL, Moss B (1985) Influenza A virus nucleoprotein is a major target antigen for cross-reactive anti-influenza A virus cytotoxic T lymphocytes. Proc Natl Acad Sci USA 83:1785-1789.
Zinkernagel RM, Anthage A (1977) Antiviral protection by virus immune cytotoxic T cells. Infected targets are lysed before infectious virus progeny is assembled. J Exp Med 145:644-647.
Zinkernagel RM, Doherty PC (1979) MHC restricted cytotoxic T cells: Studies on the biological role of polymorphic major transplantation antigens determining T cell restriction specificity, function and responsiveness. Adv Immunol 27:701-702.
Zinkernagel RM, Rosenthal KL (1981) Experiments and speculation and anti-viral specificity of T and B cells. Immunol Revs 58:131-155.

DISCUSSION

Burnette G (Burnette Labs, Buena Park, CA):

My comment is in regards to the recent papers published about the vaccines for malaria that somehow share with the viruses the intracellular parasitic mechanism. I understand they have been very successfull in using polypeptides in the case of immunization to malaria. I wonder if you care to comment about those particular vaccines.

Rouse B:

My understanding of the malaria vaccine field differs from your own. My understanding is that subunit malaria vaccines produce responses in some strains of mice and not others, and when used to immunize humans, gave variable and often negative responses. I think I would have used malaria to indicate the problems of using subunits to induce adequate immunity although immunity to malaria may not involve cytotoxic T lymphocytes.

Trachtenberg A (California Department of Health Services, Berkeley, CA):

First a comment, I would suggest that you not dispair in regards to live virus and their development and potential use. The problem of liability is recognized even among legislators, which are mostly made up of lawyers. In California, at least, both funds and committees to start addressing the problem now have been legislate mandated. There is still hope for the social possibility of live viruse vaccines in the Western World. A question, specifically in regards to cytotoxic T cell and potential vaccines for HIV, either treatment or prophylactic vaccines. Can you give me some sense of just how dependent the cytotoxic T cell response is on the helper cell, and what, over the course of exposure and disease with HIV, might that affect its action in someone who is infected?

Rouse B:

In some viral systems, helper cells need to be co-stimulated in order to induce CTL, but in others this is unnecessary. It may depend upon the dose of virus in the host genotype, and it may depend upon the individual viral antigen. It could be, for instance, that if you looked at a virus which had five target proteins, helper cells dependence would be necessary for four of these, but the fifth may be helper cell independent. How you would predict which one is which, I don't know. As yet, we lack algorithms which adequately predict CTL recognizable epitopes, whereas certain algorithms are helpful to predict helper cell determinants in some situations. As regards giving vaccines to the already infected, that is something which fascinates me, but I see little cause for optimism at present. The topic was discussed recently by Jonas Salk in Nature and I believe the article has inspired some testing of the concept in HIV infection.

ABSTRACTS

USE OF WHOLE BLOOD IN THE DOT-IMMUNOBINDING ASSAY

S.S. KALTER and R.L. HEBERLING

Southwest Foundation for Biomedical Research, NIH and WHO Collaborating Center for Reference and Research in Simian Viruses, San Antonio, Texas, USA

Recent advances in laboratory diagnostic technology has shifted from classical procedures such as serum neutralization and complement fixation, to procedures that are more rapid, specific, sensitive and generally less expensive. Accordingly for routine serologic diagnosis in the virus laboratory, ELISA-EIA has replaced for practical purposes, those first generation procedures. Here too, modifications of the original procedure have resulted in simplified methodologies that have practically eliminated the need for laboratory equipment. Development in this laboratory of a dot-immunobinding assay (DIA) for the rapid detection of viral antibody is one such method. Application of this DIA to a variety of different viral antigens has clearly demonstrated its usefulness in the field or office setting. One area of the test requiring some laboratory support was that of using patients' serum which necessitated use of a syringe and generally a centrifuge. That whole blood could be used rather than serum and, further, applied directly onto the DIA test system was recently demonstrated. Blood collected on filter paper precut to fit the test following a finger stick has been successfully applied to the DIA. Further, blood collected in this fashion may be shipped via the mail and remains stable for relatively long periods of time. DIA results using whole blood or serum were identical. No increased nonspecific reactions or interfering background was observed.

EFFECT OF ACETONE AND METHANOL FOR FIXATION ON FLUORESCENCE INTENSITY OF HSV-INFECTED CELLS IN SHELL VIAL MONOLAYERS AND IN TUBE CULTURES

M. BANACH[1], B. JUDSON[1] and C. GLEAVES[2]

Syva Company, Palo Alto, California[1];
Fred Hutchinson Cancer Research Center, Seattle, Washington[2], USA

Two 1-dram shell vials and one 16 x 125 mm tube containing MRC-5 cells were each infected with 0.2 ml of previously frozen inoculum from HSV-positive clinical specimens. Thirty-two specimens were known HSV-1 positive, 21 were known HSV-2 positive. Paired shell vials were incubated at 37°C for 20 hr, the monolayers rinsed with PBS, and fixed with either 2 ml acetone or 2 ml methanol for 10 min. After fixation, the monolayers were rinsed with distilled water and stained with the appropriate Syva MicroTrak® HSV-1/HSV-2 Culture Identification-/Typing Reagent. An additional 16 specimens were each inoculated into 4 shell vials - 2 were fixed with acetone, 2 with methanol - and these were stained with both the MicroTrak HSV typing and the MicroTrak HSV Culture Identification Test (non-typing). Culture tubes were scraped into 200 µl of 50% PBS/50% ethanol when CPE involved at least 50% of the monolayer, and 30 µl was spotted onto four 8 mm slide wells. Two slide wells were fixed with acetone for 10 min and 2 with methanol for 10 min. One acetone-fixed specimen was stained with the appropriate Syva tying reagent, the second with the Syva HSV non-typing reagent. The 2 methanol-fixed wells were also stained with either the typing or non-typing reagent. Although all cultures were correctly diagnosed, the staining intensity with all three reagents (HSV-1, HSV-2 and non-typing) was reduced after methanol fixation compared to acetone fixation in the majority of specimens. In some specimens, particularly with the non-typing reagent, staining intensity was reduced from 3+ (bright) to 1+ in both shell vials and tube culture cell suspensions. These data strongly indicate that methanol should not be used for fixation of HSV specimens, as weakly fluorescing specimens may be misdiagnosed, and that acetone, as stated in the package insert, is the fixative of choice. The possibility also exists that there may be HSV strains where the specific staining is completely destroyed by methanol fixation.

RAPID IDENTIFICATION OF HERPES SIMPLEX VIRUS (HSV) WITH A MEMBRANE ENZYME IMMUNOASSAY IN 16-HOUR MRC-5 SHELL VIAL CULTURES

C. FRASER, R. KAMIBAYASHI, M. KAWULA,
M. HAY-KAUFMAN and R. FRASER

Monoclonal Antibodies, Inc., Mountain View, California, USA

The RAMP HSV Culture Confirmation Test (Monoclonal Antibodies, Inc., Mountain View, CA) is a visual end-point membrane enzyme immunoassay for the detection of HSV in cell culture. The RAMP HSV test was compared to a fluorescent antibody (FA) test (Syva Co., Palo Alto, CA) for detection of HSV in shell vial cultures. Conventional tube culture isolation was used as a reference. Fifty-five clinical specimens taken from various anatomical sites were inoculated into a single tube culture and triplicate shell vials. Both culture methods used the MRC-5 cell line. Shell vials were centrifuged for 40 min at 700 x g and incubated at 36°C for 16 h before testing by the RMAP and FA methods. Tube cultures were processed using standard HSV isolation procedures. The RAMP test correctly identified 20 out of 25 positive specimens (80% sensitivity) as compared to tube culture. The FA test identified 20 out of 24 positive specimens (83% sensitivity). Both RAMP and FA identified all 30 negative specimens (100% specificity). One positive specimen was unreadable by FA as no cell layer remained for evaluation. The two HSV detection tests have shown comparable sensitivity and specificity in detecting HSV in the shell vial culture method. The FA test requires many steps, highly trained personnel, expensive instrumentation and lengthy handling time before results can be obtained. The RAMP HSV Culture Confirmation Test requires less than 10 min to perform and gives an immediate visual result.

DETECTION OF HERPES SIMPLEX VIRUS (HSV) USING A549 CELLS IN CENTRIFUGATION CULTURE WITH A RAPID MEMBRANE ENZYME IMMUNOASSAY (RAMP)

D.A. HURSH, S.F. WENDT, J.D. MEYERS and C.A. GLEAVES

Fred Hutchinson Cancer Research Center, Seattle, Washington, USA

The evaluation and performance of the RAMP HSV Culture Confirmation Test (Monoclonal Antibodies, Inc., Mountain View, CA) was compared to immunofluorescence (IF) staining with a non-typing HSV monoclonal antibody reagent (Syva Co., Palo Alto, CA) for the detection of HSV from clinical specimens in centrifugation culture. Clinical specimens were inoculated at a volume of 0.25 ml into one each of MRC-5 and A549 cell culture tubes for standard viral isolation, and into two A549 shell vials for centrifugation culture. A549 shell vials were centrifuged at 700 x g for 40 min, after which 1.0 ml of MEM containing 2% FBS was added to each vial and incubated at 36° for 16 to 18 h. Standard cell cultures were read daily for 6 to 7 days and observed for typical HSV CPE. All HSV isolates in standard culture were confirmed using the RAMP test and typed using HSV typing monoclonal antibodies (Syva Co.). Following incubation, one shell vial was fixed in acetone and stained by IF for HSV, and the other was subjected to the RAMP test as follows: One ml of harvesting buffer was added to the vial and vortexed for 15 sec, after which 0.05 ml of the solution was added to the matrix pad. Two drops of a specific HSV antibody linked enzyme conjugate was added for 6 min followed by 4 drops of substrate solution for 3 min, after which a stop reagent was added. A blue dot represented a HSV positive sample. In centrifugation culture, 34 specimens (95%) were HSV positive by IF and 31 specimens (86%) were HSV positive by the RAMP test. Two specimens positive by standard culture were negative by both IF and RAMP. Four specimens were HSV positive by IF, but negative by RAMP. One specimen was positive only by the RAMP test. All 4 of the RAMP negative specimens were "low positive" samples, taking longer than 3 days before HSV CPE was observed. Compared to IF, in centrifugation culture RAMP had a sensitivity of 88% and a specificity of 99%. The RAMP test can be performed in less than 15 min and gives an immediate visual result which compares favorably with IF staining in centrifugation culture from specimens containing moderate to high levels of HSV.

CENTRIFUGED MINK LUNG SHELL VIAL CULTURE VS. ENZYME IMMUNOASSAY (EIA) FOR RAPID DIAGNOSIS OF HERPES SIMPLEX VIRUS (HSV)

W.E. TAGGART, B.R. KENYON, V.C. SALMON and J.C. OVERALL

University of Utah School of Medicine, Salt Lake City, Utah, USA

Sensitive and specific rapid antigen detection tests for HSV are needed for timely decision concerning antiviral therapy and for optimal obstetrical management. A comparison was made of the shell vial culture (SCV), using centrifugation and immunoperoxidase (IP) staining for confirmation of HSV as previously reported from this laboratory, and the Ortho EIA for rapid detection of HSV in 611 fresh clinical specimens, SVC consisted of 2 dram shell vials containing mink lung cells, which were inoculated, centrifuged at 3,500 x g for 15 min at room temperature, and incubated overnight. IP staining at 16 h confirmed 131 positive (21.4%) in SVC. Ortho EIA detected 112 positives (18.3%), and failed to detect 38 positive SVC. Of 19 EIA positive/SVC negative samples, 7 were true positives (by blocking antibody assay) and 12 were false positives. There were 138 total true positive tests by both methods, (131 SVC and 7 true positive EIA), and 100 total true positive EIA tests (93 EIA positive and SVC positive plus the 7 EIA true positives). Sensitivities were 95% for SVC and 72% for EIA with specificities of 100% and 97%, respectively. Antigen detection methods such as EIA have faster turnaround times and may pick up some positive which are culture negative due to antiviral therapy, suboptimal specimen collection and/or transport, or other factors affecting viral infectivity. However, the EIA sensitivity of only 72% as compared with the centrifuged SVC method, suggests that cultures should be performed on EIA negative specimens. In addition, the fact that 10.7% of the positive EIA test were false positive raises concern about reporting of EIA positive/culture negative results without confirmatory blocking antibody, particularly since most of the specimens are from genital sites and results have implications for a sexually transmitted disease.

EVALUATION OF A CENTRIFUGATION CULTURE ENHANCED ENZYME-LINKED IMMUNOSORBENT ASSAY (ELISA) FOR THE DETECTION OF HERPES SIMPLEX VIRUS ANTIGEN

E.L. SHELDON and D. ANDRESEN

Viral Laboratory, Laboratory Services, San Jose, California, USA

A comparison study for the detection of herpes simplex virus (HSV) was performed to evaluate a 36 h post-inoculation ELISA centrifugation culture enhanced assay (Ortho Diagnostic Systems, Inc., Raritan, NJ) with a conventional culture method. A total of 364 clinical specimens primarily from genital sites were inoculated into conventional tubes containing MRC-5 and A549 cells and specially designed flat-bottom tubes containing MRC-5 and PRK cells. The flat bottom tubes were centrifuged for 30 min at 2,500 x g, incubated at 36 h, and processed for HSV detection by ELISA. The conventional cell culture tubes were observed daily for one week. HSV was identified by conventional cell culture in 106 (29%) of 364 specimens and by the ELISA spin amplification technique (ELISA-SAT) in 98 (27%) of 364 specimens. Two specimens were positive by the ELISA-SAT method alone, compared with eight specimens positive by conventional culture only. The sensitivity, specificity and positive and negative predictive values of the ECCE method for detection of HSV in clinical specimens were 92, 99, 97 and 97%, respectively. The ELISA-SAT method has good sensitivity and specificity, is easy to perform, and significantly reduces the culturing time required to detect HSV for high volume laboratories.

EVALUATION OF A NEW HERPES SIMPLEX VIRUS (HSV) TYPING REAGENT FOR TISSUE CULTURE CONFIRMATION

S.L. AARNAES, L.M. DE LA MAZA and E.M. PETERSON

University of California Irvine Medical Center, Orange, California, USA

Monoclonal antibody typing reagents for HSV-1 and HSV-2 from California Integrated Diagnostics, Inc. (Berkeley, CA) were compared to two other commercially available HSV monoclonal antibody typing reagents from Syva Corp. (Palo Alto, CA) and Kallestad Diagnostics (Austin, TX). One hundred and five specimens submitted for an HSV culture were included in the study. Specimens previously frozen at -70°C (n=61) as well as fresh specimens (n=44) were tested. All specimens were inoculated into conventional tissue culture of MRC-5 cells. Upon detection of CPE cells were scraped, pelleted and washed with phosphate buffered saline. Cells were then placed on glass slides, fixed with acetone and stained with the three reagents. All three typing reagents identified 34 HSV-1 isolates and 47 HSV-2 isolates. In addition, 24 cell cultures that did not exhibit CPE were negative by all 3 typing reagents as were the 4 cytomegalovirus, 3 enterovirus, 1 adenovirus, 4 *Ureaplasma urealyticum* and 2 *Mycoplasma hominis* isolates that were included for specificity testing. To confirm the typing results, 27 HSV-1 and 34 HSV-2 were subjected to restriction endonuclease analysis and results agreed with the monoclonal typing reagents. In summary, all three typing reagents agreed with 100% sensitivity and specificity.

IDENTIFYING SUITABLE ANTIGEN SUBSTRATES FOR A HUMAN HERPESVIRUS TYPE 6 (HHV6) (HBLV) INDIRECT FLUORESCENT ASSAY (IFA)

F.K. MUNDON, J.M. BREWER, P.P. BREWER and W.E. BARTON

Steller Bio Systems, Inc., Columbia, Maryland, USA

We found the use of freshly isolated lymphocytes from human cord blood for infection with HHV6 and its' subsequent use as an IFA antigen substrate not to be feasible for the following reasons: variability for permissive infection, and ratio of infected to uninfected cells between lots; increasing difficulty in obtaining a constant source of fresh cord bloods. This study was undertaken to identify a continuous human cell line which is permissive for HHV6 infection that would express cell surface antigens in a sufficient number of cells to provide a useful antigen substrate for an IFA system. HHV6 infections were attempted in six human continuous cell lines: two T-cell lines, Molt 4 and RPMI 8226; two B-cell lines, Ramos and Daudi and two fibroblast-like, human glioblastoma cell lines (ATCC, Rockville, MD). Five ml of cell-free supernate from HHV6 infected cord blood lymphocytes containing approximately one $TCID_{50}$/ml were inoculated into one flask of each of the suspension cultures containing 10^5 cells/ml. The same amount of inoculum was used for one flask of each monolayered glioblastoma cultures. Co-cultivation of the infected cord blood lymphocytes with these two cultures was also used to attempt cross-infection since it was shown the media for these cultures would not support the continuous growth of the lymphocytes. Cells were harvested from each of the cultures at 10-day intervals, washed, applied to paint delineated glass wells of microscope slides, dried and fixed in acetone. A standard IFA was performed: dilutions of HHV6 antibody positives and antibody negative sera were applied to the wells and the slides were incubated for 30 min. at 37°C. After washing and drying the FITC conjugated anti-human IgG was applied over mounting media and the slides were viewed under a microscope equipped with fluorescence optics.

COMPARISON OF DIFFERENT MONOCLONAL ANTIBODIES FOR EARLY DETECTION OF CYTOMEGALOVIRUS IN CLINICAL SPECIMENS BY SHELL VIAL CULTURE

P.C. DE GIROLAMI, J. DAKOS, J. SCHIFF, S. BOLIVAR and L. MERRILL

New England Deaconess Hospital, Boston, Massachusetts, USA

A new indirect immunofluorescence (IFA) test for cytomegalovirus (CMV) early detection in centrifugation enhanced shell vial cultures, using a mixture of two monoclonal antibodies to CMV early antigens (E3 and H5, Syva, Palo Alto, CA) was compared to a previously evaluated IFA method using a single monoclonal antibody to an early protein of CMV (2H2.4, DuPont, Wilmington, DE) and to conventional tube cell cultures. A total of 206 clinical specimens including 66 respiratory, 77 urine and 63 tissue samples were inoculated into two tubes and four shell vial cultures of human lung fibroblasts (MRC-5). Tube cell cultures were observed for cytopathic effect for 28 days. Two shell vial cultures were stained at 22 h incubation with DuPont and two were stained with Syva reagents (one at 22 and one at 46 h incubation). Of a total of 47 samples positive by any one method, 43 (91%) produced cytopathic effects in tube cell culture, 38 (81%) were positive by Syva and 36 (77%) by DuPont methods ($p>0.05$ by the χ^2 test). Technically, the quality of stained monolayers was comparable, with Syva producing slightly brighter specific fluorescence than DuPont. We conclude that the Syva test for early detection of CMV in shell vial cultures approaches the sensitivity of conventional culture and performs at least as well as an established similar rapid method.

LABORATORY PREPARED AND COMMERCIALLY OBTAINED MRC-5 SHELL VIALS FOR THE DETECTION OF CYTOMEGALOVIRUS IN CENTRIFUGATION CULTURE

C.A. GLEAVES, S.E. CURTIS, C.F. LEE and J.D. MEYERS

Fred Hutchinson Cancer Research Center, Seattle, Washington, USA

The sensitivity of MRC-5 shell vials commercially prepared in the laboratory was compared simultaneously to MRC-5 shell vials commercially obtained for the detection of CMV in centrifugation culture (CC) and standard culture. MRC-5 cells used to prepare laboratory shell vials and 16 x 125 mm culture tubes were originally obtained from ViroMed Laboratories (Minneapolis, MN) in 175 cm^2 flasks and subcultured weekly with MEM with 10% FBS into shell vials. Commercially prepared MRC-5 shell vials were also obtained from ViroMed Laboratories. Laboratory prepared vials were used from day 4 to 9 after seeding and commercially obtained vials were used from day 8 to 14 following the seeding date. Specimens were inoculated at a volume 0.25 ml into two of each type of vials. Vials were then centrifuged at 700 x g for 40 min after which 1.0 ml of MEM with 2% FBS was added back to each vial and incubated at 36°C. One vial of each type was fixed in cold acetone and stained by indirect immunofluorescence at 16 and 40 h post-inoculation (pi) using CMV monoclonal antibodies (Syva Co., Palo Alto, CA). The coverslips were then examined under a fluorescent microscope at a magnification of 250x. Of 158 specimens tested (45 urine, 40 tissue, 38 throat, 18 blood, 13 BAL, 4 other) 46 (29.1%) were positive for CMV, 44 by CC. In laboratory prepared vials, 42 of 46 (91%) positive specimens were detected at 16 h pi compared to 27 of 46 (59%) positive specimens detected at 16 h pi in commercially obtained vials. At 40 h pi, 44 of 46 (96%) positives were detected in laboratory prepared vials, and 38 of 46 (83%) positives were detected in commercially obtained vials. Five positive specimens were detected only in laboratory prepared vials. One positive specimen was positive at 16 h pi in commercial vials and negative at 40 h pi in commercial vials, although positive at both 16 h and 40 h in laboratory prepared vials. The data show that laboratory prepared MRC-5 shell vials are more sensitive than commercially obtained MRC-5 shell vials for the rapid detection of CMV at both 16 and 40 h pi.

DETECTION OF CYTOMEGALOVIRUS FROM BLOOD LEUKOCYTES SEPARATED BY SEPRACELL-MN AND FICOL-PAQUE/MACRODEX METHODS

C.V. PAYA, A.D. WOLD and T.F. SMITH

Mayo Clinic and Foundation, Rochester, Minnesota, USA

Sepracell-MN (Sep), a colloidal silica medium, which can separate polymorphonuclear (PMN) and mononuclear fractions of blood in a single step, was compared in a prospective study to the conventional, two-step Ficol-Paque/Macrodex (F-P/M) procedure for the detection of cytomegalovirus (CMV) from blood specimens. Of 510 blood specimens processed by both methods, 76 (14.9%) yielded CMV. Sixty-six (87%) and 65 (86%) of the 76 positive specimens were processed by F-P/M and Sep methods, respectively. Seventy-two of the 76 (95%) CMV-positive blood specimens were detected in shell vial cell cultures, but only 42 (55%) in conventional tube cell cultures. The time for recognition of specific CPE due to CMV in tube cell cultures (8.0 d vs. 7.1 d), the number of fluorescent foci in each positive shell vial culture (19.3 vs. 20.1), and the cost of the reagent ($3.50 vs. $2.80) were similar and independent of the leukocyte separation method (F-P/M vs. Sep). The Sep method is a rapid and sensitive method for the detection of CMV from blood specimens and is recommended as a replacement for the more tedious and time consuming procedure.

WESTERN BLOT (WB) ANALYSES OF THE CYTOMEGALOVIRUS (CMV) SPECIFIC IgM RESPONSES IN CARDIAC TRANSPLANT PATIENTS

C. COZZA, H. ABBO, M. ALLEN and R. ASHLEY

University of Washington - Children's Hospital Medical Center, Seattle, Washington, USA

CMV IgM WB was used to monitor six CMV seronegative cardiac transplant recipients ("CMV-N") for the development of primary CMV infections and nine CMV-seropositive recipients ("CMV-P") for CMV reactivation. Sera were tested weekly for 4 months post-transplant (PT), every 2 weeks for another 2 months, and monthly, thereafter. CMV cultures were performed for the first 2 weeks post-transplant and during febrile or rejection episodes of when lymphocytopenia occurred. Twelve of 15 patients (80%) developed CMV IgM. IgM was detected in 7 CMV-P and 5 CMV-N recipients at 4 to 11 weeks PT (median 6 weeks). Eleven remained IgM-positive for the duration of follow-up (12-104 weeks; median 28 in CMV-N and 4-78 weeks; median 31.5 in CMV-P), while 1 had intermittent IgM for 100 weeks. CMV was isolated from 8/12 IgM-positive patients within 6 weeks of IgM seroconversion. Two patients remained well and were not cultured until 12 and 56 weeks after IgM seroconversion; at which times CMV was isolated. Two CMV-P recipients with persistent IgM were currently asymptomatic at 8 and 40 weeks. CMV IgM WB profiles were unremarkable in 6 patients (the typical profile has prominent bands at 50kD, 40kD and 35kD plus minor bands of higher molecular weight). However, 6 had atypical profiles consisting of 50kD only (N=2), 50kD and 40kD (N=2), 50kD = minor bands (N=1) or minor bands only (N=1). Atypical profiles occurred with both primary and reactivation infection. In summary, 80% of cardiac transplant patients develop CMV IgM which persisted for up to 2 years. Patients with primary versus reactivation CMV were similar in their time to IgM seroconversion, duration of IgM persistence, and WB pattern. The unusual IgM WB profiles seen in 6/12 patients and the prolonged IgM production in all IgM-positive patients may be related to immunosuppressive therapy or other host factors yet to be determined.

VARIATION IN CYTOMEGALOVIRUS (CMV) ANTIBODY (AB) STATUS OF IMMUNOCOMPROMISED CHILDREN (IC)

M.T. BRADY and J.F. CUARTAS

The Ohio State University and Children's Hospital,
Columbus, Ohio, USA

Detection of CMV AB can be utilized to document prior infection to CMV. Newer methods of CMV AB determination (ELISA, latex agglutination (LA), indirect hemagglutination assay) have reduced some problems of sensitivity, specificity and reproducibility noted with the complement fixation method. From July, 1983 to June, 1986, serial CMV AB determination (2-8 specimens) were performed by ELISA and LA on 105 pediatric oncology patients on immunosuppressive chemotherapy. Forty-eight (46%) were initially CMV-seropositive (POS) and 57 (54%) were initially CMV-seronegative (NEG). Thirty-seven (77%) of those initially POS remained POS on follow-up. Thirty (53%) of those initially NEG remained NEG on follow-up. Of the 27 NEG IC who were not consistently NEG, 15 seroconverted from NEG to POS by both methods (seroconversion), 9 had single ELISA-positive samples not confirmed by LA (false positive ELISA), and 3 varied without explanation. Of the 11 initially POS IC who were not consistently POS, 4 were POS in all samples by LA but had one or more NEG results by ELISA (false negative ELISA), 3 varied from POS to NEG (both methods), 2 varied from POS to NEG and back to NEG (both methods) and 2 had results that were not consistent between the two methods. Of the 38 IC with variable serostatus, 28 could be explained (15 seroconversion; 9 false positive ELISA; 4 false negative ELISA). For the others, no specific explanation was made. POS results could not be accounted for by recent infusions of blood products. Since results for IC were similar if not identical using both methods, the variation could be explained by variability in the ability to produce CMV AB as a result of immunosuppressive therapy or the intermittent presence of a cross-reacting substance in the blood of these IC. Our results suggest that single samples for CMV AB determination may not be adequate to document serostatus of IC. Also, the LA method appears to be more sensitive and more specific than ELISA for CMV AB determination in IC.

MEASUREMENT OF HUMAN IgM ANTIBODY TO CMV USING FLUORESCENCE CONCENTRATION ANALYSIS

R.C. MINER, N. RABELLA and W.L. DREW

Mount Zion Hospital and Medical Center, Diskind Pathology Research Laboratory, San Francisco, California, USA

We have developed a method for measuring human IgM antibody directed against cytomegalovirus using fluorescence concentration analysis. The method involves attaching viral antigens to sub-micron polystyrene beads and quantitating fluorescence using a Pandex FCA Instrument. Related and essential to this IgM assay is a bead assay for measurement of rheumatoid factor to ensure against false positives. The presented FCA data are compared to a standard slide immunofluorescence assay both qualitatively and quantitatively.

Screening of CMV IgG Positive Sera for the Presence of CMV Positive IgM Antibody

IFA Results	Pandex Results +	Pandex Results -
+ $\geq$1:20	38	0
- <1:20	0	53

Numbers of sera tested by each method. Rheumatoid factor removed by adsorption with aggregated IgG

Measurement of Rheumatoid Factor by FCA

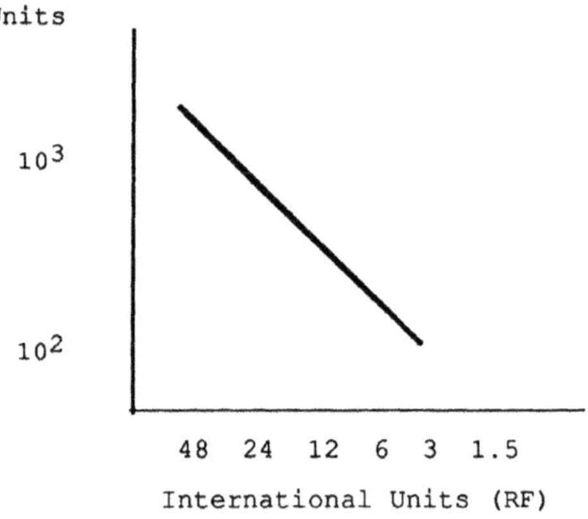

DETECTION OF VARICELLA-ZOSTER VIRUS GENOME BY *IN SITU* HYBRIDIZATION USING CLONED BIOTINYLATED DNA AND RNA PROBES

B. FORGHANI, G-J. YU and J.W. HURST

Viral and Rickettsial Disease Laboratory, California State Department of Health Services, Berkeley, California, USA

We describe here a general method for production of biotinylated DNA and RNA probes for detection of varicella-zoster virus (VZV) genome by *in situ* hybridization. Human fetal diploid lung (HFDL) cell cultures were infected with the Caque strain of VZV, and DNA was extracted from the purified viral nucleocapsids. The DNA was digested with restriction enzyme (RE) BamHI and the generated fragments were isolated. These fragments were ligated with BamHI digested plasmid pBR322 using the standard vector-insert procedure, and were then used to transform competent *E. coli* cells; recombinants containing VZV fragments were identified. Using this technique, we have cloned over 90% of the VZV genome and obtained 24 recombinants. Plasmids containing the B fragment of VZV DNA were selected, produced in large quantity and purified (the BamHI-B fragment contains the VZV immediate early genes IcPO and ICP4). A portion of purified plasmid was labeled by the nick-translation method using biotin-11-dUTP as dTTP analog. In addition, the purified plasmid was digested with RE AvaI and subcloned in plasmid pGem transcription vector, which has two promotors, T7 and SP6, between multiple cloning sites (Promega, Wisconsin). Recombinants containing VZV fragments were identified and prepared in large quantity. One sample of purified plasmid was linearized with RE PstI, and another with RE EcoRI. RNA was transcribed according to the Promega specifications, using T7 or SP6 polymerase, but substituting allylamine-UTP as UTP analog. The transcribed, highly purified RNA was treated with DNase to remove the templates and then labeled with CAB-NHS-ester(ε-caproylamidobiotin-N-hydroxysuccinimide ester). The DNA and RNA probes were used under full stringency conditions (50% formamide) for *in situ* hybridization, with alkaline phosphatase as detector and 5-bromo-4-chloro-3-indolyl phosphate/nitroblue tetrazolium (BCIP/NBT) as substrate. Tested under comparable conditions, the RNA probe was 5-10 times more sensitive than the DNA probe; both probes showed homology only to VZV infected cells.

COMPARISON OF EPSTEIN-BARR VIRUS (EBV) SEROLOGIES WITH OX-CELL HEMOLYSIN TEST, HETEROPHIL ANTIBODY TEST, AND MONOSPOT FOR THE LABORATORY DETERMINATION OF THE MONONUCLEOSIS SYNDROME

C. SIEGEL and S. GOODREAU

Bellin Memorial Hospital, Green Bay, Wisconsin, USA

Epstein-Barr Virus (EBV) is the cause of heterophil-positive and heterophil-negative infectious mononucleosis. Distinct EBV related antigens and their corresponding antibodies have been differentiated by various serologic techniques. Antibodies to viral capsid antigen (VCA) and early antigen (EA) are detected early after onset of EBV-associated infectious mononucleosis. Antibodies to EBV nuclear antigen (EBNA) are detected several weeks or months after onset. The purpose of this study is to show the advantage of using EBV serologies in place of, or as a supplement to the ox-cell hemolysin test, heterophil antibody test, and the Monospot for the laboratory detection of the mononucleosis syndrome. A total of 101 patients were tested for the mononucleosis syndrome in this study. Also patients were tested for the four EBV markers (IgG, specific IgM, EA and EBNA). In addition, patient sera were also tested for ox-cell hemolysis titers, heterophil titers, and the monospot test. Of the 101 patients studied, we detected 11 patients with positive IgM specific antibodies to VCA. Of the 11 positive patients with IgM specific antibodies detected to EBV VCA, 7 of the patients had negative monospots, 6 of the patients had negative ox-cell hemolysin titers, and 5 of the patients had heterophil antibodies with titers equal to or less than 56 dilutions. These results agree with other investigators that have shown that EBV serologies (IgG, IgM specific, EA, and EBNA) are more sensitive for the detection of the mononucleosis syndrome than either the monospot, ox-cell hemolysin titer, or the heterophil titer. Therefore, it is our recommendation that when clinical symptoms of mononucleosis are present, sera of patients with negative monospots should be tested for EBV using VCA reagents (IgG and IgM), early antigen reagents and EBNA reagents.

A DOUBLE-LABEL IMMUNOFLUORESCENCE METHOD FOR SIMULTANEOUS DETECTION OF ADENOVIRUS AND HERPES SIMPLEX VIRUS FROM THE EYE

P. WALPITA and S. DAROUGAR

University of California, San Diego School of Medicine, La Jolla, California, USA and Institute of Ophthalmology, London, England

The development and application of a shell vial/double-label immunofluorescence method for the simultaneous detection of adenovirus and herpes simplex virus is described. A total of 1,141 ocular specimens were inoculated in shell vials, centrifuged at 15,000 x g for 1 h, incubated at 37°C for 48 h, and fixed in methanol at room temperature for 15 min. The viruses were detected by staining with a double-label indirect immunofluorescence (IFA) procedure using mixtures of appropriate first antibodies, followed by fluorescein and rhodamine conjugated second antibodies. Each specimen was also inoculated in parallel by the conventional virus isolation method. The sensitivity and specificity of the shell vial/double-label IFA procedure was at least as high as the conventional method and took only 48 h to complete. The test offers a rapid and simple single vial test for the simultaneous detection of the two most common viral pathogens of the eye. The double-label rapid test described here could be adapted for simultaneous detection of multiple pathogens at other sites. Simultaneous isolation of more than one pathogen from individual genital specimens has been described previously, although we have not been able to detect adenovirus and herpes virus from the same specimen in the ocular specimens tested to date in this series.

EVALUATION OF AN IMMUNOFLUORESCENT ASSAY (IFA) FOR THE DETECTION OF ADENOVIRUS ANTIGEN IN CELL CULTURE ISOLATES

D. LEOMBRUNO[1], M.J. AUGUST[2], K. BROMBERG[3] and L.T. PIERIK[4]

Cambridge BioScience Corp.[1], Worcester, Massachusetts; Smith Kline BioScience[2], Van Nuys, California; Kings County Hospital[3], Brooklyn, New Jersey; Children's Hospital[4], Boston, Massachusetts, USA

Human adenovirus infections occur worldwide and are predominantly associated with respiratory, ocular and, recently diarrheal disease symptoms. Laboratory diagnosis has conventionally relied on the recognition of cytopathic effects in cell culture with subsequent confirmation techniques such as cell passage, enzyme immunoassay and IFA which used only polyclonal antibody reagents. We evaluated a commercially available indirect IFA (Adenoclone-IFA, Cambridge BioScience Corp.) which utilizes a well-characterized group specific monoclonal antibody. This monoclonal antibody is against the group reactive hexon antigen and has been shown to react with all 41 human adenovirus types. Three independent clinical laboratory sites in New York, California and Massachusetts participated in this study. A total of 231 patient specimens were evaluated. Specimen sources consisted of respiratory, fecal, urine, ocular, CSF and biopsy material from lung, rectum, liver and kidney. Each sample was inoculated into appropriate cell cultures and monitored for the development of characteristic cytopathic effects for adenovirus. All adenovirus isolates were identified by conventional cell culture isolation methods and in some cases by polyclonal and monoclonal IFA. Of the 231 culture isolates, 105 were positive and 126 were negative for adenovirus by conventional cell culture isolation. The Adenoclone monoclonal antibody test correctly identified all positive adenovirus cultures (sensitivity, 100%) and negative adenovirus cultures (specificity, 100%). In addition, no cross-reactivity was observed with the following viral isolates: respiratory syncytial virus, herpes simplex virus, cytomegalovirus, varicella zoster virus, influenza A and B, parainfluenza 1, 2 and 3, echovirus and coxsackie virus. In summary the Adenoclone monoclonal antibody test offers a rapid, sensitivity and specific method for the confirmation of adenoviral isolates in cell culture.

SHELL VIAL CENTRIFUGATION CULTURE METHOD (SVCCM) FOR EARLY DETECTION OF RESPIRATORY SYNCYTIAL VIRUS

M. BAHN, A. ROBERTO, B. YEN-LIEBERMAN, and M.R. PROFFITT

Cleveland Clinic Foundation, Cleveland, Ohio, USA

Forty-two specimens were received in our clinical virology laboratory for the isolation of respiratory syncytial virus (RSV). Specimens consisting of bronchial aspirates and nasal washings were processed and inoculated into tissue culture tubes containing HEp-2 cells and into shell vials containing MRC-5 cells on glass coverslips. Prior to inoculation, culture medium was aspirated from tubes and vials, then specimen (0.25 ml) was added. The vials were centrifuged at 700 x g for 45 min at 30°C. Tubes were incubated for 1 h at 36°C. After adsorption, inoculum was aspirated and both tubes and vials were fed with EMEM containing 2% fetal bovine serum. The cultures were incubated at 36°C. On day three of culture (prior to detectable CPE) the MRC-5 cell coverslips were washed with PBS, and the cells were fixed with cold methanol. Immunofluorescent (IF) staining was done using FITC conjugated anti-RSV monoclonal antibodies (Whittaker Bioproducts). Of the 42 specimens, 13 stained positive for RSV. The corresponding HEp-2 cell tube cultures were observed daily for cytopathic effect (CPE). Ten specimens developed CPE typical of RSV at an average of 5 days. These were confirmed positive by IF staining of HEp-2 cells recovered from the tubes. Three isolates detected by SVCCM did not develop CPE in HEp-2 cells for up to 14 days of culture. Cells recovered from these tubes were negative for RSV by IF staining. Our data indicate that the SVCCM for RSV detection has a sensitivity of 100% and a specificity of 91% relative to conventional culture (HEp-2) and can be reported earlier. The SVCCM is rapid and cost effective for the clinical virology laboratory. On-going testing will determine the feasibility of a shorter (24-48 h) turnaround time for reporting results using the SVCCM for RSV in MRC-5 cells.

THE USE OF RHABDOMYOSARCOMA CELLS IN A SHELL VIAL ASSAY FOR THE LABORATORY CONFIRMATION OF INFLUENZA VIRUS INFECTION

M.J. ESPY, E.J. JESPERSEN and T.F. SMITH

Mayo Clinic and Mayo Foundation, Rochester, Minnesota, USA

Shell vials seeded with rhabdomyosarcoma (RD) or with rhesus monkey kidney (RMK) cells were compared for the laboratory confirmation of influenza virus infections. Respiratory tract specimens were initially inoculated into conventional tube (RMK) cell cultures and examined for hemadsorption (HA) after 7 days. Of 325 specimens, 36 (11%) were found to be positive by HA and subsequently inoculated into shell vials seeded with either RMK or RD cells. The shell vials were then stained 24 h post-inoculation with monoclonal antibodies specific for influenza virus types A or B in an indirect immunofluorescence assay. All 36 were confirmed influenza virus type A by both the RMK and the RD seeded shell vials. RD cells appear to be as sensitive as RMK cells for the confirmation of influenza virus infections. In addition, the continuous RD cell line allows easier and less expensive preparation of shell vial cell cultures compared to RMK.

PRODUCTION AND CHARACTERIZATION OF MONOCLONAL ANTIBODIES AGAINST B19 PARVOVIRUS

D.A. KENNEDY, R.A. HILLS, D. DICK, M.T. MONETTE and G. BUTLER

Laboratory Centre for Disease Control,
Tunney's Pasture Ottawa, Ontario, Canada

B19 virus has been implicated as the causative agent of the fifth disease (erythema infectiosum); in addition it is responsible for polyarthralgias in adults, causes transient aplastic crises in persons with chronic hemolytic anemias and infection during pregnancy may lead to hydrops fetalis. With the aim of producing diagnostic reagents, we undertook the production of monoclonal antibodies. B19 virus was identified in the plasma of a Canadian blood donor by DNA hybridization. This virus was purified and used to immunize Balb/C mice, resulting in the generation of over one hundred hybridoma cultures secreting virus-specific antibody. A selection of the antibodies was characterized with respect to antibody class and subclass as well as reactivity with the homologous virus. A panel of these antibodies was further tested by radioimmunoassay against virus "isolates" from Canada, the United States, France, the United Kingdom and Japan; the antibodies reacted with all of the isolates examined, suggesting that the target epitopes are conserved.

A LABORATORY QUALITY ASSURANCE PROGRAM TO IMPROVE HIV-1 ANTIBODY TESTING

K.M. PEDDECORD, A.S. BENESON, L.K. HOFHERR, M.S. ASCHER and R. ELWELL

San Diego State University, Graduate School of Public Health, San Diego, California;
California Viral and Rickettsial Laboratories, Berkeley, California;
Centers for Disease Control, Atlanta, Georgia, USA

The SDSU/CDC Laboratory Assurance Program (LAP) is developing and evaluating methods to improve the performance of clinical laboratories, initially using HIV-1 antibody testing as the model. Serum samples with known HIV-1 content are submitted as clinical specimens by physicians, clinics, and hospitals to their usual testing laboratory; we reimburse the submitter for the cost of the test and provide a handling fee. The sera come from donor units and are rechecked by enzyme immunoassay (EIA) using multiple reagents, by Western blot and by immunofluorescence assay (IFA). In our initial survey, three sera (one a relatively strong positive, one testing false positive with some EIA systems, and one negative for HIV-Ab) were submitted to 15 laboratories for screening and confirmatory testing. Results in the EIA tests were generally acceptable, but one laboratory reported a positive result with no confirmatory testing carried out or recommended. With the Western blot, however, there was considerable variability with the report varying from the statement "positive" or "negative", to a listing of the bands detected followed by an explanation of their significance and different band patterns listed as the criteria for positivity. One laboratory reported band patterns but left it to the physician for interpretation. Different laboratories reported differing band patterns; when the same specimen was submitted to one laboratory from different submitters, different band patterns were reported. Our program calls for visits to the laboratories and review of the procedures to correct the source of discrepancies. A systems analysis is being carried out to identify "critical performance factors", identifying steps with greater potential for error in the process from the venipuncture to the final report back to the physician, including how the information is used to benefit the patient. The preliminary study has indicated the great need for this program.

SEROLOGICAL REACTIVITY OF HTLV-I ENVELOPE PROTEINS EXPRESSED IN *E. COLI*

D. PARKES, C.M. SMITH, J. BRANDIS and S.R. COATES

Triton Biosciences, Inc., Alameda, California, USA

The HTLV-I virus genome codes for an envelope protein that is processed into a 46-kDa external glycoprotein (gp46) and a 21-kDa membrane-bound protein (p21e). Two plasmids have been constructed that express portions of these proteins in *E. coli*. Plasmid env-1 contains a fragment of the envelope gene expressed as a fusion protein. Expression is regulated by the lac promoter. The env-1 protein expressed contains a substantial part of the gp46 domain and a majority of the p21e domain. Plasmid env-2 contains virtually all of p21e fused to the *E. coli* trpE leader sequence. Expression is regulated by the lambda P_l promoter. The recombinant env-1 and env-2 proteins were used in immunoblot assays to test for reactivity with human sera. Fifty sera known to have antibody to HTLV-I viral lysate by ELISA and fifty sera from normal individuals were evaluated for antibody to viral gp46 and p21e by immunoblot. Antibody to env-1 and env-2 was detected in over 95% of the patient sera known to contain antibody to viral gp46 and p21e. No antibody to these recombinant proteins was detected in sera from normal individuals. These studies demonstrate the utility of these recombinant HTLV-I envelope proteins in detecting antibodies to HTLV-I virus.

SEROLOGICAL REACTIVITY OF A HTLV-I GAG p24 GENE PROTEIN EXPRESSED IN *E. COLI*

A.J. HARRIS, M. FERRER, C.M. SMITH, D. DI DITULLIO, J. BRANDIS, M. SLIWKOWSKI and S. COATES

Triton Biosciences, Inc., Alameda, California, USA

The HTLV-I virus genome codes for a gag p55 precursor protein which is processed into three structural proteins, p24, p15 and p19. A plasmid has been constructed which directs the expression of a fusion protein containing the full length p24 protein and portions of p19 and p15. The recombinant protein was initially purified by differential extractions of contaminating proteins using a combination of detergents. The protein was further purified by solubilization in a chaotrophic solvent, followed by chromatography. An immunoblot assay was developed to evaluate the ability of a recombinant p24 protein to react with human sera. Fifty sera known to have antibody to HTLV-I viral lysate antigen by ELISA and fifty sera for normal individuals were evaluated for antibody to viral p24. No antibody to recombinant p24 was found in the sera from fifty normal individuals. These studies demonstrate the utility of this recombinant HTLV-I gag antigen in detecting antibody to HTLV-I virus.

SEROLOGICAL REACTIVITY OF A HTLV-I p19 GENE PROTEIN EXPRESSED IN *E. COLI*

C.M. CHRISTIAN, C.M. SMITH, J. BRANDIS and S. COATES

Triton Biosciences, Inc., Alameda, California, USA

A segment of the gag gene, p19 of the HTLV-I virus has been cloned in an *E. coli* expression vector. Expression was regulated by the lambda P_L promoter. Recombinant p19(rp19) was expressed as a fusion protein with the *E. coli* trp E leader sequence and was found in insoluble inclusion bodies in induced cells. An immunoblot assay was developed using this partially purified rp19 to assay human sera for antibody reactivity. Ten patient sera known to have antibody to viral p19, as assessed by immunoblot assay with SLB viral lysate antigen, and ten normal patients were tested with this rp19 immunoblot assay. One hundred percent of the positive patients' sera demonstrated positive reactions in this assay and none of the normal patient sera showed any reactivity. This study demonstrates the potential utility of this recombinant antigen in detecting antibodies to HTLV-I virus.

SEROLOGICAL REACTIVITY OF HTLV-I TAX GENE PROTEIN EXPRESSED IN *E. COLI*

E.L. KEITELMAN, S.A. WALLINGFORD, C.M. SMITH, M. FERRER, R. AKITA, A. HARRIS, H.-L. LIU, M.X. SLIWKOWSKI, M.W. BRANDIS and Y.A. TERAMOTO

Triton Biosciences, Inc., Alameda, California, USA

The HTLV-I virus genome includes a gene which encodes a 40,000 dalton transcriptional trans-activator protein recently designed tax (tat/pX). Full-length tax was cloned and expressed in *E. coli* as a fusion protein with trpE. The recombinant protein was purified to homogeneity by sequential detergent extractions and chromatography. The purified recombinant tax protein was used to develop a microtiter plate ELISA. Two thousand human sera, from a population known to be at risk for HTLV-I infection were tested by ELISA for antibodies to recombinant tax antigen. Of these sera, 155 (7.8%) were found to have antibodies to recombinant tax. Almost half of these sera (71 of 155) scored negative when tested with commercially available viral lysate ELISAs. These 71 sera were retested by radioimmune precipitation assays and by immunoblotting. Some of these sera (41 of 71) were found to have antibody to either gag or env proteins in addition to tax by radioimmune precipitation or immunoblot assays, despite scoring negatively in the commercial ELISAs. Furthermore, some serum samples appeared to have a humoral immune response limited only to the tax protein. These studies demonstrate the utility of the tax antigen in detecting antibody responses to HTLV-I infection.

DEVELOPMENT AND CHARACTERIZATION OF MONOCLONAL ANTIBODIES TO THE HTLV-I TAX (P40X) PROTEIN

B.C. LANGTON, M. SLIWKOWSKI, K.V. TRAN, S. KNAPP, E. KEITELMAN, C. SMITH, S. WALLINGFORD, H.-L. LIU, J.S. RALSTON, J. BRANDIS and S. COATES

Triton Biosciences, Inc., Alameda, California, USA

The 40 Kd tax (p40X or tat) protein from HTLV-I, known to exhibit transcriptional transactivation activity has been cloned and expressed in *E. coli* as a trpE fusion protein. This recombinant protein was purified to homogeneity and used as an immunogen to develop several monoclonal antibodies (MABs) to tax, of the IgM and IgG_{2a} isotypes. Spleen cells from immunized Balb/c mice were fused with P3X63Ag8 myeloma cells, screened on an ELISA assay to the recombinant protein, cloned by limiting dilution and ascites prepared and purified by HPLC. Selected MAbs were further characterized by several methods including Western blot, radioimmunoprecipitation, competition ELISA assay and Geysen epitope mapping. All of the MAbs recognize the recombinant tax protein and several of the MAbs also recognize the native tax from a lysate of an HTLV-I transformed cell line, SLB-1. This binding to native tax can be competed by the recombinant protein. Competition ELISA assays with patient sera, known to contain antibody to the tax protein, reveal that the MAbs and some patient sera may recognize similar epitopes on the protein. In an attempt to localize binding sites on tax, mapping by Geysen technology has revealed a few potential epitopes for selected MAbs. Work is in progress to complete this mapping and determine if these sites are relevant binding sites for tax reactive antibodies in patient sera. In conclusion, anti-tax MAbs have been developed which are useful both for the detection and quantitation of native and recombinant tax. These MAbs are also being used to localize binding sites on the tax protein.

DEVELOPMENT OF AN ENZYME-LINKED IMMUNOSORSBENT ASSAY FOR SCREENING HUMAN T-CELL LEUKEMIA VIRUS TYPE I ANTIBODIES

C. CHETTY, J.W. NORTHING, P.D. SCRUGGS,
P.L. TAN, T.S. FLOWERS and J.W.D. KAY

Organon Teknika Corp., Durham, North Carolina, USA

An enzyme-linked immunosorbent assay (ELISA) for detection of human T-cell leukemia virus type I (HTLV-I) antibodies was developed using the microelisa plate system. The assay employed viral lysate coated wells to capture HTLV-I antibodies in plasma or serum. Captured antibodies were detected by addition of peroxidase-conjugated goat anti-human IgG and an o-phenylenediamine (OPD) substrate. The assay, which could be performed within two hours, was designed for screening HTLV-I antibodies in blood donors as well as in high risk groups. Greater than 99.6% of samples from blood donors gave signal to cut-off (S/CO) ratios less than 0.95 with a mean S/CO of 0.24. The mean S/CO of potentially problematic samples, e.g., those positive for RPR, antinuclear antibody, rheumatoid factor, cytomegalovirus, rubella, herpes, etc., was 0.23. Of HTLV-I positive samples from high-risk groups 88.8% gave S/CO ratios >2.0. The S/CO of the remaining positive samples ranged from 1.01 to 1.99. These results indicated a good separation of the S/CO ratios between the HTLV-I negative and the HTLV-I positive samples and suggested the feasibility of using this HTLV-I microelisa for the screening of HTLV-I antibodies.

DETERMINATION OF IMMUNE STATUS REQUIRED FOR RAPID VIRUS CLEARANCE IN A MURINE MODEL OF HSV-1 INFECTION

R.R. McKENDALL and W. WOO

University of Texas Medical Branch, Galveston, Texas, USA

Human recurrent herpes labialis is characterized by a period of virus shedding lasting 3-5 days. No humoral or cellular immunologic defect has been consistently demonstrated to explain the long period of virus shedding. To determine the immune status required to rapidly clear virus, we studied virus clearance from footpad in five groups of mice. 1) Mice hyperimmunized with 4 doses of HSV-1. 2) Mice immunized with one dose of virus. 3) Mice adoptively immunized with 8 x 10^7 spleen cells from immune donor mice. 4) Mice passively immunized with immune mouse serum (neutralization level = 1:32 post-transfer) and 5) Mice given both immune whole spleen and immune serum. Control mice were given either spleen from non-immune donors or non-immune serum. We defined rapid virus clearance as complete clearance of virus from the footpad by 48 h after inoculation into the footpad. We found that rapid virus clearance occurred only in the hyperimmunized group and in the group receiving both immune serum and immune spleen. In the group receiving immune serum alone, virus clearance occurred at 3-5 days, whereas it took 7 days to clear virus in the group receiving immune whole spleen. The illness rate was 0% in all groups compared to 80% in controls receiving non-immune serum. In the group given immune spleen alone, antibody appearance measured by RIA, neutralization assay and immunoblotting detected antibody to gD appeared earliest on day 3 after infection. Antibody to gB, gC, gE, the major capsid protein and the nucleocapsid complex proteins of 42K and 44K appeared later during the time of viral clearance. In separate studies, antibody to IFN-alpha/beta was used to block the IFN effect. In animals treated with both immune anti-HSV-1 serum and anti-IFN alpha/beta, virus tissue titers were higher compared to animals given antiviral antibody alone. These studies suggest that maximum virus clearance required contributions from antibody, cell-mediated immunity and IFN. Since increasing and maintaining the level of anti-HSV-1 immunity in humans with recurrent herpetic disease is impractical, immunomodulant therapy should be directed at strategies to enhance the immune system components which have access to recrudescent lesions early in the course of disease - namely macrophages and some other cellular elements and antibody.

RHEUMATOID ARTHRITIS: EVIDENCE FOR A RETROVIRUS IN PERIPHERAL MONOCYTES

A.M. AL-SUMIDATE and J.C. WOODROW

University of Liverpool, Liverpool, England

We report here the regular findings of a particle with the size, form and enzymatic activity of a retrovirus in monocytes from patients with rheumatoid arthritis (RA). Monocytes from 24 patients with RA and from 36 age-matched healthy controls with no evidence of RA or other connective tissue diseases were incubated in Eagle's medium supplemented with 10% fetal calf serum and 5'azacytidine for six days. After incubation the cells were separated by centrifugation at 600xg. The supernatant was filtered (220 nm) and centrifuged at 100,000xg. The pellet was assayed for reverse transcriptase (RT) and negatively stained and examined by electron microscopy. Taking a cut-off for positivity of 18 picomoles of dCTP incorporated/1 x 10^6 monocytes, RT activity was detected in 21 out of 24 patients with RA (276$\pm$54; SEM), but in only 1 out of 36 controls (14$\pm$2; SEM) ($p<0.001$, Wilcoxon rank sum test). Negative stain electron microscopy of the pellet revealed the presence of nucleocapsids (100-120 nm in diameter) with fringed surface similar to HIV-1, and the retrovirus present in cultured monocytes from patients with breast cancer. These findings strongly suggest the presence of a retrovirus in monocytes from patients with RA. The role of this virus in the pathophysiology of the disease remains to be established.

HIV-1 INFECTION OF MULTIPLY-TRANSFUSED PEDIATRIC ONCOLOGY PATIENTS (MTPOP) IN LOW PREVALENCE AREA

M.T. BRADY and J.F. CUARTAS

Children's Hospital, Columbus, Ohio, USA

A few reports of transfusion-associated (TA) HIV-1 infection of multiply-transfused oncology patients and neonates from New York and California have enhanced concerns of these and other patient populations who received blood products before donor screening about their risks of acquiring HIV-1 infection. We attempted to determine the occurrence of TA HIV-1 infection in MTPOP from an area with blood donors whose prevalence of HIV-1 antibody is lower than that of the previously mentioned areas (Central Ohio - 2/10,000 donors are HIV-1 antibody (Ab)-positive with Western blot confirmation). We identified 85 MTPOP (initially HIV-1 Ab negative) who had received 4,346 transfusions (968 red blood cells, 3,192 platelets, 137 fresh frozen plasma; 49 leukocytes) prior to donor screening (HIV-1 antibody) in March, 1985. Only one (1.2%) of these MTPOP seroconverted following transfusion. In January, 1985, the MTPOP who seroconverted had received packed red blood cells from a donor who was subsequently identified as being HIV-1 Ab+ through the "Lookback" program of the American Red Cross. TA HIV-1 infection in our MTPOP was infrequent despite marked exposure to blood products (average of 51.2 donor exposures/MTPOP) and occurred at a frequency consistent with the prevalence of HIV-1 infection in our donor population. While our findings may not be unexpected, they offer documentation of a lower risk of TA HIV-1 infection in our blood recipients than in some previous studies conducted in areas where donor HIV-1 Ab prevalence is higher. This might lessen the anxiety of MTPOP from areas of low HIV Ab prevalence who were transfused prior to March, 1985. We also document that the effort associated with performing "Lookback" evaluations (contacting recipients of blood products from donors subsequently identified as being HIV-1 Ab+) are valuable since our only HIV-1-infected MTPOP was identified in this manner.

CYTOMEGALOVIRUS INFECTION OF HUMAN HEART MUSCLE CELLS

L. PULLIAM, N. TANG and P. SIMPSON

Veterans Administration Medical Center and the University of California, San Francisco, San Francisco, California, USA

Cytomegalovirus (CMV) is a major source of morbidity and mortality in cardiac as well as other organ transplant recipients. Research in this area has been hindered by the lack of a human heart model. We have established a human heart cell culture system to study viral infection of heart cells. This system entails repeated trypsinization of human fetal heart tissue. Cultures yield a high concentration of heart muscle cells (HMC) (up to 80%) as determined by immunocytochemistry using a monoclonal antibody to myosin heavy chain. The few remaining non-muscle cells (NMC) appear morphologically to be fibroblasts. These primary heart cultures survive at least 3 weeks in culture with the percentage of HMC remaining stable over time in culture. Infection with the Towne strain of CMV at a MOI of 1 resulted in rapid infection of NMCs by demonstration of intranuclear inclusions and CMV viral antigens; lytic infection of HMC was observed several days after the NMC lysis. Syncytial formation was also observed during infection. Intracellular CMV DNA was assayed using a ^{125}I-DNA-DNA hybridization technique. Peak CMV DNA recovery was 9 to 14 days after infection. Rapid CMV DNA replication correlated with cultures containing increased numbers of NMCs. We conclude that CMV infects both NMC and HMC in a lytic fashion with resulting syncytia. The kinetics of infection differ between these two cell types.

SEROLOGIC TESTING AND RISK ANALYSIS FOR HUMAN IMMUNODEFICIENCY VIRUS TYPE 1 (HIV-1) INFECTION IN INDIVIDUALS WITH RECENT OR COEXISTING SYPHILIS

R.J. RICE, M. NEFF-SMITH and F.E. CHURCHILL

The Medical College of Virginia and The Richmond City Health Department, Richmond, Virginia, USA

Genital ulcer disease (syphilis, herpes, chancroid) may increase the risk of HIV-1 infection. To define HIV-1 seroprevalence and associated behaviors among persons with syphilis, we analyzed risk behavior data and HIV-1 serologic testing for 115 cases of early syphilis diagnosed by clinical criteria, darkfield microscopic exam, and syphilis serology. Twenty-eight percent (32/115) of the syphilis cases were HIV-1 antibody positive as confirmed by FDA-approved ELISA and Western blot. Of the 32 HIV-1 positive cases, 60% were gay/bisexual males who used intravenous drugs (heroin or cocaine) or exchanged sex for drugs or money. Despite a reported decrease in rectal sex, no change was reported in oral-genital sex by this group. At least 20% of their elicited sexual partners were HIV-1 antibody positive. The most common risk behavior reported by syphilis and HIV-1 positive heterosexuals was exchange of sex for drugs or money. Of the 83 HIV-1 antibody negative syphilis cases, 40% were younger women (16% were aged 12-19 years) with first episode syphilis who reported exposure to male sexual partners who, in turn, reported multiple anonymous sexual partners or older prostitute contacts. Despite reported sexual behavior changes and a decline in anorectal sex by gay/bisexual men, those who abuse intravenous drugs and women who exchange sex for drugs or money may represent an expanding but inadequately-studied reservoir for syphilis and HIV-1.

USE OF A COMMUNITY ASSESSMENT SURVEY MODEL TO PREDICT AIDS MORBIDITY IN AT RISK POPULATIONS

M. NEFF-SMITH, R.J. RICE and THE COMMUNITY SURVEY TEAM

The Medical College of Virginia, Richmond, Virginia, USA

Population-specific data on risk of AIDS are needed to predict future trends, define shifts in morbidity, and assist public health efforts to contain the epidemic. Low income, indigent, and homeless populations may be at higher risk of AIDS due to more limited access to medical care and AIDS prevention education. To define predictive indicators of AIDS morbidity in these populations, we analyzed data obtained from systematic community assessment surveys to ascertain sociodemographic, behavioral, and environmental risk factors for AIDS. Surveys were conducted over a two year period using the Genesis Model of Ethanographic Community Assessment to collect data from key informants. Computer-assisted analysis of survey and regional prevalence data on sexually transmitted diseases, AIDS, adverse pregnancy outcome, violent crime, prostitution, poverty, and drug abuse correctly predicted the current 200% increase in AIDS and syphilis in the survey population that was not predicted by traditional health care information systems. Excluding same sex sexual exposure, the variables of poverty, early sexual activity or sexual abuse, early exposure to illicit drugs including "crack", and prior syphilis or gonorrhea were most highly predictive of AIDS in the survey population. Careful, systematic collection and analysis of data using the Genesis Model may assist in predicting shifting trends in AIDS morbidity, define population foci at highest risk for AIDS, and direct public health control efforts to control the epidemic in emerging population risk groups.

CLINICAL AND OBJECTIVE EVIDENCE OF A FAVORABLE RESPONSE TO ACEMANNAN (ACE-M) TREATED HIV-1 PATIENTS

H.R. McDANIEL, T. PULSE, B.H. McANALLEY and R.H. CARPENTER

Dallas-Fort Worth Medical Center Research Laboratory, Grand Prairie, Texas, USA

In three open clinical pilot studies numbering 41 patients and spanning 1985-1988, ACE-M was given p.o. up to 250 mg qid. Retrospective analysis of initial 2 pilot groups suggested that the patients could be divided into two groups of response, patients with an Ab T-4 lymphocyte count of $>150/mm^3$ and a p-24 core antigen (Abbott Diagnostics) of <300 Pg/dL were noted to do well. Patients with T-4 <150 and p-24 >300 tended to do poorly. Good response was defined as progressive lowering of patient clinical scores, serial p-24 reductions and increases in T-4 counts. In a 13 patient grouping the T-4 counts before therapy (Rx) averaged 340, 360 at 8 weeks, 450 at 16 weeks and 534 at 32 weeks. Only two patients' T-4 were <500 at 32 weeks. A prospective projection of response to ACE-M for 26 patients was made using the above criteria in a 1987-88 grouping. Seventy-seven percent responded to ACE-M as predicted. Thirteen percent had a mixed response of meeting only 2 of our 3 criteria. Combining the 3 pilots, 24/41 or 59% met criteria for a "good" response. Patients on Acemannan in excess of 800 days have maintained improvement and been able to live active, productive lives. We detected no signs of hematological, biochemical or coagulation toxicity. In a 16 patient grouping, the average Hgb was 145 g/dL pre-Rx and 147 at 90 days. Twelve out of sixteen patients had pre-Rx leukopenia (<5,000 WBC/mm3). At 90 days, 2 patients had a small decrease in WBC, 11 had a progressive rise, with 5 over 5,000 at 90 days. Pre-Rx platelets average 178,000 and 182,000 at 90 days. Abnormal biochemistry values prior to Rx normalized on receiving ACE-M. The observed spectrum of effectiveness and lack of toxicity was predicted based on the molecular homology and deletions found in comparison of ACE-M to bacterial cell-wall LPS and dextran sulfate. In conclusion, prediction of HIV-1 pt response to ACE-M Rx was made with high accuracy. In 41 patients, 59% met the full criteria of improvement using clinical scoring, p-24 and Ab T-4 values. Side effects and toxicity attributed to ACE-M were not detected in this limited exposure to ACE-M in three open clinical pilot studies involving 41 patients. A formal, placebo controlled, random selection based investigation was recommended as justified and is currently in progress.

INHIBITION OF HIV-1 REPLICATION BY A VIRUS-ACTIVATED RETROVIRAL VECTOR

D.P. BEDNARIK, J.D. MOSCA, N.B.K. RAJ and P.M. PITHA

The Johns Hopkins University Oncology Center, Baltimore, Maryland, USA

The aim of this study was to determine the molecular mechanism by which interferons (IFNs) inhibit HIV-1 replication, and to employ a novel, site-directed approach to IFN production. Retroviral expression vectors, in which expression of HIV-1 was limited to HIV-infected cells were constructed with the HIV-1-LTR directing α2 IFN. Stable cell lines containing the constructs were established, infected with HIV-1 and compared to companion cells receiving exogenous α2 IFN. The levels of viral mRNA and protein were quantitated and compared to control cultures. Addition of recombinant α2 IFN (250 units/ml) prior to HIV-1 infection caused a 35% reduction in RT activity after 11 days post-infection. The overall reduction in viral mRNA and viral proteins, in IFN-treated cells, was observed. Several cell lines containing the α 2 IFN retroviral construct produced low levels of IFN constitutively, however, synthesis was greatly enhanced after infection with HIV-1, and cells containing these constructs are presently being studied. HIV-1 replication is sensitive to α2 IFN treatment. Molecular studies indicate that the mechanism of this inhibition may occur at several steps of the virus life-cycle. HIV-1-activated, *in situ* synthesis of α2 IFN as a new concept directed against HIV-1 replication is proposed. Results of these studies demonstrate that HIV-1-induced IFN synthesis can nearly abolish replication of HIV-1 in cell lines containing the novel retroviral vector.

EFFECT OF 6-DIAXO-5-OXO-L-NORLEUCINE (DON) ON HERPES SIMPLEX VIRUS TYPE 1 (HSV-1) REPLICATION

G. GOLDSTEIN

Ohio Wesleyan University, Delaware, Ohio, USA

The effects of DON on the replication of HSV-1 in cultured buffalo green monkey kidney cells were examined. DON is a structural analogue of the amino acid glutamine and inhibits the amino transferase reaction which converts fructose-6-phosphate into glucosamine. DON also inhibits the *de novo* synthesis of ribonucleotides by preventing the transfer of amino groups from glutamine to ribonucleotide precursors. Treatment of BGMK cells with cell culture medium containing DON, at a concentration of 20 μg/ml (0.117 mM) for 1 h prior to and during infection with HSV-1, reduced virus replication *in vitro* 99.4%. Inhibition of virus replication is initially seen at 8 h post infection when assembly of infectious virus particles begins. At a concentration of DON of 20 μg/ml, BGMK cells grown in the presence of DON show no statistical difference in their growth rate compared to untreated control cells cultured over a period of 36 h. A dose response curve shows that the concentration of DON at which HSV-1 replication is inhibited 50% is about 0.17 μg/ml or less than 1 mM DON. Additional experiments demonstrate that the decrease in virus synthesis was not due to virus inactivation by DON or by inhibition of virus adsorption to susceptible host cells. The inhibitory activity of DON on virus replication was also shown to be able to be partially reversed by the presence of excess glutamine in the medium. Treatment with DON may provide to be an effective therapy for HSV-1.

RAPID SCREENING METHOD WITH A SHELL VIAL ASSAY FOR ANTIVIRAL ACTIVITY AGAINST CYTOMEGALOVIRUS (CMV)

A. TELENTI and T.F. SMITH

Mayo Clinic and Mayo Foundation, Rochester, Minnesota, USA

An assay based on the shell vial cell culture (SVA) was devised to assess differences in fluorescent patterns of 8 CMV strains incubated in the presence or absence of three antiviral substances. The drug concentrations selected for *in vitro* testing were in the range of peak serum levels achievable in humans (ganciclovir 20 μM). Four days after inoculation of the shell vials, the cell monolayer was fixed, stained with a mixture of monoclonal antibodies to early and late antigens of CMV and reacted with anti-mouse fluorescein-labeled globulin. Twenty μM of ganciclovir in the media inhibited the formation of plaque-like foci and specific cytoplasmic fluorescence of laboratory strain AD169 and four clinical isolates. Two documented ganciclovir resistant strains exhibited partial inhibition at 20 μM and a clinical isolate exhibited incomplete inhibition in 3 of 4 occasions. None of the strains was fully inhibited by 80 μM of acyclovir; however, different degrees of partial inhibition were observed. All isolates were inhibited by 300 μM of phosphonoformic acid. The SVA can be used as a rapid screening test for anti-CMV substances and as a screening of CMV resistance to ganciclovir prior to quantitative testing by the more complex plaque reduction assay.

EFFECT OF A NON-STEROIDAL ANTI-INFLAMMATORY AGENT (DICLOFENAC) ON HERPES KERATITIS

M.D. TROUSDALE[1], W.E. BARLOW[1], L.J.B. McGUIGAN[2]

Doheny Eye Institute, Los Angeles, California[1] and
Ciba-Geigy Corp., Summit, New Jersey[2], USA

The initial consideration in the treatment of ocular herpes infection is the selection of an antiviral agent, but there is often the need for an additional drug/agent to control the accompanying inflammatory reaction. Steroids, however, can cause deleterious effects. Diclofenac (0.1%), a non-steroidal anti-inflammatory agent, was compared with 1% prednisolone phosphate (Inflamase Forte), 0.03% flurbiprofen (Ocufen), and a saline placebo in rabbit eyes with acute herpetic keratitis in a double blind study. Twenty-four New Zealand white rabbits were infected with 2 x 10^5 pfu/eye of herpes simplex virus type 1, and divided into four groups of six on day three post-inoculation; topical therapy 4x/day was then initiated for ten days. Corneal epithelial involvement (CEI), conjunctivitis, clouding, virus shedding and histopathology of the cornea were evaluated and compared for each group. Maximum CEI was observed in each group on day six post-inoculation, but in eyes treated with prednisolone phosphate the CEI appeared to be more severe and to resolve more slowly. Conjunctivitis and corneal clouding peaked on day seven for all treatment groups; this was most severe in the placebo treated eyes, followed closely by those treated with prednisolone phosphate. The duration of virus shedding was the same for placebo, flurbiprofen and diclofenac treated groups (50% or more of eyes were virus negative by day 10 or 11). Only prednisolone phosphate treated eyes had an extended period of virus shedding, and the mortality rate in this group was slightly higher. It was found that topical diclofenac does not exacerbate acute herpes keratitis; diclofenac-treated eyes displayed less or at least no more severe disease than did eyes treated with the other anti-inflammatory agents tested.

FLOW CYTOMETRIC ANALYSIS OF CELL CYCLE CHANGES DUE TO REPLICATING BOVINE HERPES-VIRUS-1 (BHV-1)

T.E. TOTH and R.E. HAND

College of Veterinary Medicine, Virginia Tech, Blacksburg, Virginia; Oakridge National Laboratory, Oakridge, Tennessee, USA

Bovine turbinate (BT) cells were inoculated by adsorbing, for defined periods of time, defined quantities of BHV-1. At predetermined post-inoculation times cytopathic effects were observed, culture fluids harvested for viral-infectivity determination, cells were harvested, fixed in 70% alcohol, treated with RNase, and the DNA stained with propidium iodide. These, and uninoculated control cells, were analyzed for cell phase by exciting with an Argon ion laser at 488 nm, and measuring fluorescence emittance at 630 nm in an Ortho 50H cytofluorograph coupled with an Ortha Data Analysis system. Replicating BHV-1 induced non-synchronized BT cells to enter the S phase of the cell cycle. This effect appears to be restricted to a single round of DNA synthesis. Cells inoculated with highly productive levels of BHV-1 did not continue into the next phase as indicated by the decreasing G2+M proportions among these cells. The appearance and degree of the effect positively correlated with the multiplicities of infection, length of inoculum absorption period, length of virus replication time, appearance of infectious virions in the culture fluid, and the titer of the infectious virions released into the culture. Threshold values and the degree of the effect was determined by these parameters synergistically.

EXPRESSION OF HUMAN T-CELL LEUKEMIA VIRUS TYPE I (HTLV-I) GP46 ENVELOPE ANTIGEN IN CHINESE HAMSTER OVARY (CHO) CELLS

D. LAROCCA, L.A. CHAO, M. WRIGHT, C. SMITH and S. COATES

Triton Biosciences, Inc., Alameda, California, USA

The human retrovirus, HTLV-I, is associated with several diseases, including adult T-cell leukemia, myelopathy, and tropical spastic paraparesis. After several attempts to produce the full length (gp68) or truncated (gp46) HTLV-I envelope in mammalian cells were unsuccessful, we fused a heterologous human signal peptide and N-terminus to 92% of gp46 to make the expression plasmid pGHE. The resulting fusion protein was efficiently produced in COS and CHO cells. Stable pGHE transformed CHO clones were obtained by co-transfection with pSV5 DHFR. Positive transformants were amplified using incremental increases in methotrexate concentration. The extent of large syncytia formation was much greater in the pGHE transformants than in a similarly treated negative control, suggesting biological activity. The recombinant envelope was found to be intracellular, membrane associated, and poorly secreted. In a screen of 50 HTLV-I positive sera, 86% reacted with the recombinant envelope of Western blots. None of 50 negative sera reacted with the recombinant envelope. We estimate that we can recover approximately 2.5 mg of recombinant envelope from 10^9 cells.

CHARACTERIZATION OF IgG Fc RECEPTORS INDUCED BY HUMAN CYTOMEGALOVIRUS

X. BIN, T. MURAYAMA, K. ISHIDA and T. FURUKAWA

Kanazawa Medical University, Uchinada, Ishikawa, Japan

The 65K protein of human cytomegalovirus (HCMV) is a major viral protein and all the sera from seropositive population contain the antibody for this protein. This protein had immunoglobulin G Fc receptor (FcR-IgG) demonstrated by IgG-affinity chromatography and by Western blotting using ^{125}I-labeled human IgG and Fc. FcR of the 65K polypeptide was found to mediate the binding of IgG and was also detected on HCMV virions. The 65K FcR was reactive with murine monoclonal antibodies to herpes simplex virus gE and granulocyte FcR. Immunofluorescence study using monospecific and monoclonal antibodies to the 65K protein was also carried out to monitor the expression of this protein in infected cells. Regardless of differences in the reactivity of the monoclonal antibodies, as determined by immunoblotting and immunofluorescent staining, all stained cytoplasmic inclusion bodies localized to the site of the HCMV-induced FcR, suggesting that most of the 65K major protein of HCMV co-localizes with the HCMV-induced FcR. To determine the evolution of the 65K protein expression, cells harvested at various times after infection were immunoblotted on nitrocellulose membrane and reacted anti-65K serum. The 65K protein was first detected at 48 h p.i. which increased in intensity during the late stage of infection. Infected cells incubated with PAA or tunicamycin decreased the reactivity of this antiserum with the 65K protein.

TRANSACTIVATION OF HUMAN IMMUNODEFICIENCY VIRUS BY HEPATITIS B VIRUS X GENE PRODUCT

[1]R.W. FARR, [1]P. BISHOP, [1]A. SIDDIQUI, and [2]R. GAYNOR

University of Colorado Health Sciences Center, Denver, Colorado[1]; University of California, Los Angeles School of Medicine, Los Angeles, California[2], USA

Evidence of hepatitis B virus (HBV) infection is present in 87% of American AIDS patients and 97% of African AIDS patients, and HBV has been suggested as a co-factor in human immunodeficiency virus type 1 (HIV-1) infections. In this study, we describe transactivating effect of the HBV X gene product on the HIV-1 LTR. The X gene product of HBV induces transcription of the HIV-1 long terminal repeat (LTR) as measured by the chloramphenicol acetyltransferase (CAT) assay and luciferase assay. HepG2 cells, which do not contain the X gene, and Get cells, a HepG2 cell line permanently transfected with the X gene vector, were transfected with HIV-1 LTR constructs containing the CAT gene or luciferase gene. Get cells (X gene positive) stimulated transcription of HIV-1 LTR more than 10-fold compared to HepG2 cells (X gene negative) by both CAT and luciferase assays. During transfections with HIV-1 LTR deletional mutants, little transcription was seen with HIV-1 LTR 46 and HIV-1 LTR 70. The greatest activity was with HIV-1 LTR 160 which contains the HIV-1 enhancer region. Transcription was less with HIV-1 LTR 218 which contains the HIV-1 negative regulatory element. Transfections with a point mutant of the enhancer region abolished transcription of HIV-1 LTR. Transfections with a point mutant of the negative regulatory element restored some of the transcription when compared with HIV-1 LTR 218. Our results indicate that the X gene product of HBV can transactivate the HIV-1 LTR. Since HBV DNA has been found in lymphoid cells of patients with HIV-1 infection, *in vitro* evidence of HBV transactivation of HIV-1 supports the role of HBV as a possible co-factor of AIDS.

MONOCLONAL ANTIBODY BINDING TO A VIRUS CHANGES CONFORMATION OF UNRELATED EPITOPES

A. CEPICA, C. YASON and G. RALLING

Atlantic Veterinary College, University of Prince Edward Island, Charlottetown, Canada

It has been assumed that the conformation of proteins remains unchanged after an interaction of some of the epitopes with antibodies. The issue of conformational changes of viruses induced by antibody is of importance for two reasons: because most antigenic determinants of proteins are conformational rather than sequential, and because the recent advances in genetic and protein engineering favor the development of vaccines based on isolated antigenic determinants. We reacted six monoclonal antibodies (Mabs) with potato virus X (PVX) in a blocking antigen-coated plate (ACP) enzyme-linked immunoassay (ELISA) and a double antibody sandwich (DAS) ELISA. After binding of some Mabs, diminished availability of other epitopes was observed. We concluded, that induction of conformational antigenic changes was responsible for this phenomenon. In support of this conclusion, use of the Mab with the ability to modulate unrelated epitope, as capture antibody in DAS-ELISA, led to a decrease of the sensitivity. Understanding of antigenic modulation by antibody is considered important for studies of immunity to viruses, and for a rational development of immunoassays. If the phenomenon of epitope modulation is confirmed to occur universally with protein antigens, it would have to be taken into consideration in devising vaccines, passive immunotherapy, and immunoassays.

THE ANTIVIRAL EFFICACY AND MECHANISM OF INTERLEUKIN-2 (IL-2) ACTIVATED BONE MARROW CELLS AGAINST MURINE CYTOMEGALOVIRUS (MCMV) INFECTION *IN VIVO* AND ITS POTENTIAL USE IN BONE MARROW TRANSPLANTATION SETTING

R. AGAH, P.E. TWOMEY, L. BUFFA and A. MAZUMDER

University of Southern California School of Medicine and Norris Hospital and Research Institute, Los Angeles, California, USA

No effective therapy is currently available for cytomegalovirus (CMV) infection in patients undergoing bone marrow transplantation. CMV infection is a leading cause of morbidity and mortality in these patients. Earlier results from our laboratory using the murine model of MCMV showed that IL-2 activated bone marrow cells have antiviral activity and are cytotoxic *in vitro* both against spleen and salivary gland cells infected with MCMV in the acute and chronic infection models of disease, respectively. Immunocompromised (irradiated) and intact Balb/c mice were infected with 5×10^4 plaque forming units (PFU) of MCMV. Some of the mice were subsequently infected with 5×10^6 ABM intravenously into the tail vein one day post viral infection with daily intraperitoneal injections of IL-2 (25,000 units/mouse) for 2 days. The mice were then sacrificed and their organs (spleen for acute infection model and salivary gland for chronic infection model) were measured for viral titer at different time intervals post therapy. Appropriate controls were also included in the study. Our results indicated that ABM cleared the virus from the normal reservoir of virus, namely the salivary gland in the chronic infection model in the intact mice (zero PFU/ml in treated mice vs. 4.4×10^4 PFU/ml in untreated mice) and reduced the titer of virus significantly in the immunocompromised mice (3×10^2 PFU/ml in treated vs. 1.8×10^6 in untreated group). The survival of treated immunocompromised mice was also significantly increased ($P<0.002$). The antiviral efficacy was not due to IL-2 alone since no significant antiviral efficacy was seen when IL-2 was used along with other control groups. The antiviral mechanism was also investigated indicating that macrophages are the most important immune component involved in this antiviral therapy. No significant reduction in antiviral therapy was seen when endogenous Natural Killer cell population were eliminated in animals being treated with IL-2 activated bone marrow cells along with intraperitoneal IL-2 injection.

CONTRIBUTORS

HELEN M. ARIZPE The University of Texas Health Science Center of San Antonio, San Antonio, Texas 78284-7758; USA

ANN M. ARVIN Infectious Disease Division, Department of Pediatrics and Department of Obstetrics and Gynecology, Stanford University School of Medicine, Stanford, California; USA

ANGELOS ATHANASSOPOULOS NCI-Frederick Cancer Research Facility, BRI-Basic Research Program, P.O BOX B, Frederick, Maryland; USA

SUSAN BAKER National Jewish Center for Immunology and Respiratory Medicine, Denver, Colorado; USA

WILLIAM BENNETT Harvard School of Public Health and Harvard Medical School, Boston, Massachusetts; USA

FRANCOIS BOUCHER Infectious Disease Division, Department of Pediatrics and Department of Obstetrics and Gynecology, Stanford University School of Medicine, Stanford, California; USA

CHRISTOS CLADARAS NCI-Frederick Cancer Research Facility, BRI-Basic Research Program, P.O BOX B, Frederick, Maryland; USA

NORA M. CRAPMAN The University of Texas Health Science Center of San Antonio, San Antonio, Texas 78284-7758; USA

PAMELA S. DIAZ Infectious Disease Division, Department of Pediatrics and Department of Obstetrics and Gynecology, Stanford University School of Medicine, Stanford, California; USA

BARBARA K. FELBER NCI-Frederick Cancer Research Facility, BRI-Basic Research Program, P.O BOX B, Frederick, Maryland; USA

BONINO FERRUCCIO Division of Gastroenterology, San Giovanni Battista "Molinette" Hospital Corso Bramante 88, 10126, Torino, Italy

OLIVERI FILIPPO Division of Gastroenterology, San Giovanni Battista "Molinette" Hospital Corso Bramante 88, 10126, Torino, Italy

PHYLLIS R. FLOMENBERG Department of Medicine, Veterans Administration Medical Center, 1st Avenue at E. 24th Street, New York, NY; USA

ROSINA FLORIANO Division of Gastroenterology, San Giovanni Battista "Molinette" Hospital Corso Bramante 88, 10126, Torino, Italy

DANA GALLO California Department of Health Service, Division of Laboratories, Viral and Rickettsial Disease Laboratory, 2151 Berkeley Way Berkeley, CA 94704; USA

CHARLES J. GAUNTT The University of Texas Health Science Center of San Antonio, San Antonio, Texas 78284-7758; USA

ELMER K. GODENY The University of Texas Health Science Center of San Antonio, San Antonio, Texas 78284-7758; USA

JONATHAN W. M. GOLD Infectious Disease Service, Memorial Sloan-Kettering Cancer Center, 1275 York Avenue, New York, New York 10021; USA

VICTOR DE GRUTTOLA Harvard School of Public Health and Harvard Medical School, Boston, Massachusetts; USA

MARGARITA HADZOPOULOU-CLADARAS NCI-Frederick Cancer Research Facility, BRI-Basic Research Program, P.O BOX B, Frederick, Maryland; USA

PAUL A. HENSLEIGH Infectious Disease Division, Department of Pediatrics and Department of Obstetrics and Gynecology, Stanford University School of Medicine, Stanford, California; USA

MARSHALL S. HORWITZ Departments of Microbiology-Immunology, Cell Biology and Pediatrics, Albert Einstein College of Medicine, 1300 Morris Park Avenue, Bronx, New York; USA

JAMES F. JONES National Jewish Center for Immunology and Respiratory Medicine, Denver, Colorado; USA

C. WILLIAM LUTTON The University of Texas Health Science Center of San Antonio, San Antonio, Texas 78284-7758; USA

STEPHEN MARTIN Department of Microbiology, College of Veterinary Medicine, University of Tennessee, Knoxville, TN 37996-0845; USA

JOSEPH W. McCRAY Sandoz-Forschungsinstitut GmbH, Brunnerstrasse 59 A-1235 Vienna, Austria

JOSEPH D. MOSCA Henry M. Jackson Foundation, Rockville, MD 20850; and the Department of Immunology and Infectious Diseases, The Johns Hopkins University School of Public Health, Baltimore, MD 21205; USA

GEORGE N. PAVLAKIS NCI-Frederick Cancer Research Facility, BRI-Basic Research Program, P.O BOX B, Frederick, Maryland; USA

CHARLES G. PROBER Infectious Disease Division, Department of Pediatrics and Department of Obstetrics and Gynecology, Stanford University School of Medicine, Stanford, California; USA

GEORGE E. REVTYAK The University of Texas Health Science Center of San Antonio, San Antonio, Texas 78284-7758; USA

BRUNETTO MAURIZIA ROSSANA Division of Gastroenterology, San Giovanni Battista "Molinette" Hospital Corso Bramante 88, 10126, Torino, Italy

BARRY T. ROUSE Department of Microbiology, College of Veterinary Medicine, University of Tennessee, Knoxville, TN 37996-0845; USA

MARIUS M. ROZEK The University of Texas Health Science Center of San Antonio, San Antonio, Texas 78284-7758; USA

EVAN STEINBERG Regional Virology Laboratory and Department of Pediatrics, Southern California Kaiser Permanente, North Hollywood and Los Angeles, California; USA

JOANNE STREIB National Jewish Center for Immunology and Respiratory Medicine, Denver, Colorado; USA

WAYNE M. SULLENDER, Infectious Disease Division, Department of Pediatrics and Department of Obstetrics and Gynecology, Stanford University School of Medicine, Stanford, California; USA

STEVEN M. TRACY The University of Texas Health Science Center of San Antonio, San Antonio, Texas 78284-7758; USA

ANTHONY J. VALENTE The University of Texas Health Science Center of San Antonio, San Antonio, Texas 78284-7758; USA

ANN L. WARFORD Regional Virology Laboratory and Department of Pediatrics, Southern California Kaiser Permanente, North Hollywood and Los Angeles, California; USA

GUDRUN WERNER Sandoz-Forschungsinstitut GmbH, Brunnerstrasse 59 A-1235 Vienna, Austria

CONNIE M. WRIGHT NCI-Frederick Cancer Research Facility, BRI-Basic Research Program, P.O BOX B, Frederick, Maryland; USA

AUTHOR INDEX

SUBJECT INDEX

GPSR Compliance

The European Union's (EU) General Product Safety Regulation (GPSR) is a set of rules that requires consumer products to be safe and our obligations to ensure this.

If you have any concerns about our products, you can contact us on ProductSafety@springernature.com

In case Publisher is established outside the EU, the EU authorized representative is:

Springer Nature Customer Service Center GmbH
Europaplatz 3
69115 Heidelberg, Germany

Batch number: 10761599

Printed by Printforce, the Netherlands